Acquisition of
Mathematics Concepts
and Processes

DEVELOPMENTAL PSYCHOLOGY SERIES

SERIES EDITOR
Harry Beilin

Developmental Psychology Program
City University of New York Graduate School
New York, New York

In Preparation

DIANE L. BRIDGEMAN. (Editor). *The Nature of Prosocial Development: Interdisciplinary Theories and Strategies*

EUGENE S. GOLLIN. (Editor). *Malformations of Development: Biological and Psychological Sources and Consequences*

ALLEN W. GOTTFRIED. (Editor). *Home Environment and Early Mental Development*

Published

ROBERT L. LEAHY. (Editor). *The Child's Construction of Social Inequality*

RICHARD LESH and MARSHA LANDAU. (Editors). *Acquisition of Mathematics Concepts and Processes*

MARSHA B. LISS. (Editor). *Social and Cognitive Skills: Sex Roles and Children's Play*

DAVID F. LANCY. *Cross-Cultural Studies in Cognition and Mathematics*

HERBERT P. GINSBURG. (Editor). *The Development of Mathematical Thinking*

MICHAEL POTEGAL. (Editor). *Spatial Abilities: Development and Physiological Foundations*

NANCY EISENBERG. (Editor). *The Development of Prosocial Behavior*

WILLIAM J. FRIEDMAN. (Editor). *The Developmental Psychology of Time*

SIDNEY STRAUSS. (Editor). *U-Shaped Behavioral Growth*

GEORGE E. FORMAN. (Editor). *Action and Thought: From Sensorimotor Schemes to Symbolic Operations*

The list of titles in this series continues on the last page of this volume.

Acquisition of Mathematics Concepts and Processes

EDITED BY

RICHARD LESH

MARSHA LANDAU

School of Education
Northwestern University
Evanston, Illinois

1983

ACADEMIC PRESS

A Subsidiary of Harcourt Brace Jovanovich, Publishers

New York London
Paris San Diego San Francisco São Paulo Sydney Tokyo Toronto

ACADEMIC PRESS, INC.
111 Fifth Avenue, New York, New York 10003

United Kingdom Edition published by
ACADEMIC PRESS, INC. (LONDON) LTD.
24/28 Oval Road, London NW1 7DX

Library of Congress Cataloging in Publication Data

Main entry under title:

Acquisition of mathematics concepts and processes.

(Developmental psychology series)
Companion vol. to: The Development of mathematical thinking.
Includes bibliographies and index.
Contents: Introduction -- The acquisition of addition and subtraction concepts / Thomas P. Carpenter and James M. Moser -- Proportional reasoning of early adolescents / Robert Karplus, Steven Pulos, and Elizabeth K. Stage -- [etc.]
1. Mathematics--Study and teaching (Elementary)--Psychological aspects. 2. Cognition in children.
I. Lesh, Richard A. II. Landau, Marsha. III. Series.
QA135.5.A27 1983 155.4'13 83-2845
ISBN 0-12-444220-X

PRINTED IN THE UNITED STATES OF AMERICA

83 84 85 86 9 8 7 6 5 4 3 2 1

Contents

Contributors ix

Preface xi

1. Introduction

Richard Lesh and Marsha Landau

References 6

2. The Acquisition of Addition and Subtraction Concepts

Thomas P. Carpenter and James M. Moser

Introduction 7
Research on Problem Difficulty 8
Piagetian Research 11
Current Research: A Process Approach 14
Discussion 37
References 40

3. Proportional Reasoning of Early Adolescents

Robert Karplus, Steven Pulos, and Elizabeth K. Stage

Background 45
Research Questions 50
The Lemonade Puzzle Interviews 51
Cognitive and Attitudinal Correlates 62
Conclusions 72
Implications for Teaching 79
Appendix 84
References 86

4. Rational-Number Concepts

Merlyn J. Behr, Richard Lesh, Thomas R. Post, and Edward A. Silver

A Mathematical and Curricular Analysis
of Rational-Number Concepts 92
The Rational-Number Project 98
Perceptual Cues and the Quality of Children's Thinking:
A Project Study 108
Definitions 110
Directions for Future Research 120
Reference Note 124
References 124

5. Multiplicative Structures

Gerard Vergnaud

Preliminary Analysis 128
Experiments 140
Further Analysis and Experiments 160
Conclusion 172
References 173

6. Space and Geometry

Alan J. Bishop

Meanings and Understandings 176
Abilities and Processes 181
Spatial Understanding and Processes of Students
in Papua New Guinea 186
Guidelines for Future Research 197
References 200

7. Van Hiele–Based Research

Alan Hoffer

Introduction 205
Work of the Van Hieles 206
Soviet Studies 209
Preliminary Work in the United States 210
Recent Studies in the United States 212
Other Aspects of the Van Hiele Model 219
Future Work 223
Reference Notes 226
References 226

8. Trends and Issues in Mathematical Problem-Solving Research

Frank K. Lester, Jr.

Introduction 229
The Nature of Mathematical Problem-Solving
and Mathematical Problem-Solving Research 231
Overview of Mathematical Problem-Solving Research 233
Recent Studies at Indiana University 240
Issues for Future Research 249
Looking Back 255
Reference Notes 257
References 258

9. Conceptual Models and Applied Mathematical Problem-Solving Research

Richard Lesh, Marsha Landau, and Eric Hamilton

Conceptual Models 264
Background for Current Research on Applied Problem-Solving 265
Rational-Number Task Interviews 272
Rational-Number Written-Test Results 297
Concluding Remarks 306
Appendix 9.A: Reordered Versions of the CA and OA Instruments 337
Appendix 9.B: Rational-Number Characteristics
on the CA and OA Instruments 337
Reference Notes 342
References 343

10. Episodes and Executive Decisions in Mathematical Problem-Solving

Alan H. Schoenfeld

Introduction 345
A Discussion of Antecedents 347
An Informal Analysis of Two Protocols 350
Framework for the Macroscopic Analysis
of Problem-Solving Protocols 353
The Full Analysis of a Protocol 360
Some Empirical Results 364
Discussion 369
Appendix 10.A: Protocol 1 371
Appendix 10.B: Protocol 2 375
Appendix 10.C: Protocol 3 381
Appendix 10.D: Protocol 4 389
Appendix 10.E: Protocol 5 394

Reference Notes 394
References 395

Author Index 397

Subject Index 403

Contributors

Numbers in parentheses indicate the pages on which the authors' contributions begin.

MERLYN J. BEHR (91), Department of Mathematical Sciences, Northern Illinois University, DeKalb, Illinois 60115

ALAN J. BISHOP (175), Department of Education, University of Cambridge, Cambridge CB2 1PT, England

THOMAS P. CARPENTER (7), Department of Curriculum and Instruction, University of Wisconsin, Madison, Wisconsin 53706

ERIC HAMILTON (263), School of Education, Northwestern University, Evanston, Illinois 60201

ALAN HOFFER (205), Department of Mathematics, University of Oregon, Eugene, Oregon 97403

ROBERT KARPLUS (45), Lawrence Hall of Science, University of California, Berkeley, California 94720

MARSHA LANDAU (1, 263), School of Education, Northwestern University, Evanston, Illinois 60201

RICHARD LESH (1, 91, 263), School of Education, Northwestern University, Evanston, Illinois 60201

FRANK K. LESTER, JR. (229), School of Education, Indiana University, Bloomington, Indiana 47405

JAMES M. MOSER[1] (7), Wisconsin Center for Education Research, Madison, Wisconsin 53706

THOMAS R. POST (91), Department of Curriculum and Instruction, University of Minnesota, Minneapolis, Minnesota 55455

STEVEN PULOS (45), Lawrence Hall of Science, University of California, Berkeley, California 94720

[1]Present address: Laboratoire de Psychologie, 75270 Paris, Cedex 06, France.

ALAN H. SCHOENFELD (345), Graduate School of Education and Department of Mathematics, University of Rochester, Rochester, New York 14627

EDWARD A. SILVER (91), Department of Mathematical Sciences, San Diego State University, San Diego, California 92182

ELIZABETH K. STAGE (45), Lawrence Hall of Science, University of California, Berkeley, California 94720

GERARD VERGNAUD (127), École des Hautes Études en Sciences Sociales, Centre National de la Recherche Scientifique, Laboratoire de Psychologie, Centre d'Étude des Processus Cognitifs et du Langage, 75270 Paris, Cedex 06, France

Preface

This volume is a companion to Herbert Ginsburg's *Development of Mathematical Thinking* in the Developmental Psychology Series edited by Harry Beilin. The common goal of these two books is to present some of the most promising and productive areas of current research in mathematics learning and problem solving. Leading researchers in each area have been enlisted to characterize state-of-the-art developments in their fields of specialization, explain where their own work fits into this picture, describe one project or study as an exemplar of what is current, and identify some important directions for future research.

Because a community of scholars is beginning to emerge at the interface of several branches of psychology, mathematics, and mathematics education, the editors believe that it is timely to publish a set of volumes that might further this movement. Each volume was planned for readers from both the psychology and mathematics education research communities. Ginsburg's book was intended to focus on the work of researchers who approach mathematics learning and problem solving from a psychological perspective, whereas this volume was designed to represent a mathematics education point of view.

The distinctions between a psychological and a mathematics education perspective are discussed in the introduction (Chapter 1). One characteristic of the mathematics education research reported in this volume is an emphasis on mathematics as a highly structured content domain. Another is the important role of instruction, both as a goal of research and as a methodological tool in research settings.

Carpenter and Moser (Chapter 2) trace the development of whole-number arithmetic and describe their program of research on the procedures children use to obtain answers to addition and subtraction problems. The next three chapters provide information about three different, but in many ways complementary, approaches to research on the development of rational-number ideas. Karplus, Pulos, and Stage (Chapter 3) investigate the use of proportional reasoning patterns by early adolescents, seeking data on individual differences, sex differences, grade-level differences, and the relationship between student attitudes and success on proportional reasoning tasks. Behr, Lesh, Post, and Silver (Chapter 4) conducted rational-number research that included a large-scale testing component, a small-group instructional component, and the careful observation of the effects of theory-based instruction on rational-number understandings. Vergnaud (Chapter 5) describes three categories of multiplicative structures (isomorphism of measures, product of measures, and multiple proportion) and reports results of didactic experiments involving the topic of volume in seventh-grade classrooms.

Chapters 6 and 7 deal with research on space and geometry. Bishop (Chapter 6) reviews two major thrusts in this field: (*a*) the child's understanding of spatial and geometric concepts and (*b*) spatial abilities and visual processing. Hoffer (Chapter 7) provides an overview of the research and curricular developments in the Soviet Union and the United States that have been based on the Van Hiele levels of thought and phases of learning for geometry.

The last three chapters focus on mathematical problem-solving. Lester (Chapter 8) discusses the complex nature of mathematical problem-solving and reviews the necessarily slow progress that mathematics education researchers have made in developing a stable body of knowledge concerning its acquisition. He then describes past research and some recent studies conducted at Indiana University, and raises several key issues for future research, including a call for more open dialogue among researchers in various disciplines.

Lesh, Landau, and Hamilton (Chapter 9) define and illustrate a theoretical construct—a conceptual model—central to a current research project on applied mathematical problem-solving.

Schoenfeld (Chapter 10) characterizes the "managerial" aspects of expert and novice problem-solving behaviors among college students and describes the impact of managerial or "executive" actions on success or failure in problem solving.

The mathematics content treated in this volume ranges from early number-concepts in the primary grades to complex problem-solving at the college level. Thus the book should be of interest to teachers of mathematics and to trainers of mathematics teachers in grades K–13, as well as to mathematics education researchers and graduate students in mathematics education.

The editors are indebted to Christine Duffy for her help in preparing the manuscript.

Acquisition of
Mathematics Concepts
and Processes

CHAPTER 1

Introduction

Richard Lesh and Marsha Landau

This book presents some of the most promising and productive current research in mathematics learning and problem solving to readers from both the psychology and mathematics education communities, and does so from a *mathematics education* point of view. What constitutes a distinctive mathematics education perspective? The answer includes an emphasis both on *mathematics* as a content domain, and on *education,* for example, the development of instructional materials. These dual emphases influence the nature of the research questions asked, the underlying assumptions made, the research procedures used, and the generalizations formulated.

In the past, mathematics education researchers borrowed most of their theoretical perspectives and research methodologies from other fields, largely within psychology (e.g., developmental psychology and information processing). However, the chapters in this book represent subareas of mathematics education research that have matured sufficiently so that "theory building" has replaced "theory borrowing." For example, Lester emphasizes in Chapter 8 the movement toward theory-based research in problem solving. A consequence has been that formerly useful, borrowed methodologies are frequently inconsistent with the purposes and assumptions underlying the newly emerging theoretical perspectives. Major mathematics education research projects, such as those discussed in this book, have had to engage in the development of research methodologies as well as in the generation of knowledge related to the improvement of mathematics instruction.

Many of the most promising techniques integrate some form of instructional intervention into the data gathering process. For example, Behr, Lesh, Post, and Silver's rational number research (Chapter 4) made extensive use of Soviet-style teaching experiment methodologies (Kantowski, 1978). The results of both

ACQUISITION OF MATHEMATICS
CONCEPTS AND PROCESSES

ISBN No. 0-12-444220-X

Bishop's (Chapter 6) and Hoffer's (Chapter 7) geometry research were linked to the presence–absence and quality–quantity of prior instruction. In Carpenter and Moser's early number research (Chapter 2), in Karplus, Pulos, and Stage's proportional reasoning research (Chapter 3), and in Vergnaud's multiplicative structures research (Chapter 5), the explicit assumption was made that task behavior was as much a function of internal models as of external stimuli, and that these internal models (i.e., mathematical ideas) were, in large part, the products of explicit instruction.

Care must be taken in using instruction-related research methodologies because changes are induced in the subjects that the research is intended to describe, and the research environment is modified differentially to fit individual subjects, thus raising questions about standardization. On the other hand, the entire notion of a *standardized question* may be an inappropriate construct if the theory on which the research is based assumes that two students frequently interpret a single problem situation or stimulus in quite different ways (as a function of the internal models that are selected and superimposed onto the problem). In mathematics education research, it is common to assume not only that different students may interpret a single seemingly unambiguous problem in completely different ways, but also that two responses that appear identical may be produced using completely different solution paths.

Clinical interviewing techniques have begun to reflect a diminished concern with standardization in order to obtain a more complete picture of children's developing understandings of mathematical ideas and the processes that are used to produce answers. Investigators often find it useful to tailor an interview to the subject either in a relatively spontaneous manner or by beginning with a standard series of questions and then making modifications based on the interviewee's responses. Such stochastic questioning procedures might use *sets* or *sequences* of problems to identify the *processes* used to arrive at answers, and to identify the relational–operational systems that are used to make judgments based on the underlying ideas.

The projects discussed in this book frequently used properties of formal mathematical systems to generate tasks for investigating the structural properties of the conceptual models, mathematical understandings, or internal "programs" students used to interpret and manipulate problem situations. For example, in Lesh, Landau, and Hamilton (Chapter 9), sets of tasks were generated that were presumably characterized by isomorphic structures; sources of variability across tasks were investigated. Vergnaud (Chapter 5) emphasizes the importance of attending to formal mathematical structures when studying the acquisition of interconnected concepts. His analysis and classification of multiplicative structures shaped his investigation of products and proportions in seventh-grade classrooms in France. Carpenter and Moser's chapter (Chapter 2) represents another example of a body of research that has investigated relationships between process use and content understanding. In early number research, relevant processes have included counting or imaging capabilities as well as various computational processes or problem-solving procedures.

The goals of much of the research reported in this book were (*a*) to identify students' primitive conceptualizations of various mathematical ideas and processes (e.g., rational numbers, proportions, early number concepts, and spatial–geometric concepts); (*b*) to investigate similarities and differences between students' conceptual structures associated with these ideas and the formal mathematical structures that characterize them; (*c*) to describe how these conceptualizations are gradually modified into mature understandings; and (*d*) to identify factors that influence this development. These *idea analyses* are quite distinct from the *task analyses* and *analyses of children's cognitive characteristics* that tend to be used in general (i.e., not subject-matter-oriented) psychological research. These three types of analyses are clearly interrelated. However, the research procedures that are appropriate and the generalizations that result often depend on which of these analyses is chosen as central.

Child development research tends to focus on (*a*) cognitive capabilities that are (or are assumed to be) invariant across large bodies of subject matter, (*b*) changes in general cognitive capabilities before and after major cognitive reorganizations (i.e., at approximately 2 years, 6–7 years, and early adolescence), and (*c*) ideas that most students acquire "naturally," without specific instruction.

By contrast, mathematics education research designed to trace the development of a given idea is likely to focus on one of the following: (*a*) ideas that are acquired at intermediate levels between concrete and formal stages of thought, and on idea-related variations in operational ability; (*b*) factors and processes that produce or facilitate transitions from one level of conceptualization to another for individual concepts; (*c*) processes and capabilities that are linked to content understanding; and (*d*) ideas that do not develop naturally, that is, without artificial (e.g., instructional) experiences to facilitate their acquisition. Thus, research focusing on analyses of students' cognitive characteristics tends to generate labels (e.g., preoperational, impulsive, field dependent) that are associated with a given child, or on characteristics that are assumed to be difficult or impossible to change through instruction. Variability across concepts (or across tasks based on the same concept) is frequently ignored as an uninteresting *décalage*.

Idea analyses tend to result in generalizations about children's *ideas* (i.e., about behaviors that can be expected, given a particular conceptualization of an idea) rather than about children per se. Also, investigating variability across tasks is important for idea analysis. For researchers whose ultimate purpose is to design better instructional experiences for students, it is necessary (in situations in which empirical evidence is not always available) to generate theories for predicting the appropriate sequencing of ideas, tasks, and instructional models.

Given two ideas (or two conceptualizations or models for a single idea), it is important to be able to predict which will be less complex, less abstract, less difficult, or more intuitive. For example, Hoffer's review of research based on the work of the Van Hieles explicitly focuses on intermediate stages in the development of various spatial–geometric concepts, factors that produce variability across isomorphic tasks, and mechanisms that help students progress from one stage of conceptual understanding to another (Chapter 7).

For much of the research reported here, a goal (mentioned previously) was to investigate the nature of students' primitive conceptualizations of various mathematical ideas and processes, many of which do not develop "naturally." Rational number ideas provide an important example of this and are the subject of several chapters in this book (Behr *et al.;* Karplus *et al.;* Lesh, Landau, & Hamilton; Vergnaud). Further, many of these ideas are related to rather recent creations in the history of mathematics (e.g., coordinate systems, measurement systems, probability, infinity, and infinitesimals), each of which dramatically influenced the development of mathematics. These tools give students new ways to think about everyday experiences, significantly increasing their cognitive capabilities, and potentially altering (not just accelerating) the course of cognitive development.

Most of the authors in this book are concerned not only with what a student will do on a given problem when unaided, but also with what she or he is optimally capable of doing on a *class* of problems, or what she or he could do when accompanied by minimal guidance or when following theory-based instruction. A goal is to explore the "zone of proximal development" of students' understandings (Vygotsky, 1962), both to describe the "programs" that students typically use to interpret and solve mathematical problems, and to describe how these programs change as a result of theory-based instruction. Rather than seeking only to accelerate mathematical understandings along narrow conceptual paths, the concern is with broadening and strengthening deficient conceptual systems. A further goal is to investigate the results of introducing students to powerful *cultural amplifiers* (e.g., symbol systems or conceptual models) that might increase problem-solving capabilities and alter developmental paths to more sophisticated mathematical concepts.

According to the preceding characterizations, analyses of children's cognitive characteristics produce generalizations that idea analysis researchers have considered to be too crude because there is too much variability across concepts; task analyses, on the other hand, are considered to be too fine grained. An explanatory model that is sufficient to simulate students' behavior on an isolated task may not be sufficient to explain how an idea influences the interpretation and solution of that problem; explanations that correctly predict behavior within a restricted universe of tasks often yield inappropriate responses in a slightly larger universe. For these reasons, the kinds of task analyses that characterize a large share of general (i.e., not subject-matter-specific) psychological research are quite different from the idea analyses that characterize most of the mathematics education research reported in this book. For example, in the proportional reasoning studies of Karplus *et al.* (Chapter 3), the significance of a student's behavior on an individual task is considered relevant only insofar as it sheds some light on the student's understanding of proportions. On the other hand, behavioral research based on task analyses (e.g., Klahr & Siegler, 1978; Siegler, 1981) often generates models of isolated-task behavior that ignore certain mathematical ideas that a mathematics educator would regard as fundamental to the task.

From a mathematics education perspective, it is implausible that a student's behavior on a proportional reasoning task can be explained using a model that gives no substantive role to the student's concept of proportions. Mathematics educators would want to investigate the variability of a student's behavior across a set of tasks that are characterized by related underlying ideas. In contrast, behavioral task analyses are typically based on theoretical perspectives that provide no basis for generating sets of isomorphic tasks, performance models that are parsimonious across isomorphic tasks, or explanations of performance on tasks based on related underlying ideas. Vergnaud (Chapter 5) argues that one must not only study sets of isomorphic tasks, but also study a "field" of related ideas.

Essentially, the choice of whether to focus on task analyses, idea analyses, or on analyses of children's cognitive characteristics amounts to a decision about "how far down to crank the microscope" in the analysis of students' behavior. Mathematics educators, because of their interest in substantive mathematics content and educational implications, choose to focus on students' ideas as they are used in a variety of task situations.

The idea analysis focus that has become so productive in examining the development of students' mathematical ideas is currently being adapted to problem-solving research. That is, the kinds of processes that are emerging as most important in mathematical problem solving, as in other problem solving research content domains, are not the general-but-weak variety emphasized by Polya (1957) and others; they are powerful, content-specific procedures and techniques. This point is made in Lesh, Landau, and Hamilton's work on applied problem solving (Chapter 9). A related idea is discussed in the chapters by Schoenfeld and by Lester. That is, not only is process use directly related to content understanding, but the same kind of idea analysis techniques that have been applied to rational number concepts, early number concepts, and geometry concepts can also be applied to mathematical processes. For example, Schoenfeld (Chapter 10) goes beyond investigating the influence and learnability of various heuristics and strategies to study the managerial understandings that are needed to determine when these processes might be useful, how they relate to other processes within some system, and so on. Lester (Chapter 8) also discusses research efforts that recognize that it is not productive to aim at simply getting an isolated problem-solving process into a student's head; the student must learn when to use it, what other ideas and processes are related to it, and how to retrieve it when appropriate. This research suggests that instructional principles similar to those that apply to other mathematical ideas can be extended to problem-solving processes, heuristics, and strategies, thus providing a useful direction for the dual mathematics and education components of mathematics education research.

References

Kantowski, M. G. The teaching experiment and Soviet studies of problem solving. In L. L. Hatfield (Ed.), *Mathematical problem solving: Papers from a research workshop*. Columbus, Ohio: ERIC/SMEAC, 1978.

Klahr, D., & Siegler, R. S. The representation of children's knowledge. In H. Reese & L. P. Lipsett (Eds.), *Advances in child development and behavior* (Vol. 12). New York: Academic Press, 1978.

Polya, G. *How to solve it* (2nd ed.). New York: Doubleday, 1957.

Siegler, R. S. Developmental sequences within and between concepts. *Monographs of the Society for Research in Child Development*, 1981, *46*(2, Serial No. 189).

Vygotsky, L. S. *Thought and Language*. Cambridge, Mass.: MIT Press, 1962.

CHAPTER 2

The Acquisition of Addition and Subtraction Concepts*

Thomas P. Carpenter and James M. Moser

Introduction

For almost 100 years, psychologists and educators have been investigating children's learning of arithmetic (Buswell & Judd, 1925). This research has been motivated both by an interest in using arithmetic as a foil to investigate more general psychological principles and by a desire to improve instruction in mathematics. Arithmetic provides a clearly specifiable content domain that has often been used to study general principles of learning or instruction or to provide a window on basic cognitive processing within a child or adult. On the other hand, some of the most concerted attempts to apply psychology to education have involved instruction in arithmetic. (For example, see the discussion in Cronbach & Suppes [1969] of the development of instruction in arithmetic based on the work of Thorndike.)

At certain times in the past, the work of prominent individuals such as Brownell or Thorndike has provided some coherence to research on children's learning of arithmetic, but the area has generally remained an eclectic collection of unrelated studies and sometimes conflicting results. In the last few years, however, this seems to be changing. Research in several content areas is beginning to coalesce into coherent bodies of knowledge. Researchers working in

*The project presented or reported herein was performed pursuant to a grant to The Wisconsin Center for Education Research from the National Institute of Education, Department of Health, Education, and Welfare. However, the opinions expressed herein do not necessarily reflect the position or policy of the National Institute of Education, and no official endorsement by the National Institute of Education should be inferred. Center Grant No. 08-NIE-G-80-0117

ISBN No. 0-12-444220-X

these areas are beginning to view the central problems from a common perspective, use similar research paradigms, and build theories based on common data pools.

One of the areas in which the greatest degree of coherence has been achieved is the acquisition of basic whole-number addition and subtraction concepts by young children. There are many reasons for the substained interest in addition and subtraction. The domain of problems is simple enough that differences between problems can be specified with a reasonable degree of clarity. On the other hand, the domain is rich enough to provide a variety of problems, solution strategies, and errors. By the same token, children's solution processes appear to be simple enough to provide some hope of understanding and modeling them but complex enough to be interesting.

Addition and subtraction occupy a central position in the mathematics curriculum. Vergnaud (1982) has argued that the concept of *additive structure,* of which simple addition and subtraction are the most elementary examples, underlies a large portion of mathematics and develops over an extended period of time. It has been suggested elsewhere (Carpenter, 1981) that the transition from children's informal counting and modeling strategies, developed outside formal instruction, to the use of memorized number facts and formal addition and subtraction algorithms is a critical stage in children's learning of mathematics and, further, that some of children's later difficulty in mathematics can be traced to initial instruction in addition and subtraction. Thus, there are both methodological and educational reasons for the focus of research on addition and subtraction.

The purpose of this chapter is to review recent developments in research on young children's acquisition of basic addition and subtraction concepts and skills. In order to provide a context for this analysis and to demonstrate how the current research extends earlier work, some background is needed.

Research on Problem Difficulty

Most of the early research on addition and subtraction investigated children's ability to compute and was concerned exclusively with identifying which problems children could and could not solve. The largest group of early studies attempted to rank the relative difficulty of the 100 addition and subtraction number combinations (basic facts). The motivation for this research was to provide guidelines for organizing instruction. More difficult number combinations would presumably be taught later and would be allocated more instructional time.

The general findings of these studies have been summarized elsewhere (Brownell, 1941; Buswell & Judd, 1925; Suppes, Jerman, & Brian, 1968). The only clear consistency in the rankings of different studies is that the difficulty of addition and subtraction combinations increases as the numbers get larger. Although there were relatively high correlations among the rankings of several studies (Murray, 1939), there were many conflicting results and inconsistencies (Brownell, 1941).

In addition to attempting to generate a linear ranking of the number facts, some studies attempted to identify factors that account for the relative difficulty of different number combinations. Knight and Behrens (1928) hypothesized that addition facts in which the larger addend is given first (e.g., 6 + 3) would be easier than the corresponding pair in which the smaller addend appears first (3 + 6). Their results and the results of other studies failed to confirm the hypothesis. Browne (1906) and Pottle (1937) contended that problems in which both addends were even were easier than problems in which both addends were odd, which in turn were easier than problems involving one even and one odd addend. Other studies, however, did not support this hypothesis either (e.g., Murray, 1939). In general the only consistent pattern that emerges is that doubles (e.g., 8 + 8) are easier than are other combinations with comparable-size addends.

Some of the inconsistency may result from sampling differences. Studies included subjects ranging in age from primary school children, undergoing initial instruction on number combinations, through college graduates. A related problem with most of these studies is that the rankings were based on errors or response latencies of upper-grade children or adults who had received instruction on basic facts. Consequently, the rankings were a contaminated measure of inherent difficulty of different combinations and reflected instruction and unknown amounts of practice with different combinations. To control the effects of instruction and previous practice, Knight and Behrens (1928) used children who had no previous school instruction in either addition or subtraction, and based their ranking on the amount of time required to master each number combination. Although this procedure provided a reasonably accurate measure of practice effect, the amount of time required for mastery was still potentially a function of instruction.

Brownell (1941) and Brownell and Carper (1943) severely criticized the basic assumptions underlying research on the ranking of difficulty of basic facts. They argued that there is no intrinsic order of difficulty among number combinations.

> It may be assumed that all difficulty rankings are authentic for the conditions under which they were obtained and for the techniques by which they were determined. And this is precisely why research to ascertain *the* comparative difficulty of the combinations has been unprofitable. There is no such thing as ''intrinsic'' difficulty in the number facts; their difficulty is relative, contingent upon many conditions, chief of which is method of teaching, or stated differently, the number, order, and nature of learning experiences on the part of pupils [Brownell, 1941, p. 127].

Brownell's attack seems to have dampened enthusiasm for this line of research, and after 1941 there were relatively few studies attempting to rank-order basic number-facts. Researchers have continued to be interested in the relative difficulty of different problems, but the emphasis has shifted to structural features of the problems or the conditions under which problems are administered.

One of the variables of interest has been the type of open sentence given in the problem. By varying the unknown, a total of six addition and subtraction open sentences can be generated (see Table 2.1). Studies by Beattle and Deichmann (1972), Grouws (1972), Hirstein (1979), and Weaver (1971) produced generally consistent results regarding the relative difficulty of different sentence types. The four studies all involved children in Grades 1, 2, and 3 and included numbers from the basic fact domain that resulted in whole-number solutions. The following results were consistent across the studies:

1. *Canonical* addition and subtraction sentences ($a + b = ?$) are less difficult than are *noncanonical* sentences ($a + ? = c$).
2. Canonical subtraction sentences are generally more difficult than are canonical addition sentences.
3. There are no clear differences in difficulty among the following three sentences: ($a + ? = c$; $? + b = c$; and $a - ? = c$).
4. The missing minuend sentence ($? - b = c$) is significantly more difficult than are the other five sentences.
5. Sentences with the operation on the right side of the equal sign ($c = a + ?$) are significantly more difficult than parallel problems with the operation on the left.

Consistent results have also been found regarding the availability of concrete or pictorial aids in solving simple addition and subtraction word problems (Bolduc, 1970; Carpenter & Moser, 1982; Gibb, 1956; Hebbeler, 1977; Ibarra & Lindvall, 1979; LeBlanc, 1971; MacLatchy, 1933; Marshall, 1977; Shores & Underhill, 1976; Steffe, 1970; Steffe & Johnson, 1971). Performance on problems for which concrete objects or pictures were available was consistently

TABLE 2.1

Open Sentence Types[a]

$a + b = ?$	$? = a + b$
$a + ? = c$	$c = a + ?$
$? + b = c$	$c = ? + b$
$a - b = ?$	$? = a - b$
$a - ? = c$	$c = a - ?$
$? - b = c$	$c = ? - b$

[a]Numbers are represented by a, b, and c.

higher than performance on parallel problems without aids, although the difference was less pronounced for older children. The one exception to this rule was reported by Steffe, Spikes, and Hirstein (1976). They suggested that the superior performance reported in most studies for children who used manipulative objects may result from instruction that encourages the use of physical objects and discourages the use of fingers. In other words, the use of manipulative objects rather than finger counting may not represent an intrinsically superior method of solution but may simply be more effective because it has been taught and practiced. This is essentially the same argument raised by Brownell (1941) regarding the research attempting to rank-order basic number facts.

Probably the most sophisticated attempts to account for the relative difficulty of different problems has involved word problems. Because there are potentially many more variables to be concerned about than with purely symbolic problems, this research is much more complex. The most systematic research on word problem difficulty involved the construction of linear regression models (Jerman, 1973; Jerman & Mirman, 1974; Jerman & Rees, 1972; Suppes, Loftus, & Jerman, 1969). The fundamental objective of this line of research has been to build comprehensive models to predict the difficulty of different problems. Individual studies have been able to account for as much as 87% of the variance in problem difficulty in terms of specific measures of number of words in the problem, sentence complexity, and a variety of other syntactic and structural variables. The results, however, have not been uniformly consistent across studies or even between grades in a single study (Jerman & Mirman, 1974).

In general, research that has focused narrowly on item difficulty has not produced a coherent body of knowledge that accounts for how children learn arithmetic. Early research on addition and subtraction that focused on problem difficulty was based on psychological theories that described learning as a direct product of the external environment. It is now generally acknowledged that it is necessary to take into account internal cognitive processes to explain complex human behavior. The first widespread attempt to go beyond simple analysis of problem difficulty in describing children's knowledge of arithmetic is provided by the research on early number-concepts based on the theories and research of Piaget.

Piagetian Research

Throughout the 1960s and early 1970s, the dominant theme for research on early number concepts was provided by Piaget's (1952) *The Child's Conception of Number*. At the time, Piaget's influence was so great that it led Flavell (1970) to observe, "Virtually everything of interest that we know about the early growth of number concept grows out of Piaget's pioneer work in the area" (p. 1001). The research that attempted to extend or refute Piaget's theories stood in stark contrast to the early arithmetic research on problem difficulty: Piaget was not explicitly interested in the learning of traditional arithmetic calculations. Instead, he was concerned with the development of basic logical reasoning abilities that underlie children's conception of number.

Most of the basic Piagetian research has been concerned with conservation, seriation, class inclusion, or related concepts that provide a foundation for understanding number per se; less attention has been directed at concepts underlying addition and subtraction. However, Piaget (1952) himself did consider the development of addition and subtraction concepts. He argued that an operational understanding of addition requires that a child recognizes that a whole remains constant irrespective of the composition of its parts. He found a stagewise development of this concept that paralleled the development of conservation. In the initial stage, children did not realize that a set of eight objects divided into two subsets of four objects each was equivalent to a set of eight objects divided into sets containing one and seven objects. In the second stage, children could often solve the task correctly, but only after empirical verification. It was not until the third stage, at about age seven, that children logically recognized that the composition of the set did not affect number in the set (i.e., $4 + 4 = 1 + 7$). Several other experiments, which were concerned with partitioning sets into equally numerous arrays, found similar patterns of development.

Piaget's addition and subtraction experiments have not been as widely replicated as the conservation, seriation, and class-inclusion tasks. However, a series of studies has examined children's basic understanding of addition and subtraction based on Piagetian constructs. The fundamental idea underlying this research is that certain transformations result in a change in quantity and others do not. The spatial transformations in classical conservation experiments do not change the quantities, whereas adding or removing transformations do result in change. Whole-number addition and subtraction are based on the idea that joining elements to a set increases the number of elements in the set and removing elements decreases the number of elements in the set. Starkey and Gelman (1982) reviewed a number of studies that clearly demonstrate that even preschool children understand the effect of joining elements to or removing them from a set. Complete understanding of the relationship between these operations, however, develops over an extended period of time. Children do not initially recognize that the effect of adding elements to a set can be offset by removing the same number of elements or that adding a number of elements to one of two equivalent sets can be compensated for by adding the same number of elements to the other set. These basic principles of addition and subtraction appear to develop in a pattern that is similar to but not synchronous with the development of conservation and other concrete operational concepts.

Although these studies provide some interesting insights regarding the development of basic number concepts, it is not clear how the development of these concepts affects the learning of the traditional arithmetic concepts and skills taught in school. Piaget generally did not try to draw educational implications from his work but the constructs he identified initially appeared to have profound consequences for instruction in arithmetic. Piaget was suggesting that the most elementary concepts of number were not fully developed in children until about 7 years of age, whereas instruction in addition and subtraction, which assumes knowledge of basic number concepts, typically begins at age 6. Furthermore, many of the specific constructs that Piaget identified appeared to be explicitly required to develop meaningful arithmetic operations. For example, the operation of joining two sets that is typically used to represent addition assumes that children understand that the numbers of elements in the two sets are conserved under the joining operation. Addition and subtraction also involve part–whole relationships, which assume an understanding of class inclusion.

A number of studies attempted to establish empirically the relation between the development of conservation, class inclusion, and other Piagetian constructs by using correlational techniques. (See Carpenter [1980a] and Hiebert & Carpenter [1982] for reviews of this research.) These studies have uniformly found high positive correlations between performance on tests of Piagetian tasks and various measures of arithmetic achievement. Positive correlations do not demonstrate, however, that the development of basic constructs identified by Piaget are required to learn addition and subtraction concepts and skills. In fact, children who do not pass traditional tests of conservation, seriation, or class inclusion can learn to add and subtract; and basic Piagetian measures appear to have limited use in explaining children's ability to add and subtract (Hiebert & Carpenter, 1982; Hiebert, Carpenter, & Moser, 1982).

One limitation of Piaget's (1952) theory of how number concepts develop is that he underestimates the significance of such basic quantitive skills as counting, estimating, and subitizing.[1] There is a growing body of research that suggests that the development of basic number concepts involves the integration or increasingly efficient application of such skills (Carpenter, Blume, Hiebert, Anick, & Pimm, 1981; Fuson & Hall, 1982; Gelman & Gallistel, 1978; Klahr & Wallace, 1976; Schaeffer, Eggleston, & Scott, 1974). Much of the current research pictures addition and subtraction as building upon these basic quantitive skills.

[1]*Subitizing* means directly perceiving the numerosity of something, thereby making it unnecessary to count.

Current Research: A Process Approach

Most current research on children's basic number concepts shares with the Piagetian research the concern for internal cognitive processes and attempts to describe not only what problems children can solve but also how they solve them. However, rather than focus on underlying logical abilities, such as number conservation or class inclusion, the research directly examines the processes that children use to solve arithmetic problems that are the core of the mathematics curriculum. This research also differs from Piagetian research in another important respect. In most Piagetian research, children in the early stages of the development of a particular concept are classified on the basis of certain errors or misconceptions that they exhibit. Much of the current research attributes a much higher level of understanding of basic number-concepts to young children. Rather than simply focus on the misconceptions of younger children, current research describes the development of what they can do. The development of basic addition and subtraction concepts is described here in terms of levels of increasingly sophisticated and efficient problem-solving strategies.

Although several early studies were concerned with the strategies children used to solve addition and subtraction problems (notably, Brownell, 1928; Gibb, 1956; Ilg & Ames, 1951), current research provides a much clearer specification of both problem structure and children's solution strategies than was found in earlier studies. Second, both problem structure and children's strategies are described in such a way that a clear connection can be established between different problems and how children solve them. Finally, whereas most earlier researchers were content simply to describe the overt strategies that children used, the current goal is to attempt to describe the internal cognitive processes that are manifested in the overt strategies.

Classification of Problem Structure

The classification of symbolic addition and subtraction problems was discussed in an earlier section of this chapter (see Table 2.1). Although there has not always been uniform agreement as to which distinctions are most important, the dimensions along which symbolic problems differ are reasonably clear. The characterization of addition and subtraction word problems is much more complex. Attempts to characterize word problems have focused either on syntactic variables, the semantic structure of the problem, or some combination of the two. As noted in the review of research on problem difficulty, syntactic variables such as the number of words in the problem, the sequence of information, and the presence of words that cue a particular operation significantly affect problem difficulty. However, most of the available evidence suggests that the semantic structure of a problem is much more important than syntax in determining the processes that children use in their solutions (Carpenter, Hiebert, & Moser, 1981; Carpenter & Moser, 1982). Consequently, most of the recent research on word problems has focused on semantic structure of problems rather than on syntactic variables.

The semantic structure of addition and subtraction problems has been classified and described in a number of ways. One approach distinguished among problems on the basis of whether action was involved in the problem or not (LeBlanc, 1971; Steffe, 1970). A second approach differentiated among problems in terms of the open sentences they represented (Grouws, 1972; Lindvall & Ibarra, 1980; Rosenthal & Resnick, 1974). Both approaches overlook important differences among certain classes of problems.

Recently, a number of researchers actively engaged in research on children's solutions of addition and subtraction problems have adopted a common framework to characterize word problems along dimensions that seem to be the most productive in distinguishing important differences in how children solve the different problems (Carpenter & Mose, 1982; Riley, Grenno, & Heller, 1982). This framework is generally consistent with several earlier classification schemes (Gibb, 1956; Nesher & Katriel, 1978). This analysis proposes four broad classes of addition and subtraction problems: *Change, Combine, Compare,* and *Equalize.*

There are two basic types of Change problems, both of which involve action. In Change–Join problems, there is an initial quantity and a direct or implied action that causes an increase in that quantity. For Change–Separate problems, a subset is removed from a given set. In both classes of problems, the change occurs over time. There is an initial condition at T1, which is followed by a change occurring at T2, which results in a final state at T3.

Within both the Join and Separate classes, there are three distinct types of problems depending upon which quantity is unknown (see Table 2.2). For one type, the initial quantity and the magnitude of the change are given and the resultant quantity is the unknown. For the second, the initial quantity and the result of the change are given and the object is to find the magnitude of the change. In the third case, the initial quantity is the unknown.

Both Combine and Compare problems involve static relationships for which there is no action. Combine problems involve the relationship existing among a particular set and its two, disjoint subsets. Two problem types exist: the two subsets are given and one is asked to find the size of their union, or one of the subsets and the union are given and the solver is asked to find the size of the other subset (see Table 2.2).

Compare problems involve the comparison of two distinct, disjoint sets. Because one set is compared to the other, it is possible to label one set the *referent*

TABLE 2.2

Classification of Word Problems

Change	
Join	Separate
1. Connie had 5 marbles. Jim gave her 8 more marbles. How many marbles does Connie have altogether?	2. Connie had 13 marbles. She gave 5 marbles to Jim. How many marbles does she have left?
3. Connie has 5 marbles. How many more marbles does she need to have 13 marbles altogether?	4. Connie had 13 marbles. She gave some to Jim. Now she has 8 marbles left. How many marbles did Connie give to Jim?
5. Connie had some marbles. Jim gave her 5 more marbles. Now she has 13 marbles. How many marbles did Connie have to start with?	6. Connie had some marbles. She gave 5 to Jim. Now she has 8 marbles left. How many marbles did Connie have to start with?
Combine	
7. Connie has 5 red marbles and 8 blue marbles. How many marbles does she have?	8. Connie has 13 marbles. Five are red and the rest are blue. How many blue marbles does Connie have?
Compare	
9. Connie has 13 marbles. Jim has 5 marbles. How many more marbles does Connie have than Jim?	10. Connie has 13 marbles. Jim has 5 marbles. How many fewer marbles does Jim have than Connie?
11. Jim has 5 marbles. Connie has 8 more than Jim. How many marbles does Connie have?	12. Jim has five marbles. He has 8 fewer marbles than Connie. How many marbles does Connie have?
13. Connie has 13 marbles. She has 5 more marbles than Jim. How many marbles does Jim have?	14. Connie has 13 marbles. Jim has 5 fewer marbles than Connie? How many marbles does Jim have?
Equalize	
15. Connie has 13 marbles. Jim has 5 marbles. How many marbles does Jim have to win to have as many marbles as Connie?	16. Connie has 13 marbles. Jim has 5 marbles. How many marbles does Connie have to lose to have as many marbles as Jim?
17. Jim has 5 marbles. If he wins 8 marbles, he will have the same number of marbles as Connie. How many marbles does Connie have?	18. Jim has five marbles. If Connie loses 8 marbles, she will have the same number of marbles as Jim. How many marbles does Connie have?
19. Connie has 13 marbles. If Jim wins 5 marbles, he will have the same number of marbles as Connie. How many marbles does Jim have?	20. Connie has 13 marbles. If she loses 5 marbles she will have the same number of marbles as Jim. How many marbles does Jim have?

set and the other the *compared set*. The third entity in these problems is the *difference*, or the amount by which the larger set exceeds the other. In this class of problems, any one of the three entities could be the unknown—the difference, the referent set, or the compared set. The larger set can be either the referent set or the compared set. Thus, there exist six different types of Compare problems (see Table 2.2).

The final class of problems, Equalize problems, are a hybrid of Compare and Change problems. There is the same sort of action as found in the Change problems but it is based on the comparison of two disjoint sets. Equalize problems are not commonly found in the research literature or in most American mathematics programs; however, they do appear in the Developing Mathematical Processes (DMP) program (Romberg, Harvey, Moser, & Montgomery, 1974). These problems are also present in experimental programs developed in the Soviet Union (Davidov, 1982) and in Japan (Ginbayashi, 1980). As in the Compare problems, two disjoint sets are compared; then the question is posed, "What could be done to one of the sets to make it equal to the other?" If the action to be performed is on the smaller of the two sets, then it becomes an Equalize–Join problem. On the other hand, if the action to be performed is on the larger set, then an Equalize–Separate problem results. As with Compare problems, the unknown can be varied to produce three distinct Equalize problems of each type (see Table 2.2).

The above analysis of addition and subtraction word problems is limited to simple problems that are appropriate for primary age children. It is not as complete as the framework proposed by Vergnaud (1982) that extends to operations on integers. Although this analysis does not unambiguously characterize all addition and subtraction word problems (Fuson, 1979), it has been useful to help clarify distinctions between problem types and to help distinguish problems with clearly different semantic characteristics from problems that merely differ in terminology or syntax. Furthermore, the results to be discussed in the following sections of this chapter indicate that children's solution processes clearly reflect these distinctions between problem types.

Processes of Solution

The following discussion is limited to the processes that children use to solve addition and subtraction problems with small numbers that do not require the use of formal algorithms. For a discussion of algorithmic solutions, see Brown and Van Lehn (1982) and Resnick (1982).

Identifying the processes that children use to solve simple addition and subtraction problems is not an easy task. Internal cognitive processes cannot be observed directly, and the problems are sufficiently simple that the children themselves often are not aware of how they solved a given problem. Three basic paradigms have been used to study the processes that children use to solve simple addition and subtraction problems.

The most straightforward has involved the use of individual interviews (Carpenter & Moser, 1982; Steffe, Thompson, & Richards, 1982). Children are individually presented problems and, through a combination of observation of overt behavior and probing questions, the interviewer infers how a child solved a particular problem. There are a number of limitations to the kinds of inferences that can be drawn from interviews. Most serious is that children's explanations of how they solved a problem may not accurately reflect the processes that they actually used to solve it. The interview procedures may change how a child solves a problem, or a child may have difficulty articulating the process that was really used and therefore describe another process that is easier to explain. Or a child may try to second guess what the interviewer is seeking. Another serious problem is that the inferences drawn from an interview involve a great deal of subjective judgment on the part of the experimenter.

Because of these limitations, researchers have sought alternative procedures that do not rely on children's explanations and can be based on more objective measures. One of the more popular techniques is the use of response latencies. Response latencies have been used for a number of years to assess problem difficulty (Arnett, 1905; Knight & Behrens, 1928; Smith, 1921). By hypothesizing models of the time required to solve different problems using particular strategies, researchers have been able to go beyond the simple assessment of problem difficulty and draw inferences about the processes used (Groen & Parkman, 1972; Groen & Poll, 1973; Groen & Resnick, 1977; Rosenthal & Resnick, 1974; Suppes & Groen, 1967; Suppes & Morningstar, 1972; Woods, Resnick, & Groen, 1975). The approach involves breaking operations down into a series of discrete steps. It is assumed that the time required to solve a given problem using a particular strategy is a linear function of the number of steps required to reach a solution. By finding the best fit between response latencies for subjects solving a variety of problems and the regression equations of possible solution processes, the most appropriate model can be inferred. Most of the models tested involve counting strategies that assume that a counter is set to some value and incremented or decremented in steps of one unit.

One of the limitations of most of the research using this paradigm is that the models tested are usually limited to the type of counting processes described above. Clinical studies have identified a number of processes that involve direct modeling with fingers or objects or the use of recalled number facts that do not follow this pattern. There are also a number of assumptions underlying the models that are suspect. One such assumption is that the time required for various steps is constant for different number pairs. Another questionable assumption is that children use a consistent strategy for all problems. Extreme values may result in a reasonably good fit of latency data to the regression equations of a particular strategy. This may give the impression that children consistently use a given strategy, whereas only a piece of the data really fits the model very well. For example, a reanalysis of the data from a study by Groen and Parkman (1972) indicated that Groen and Parkman's best-fitting model was much more appropriate for certain number domains than for others (Siegler & Robinson, 1981).

The third technique that has been used to infer children's solution processes is the analysis of error patterns (Lindvall & Ibarra, 1980; Riley *et al.*, 1982). For simple addition problems involving basic facts, it is usually difficult to infer what processes may be implied by errors on specific problems, as can be done for algorithmic solutions to multidigit problems (Brown & Van Lehn, 1982). However, certain general solution processes will allow students to solve some problems but not others. By examining the patterns of errors over groups of problems, attempts have been made to infer what processes children may be using. Examples of this type of analysis are provided in the following sections of this chapter. The major drawback to this procedure is that errors occur for reasons other than those identified in the models of children's general strategies. Furthermore, differences in wording or context may result in systematic differences that are unrelated to the structure of the problem or to how children solve it.

In short, all of the methods available to assess the process that children use to solve addition and subtraction problems have limitations. For all their weaknesses, individual interviews provide the most direct measure of the processes that children use. But the latency and error data provide valuable supporting evidence. In spite of the limitations of the assessment techniques, a reasonably consistent picture of children's solution processes is beginning to emerge. The processes described below have been observed in a number of studies using both interviews and analysis of response latencies (Brownell, 1928; Carpenter & Moser, 1982; Gibb, 1956; Ginsburg, 1977; Groen & Parkman, 1972; Ilg & Ames, 1951; Woods *et al.*, 1975).

ADDITION STRATEGIES

Three basic levels of addition have been identified: strategies based on direct modeling with fingers or physical objects, strategies based on the use of counting sequences, and strategies based on recalled number facts.

In the most basic strategy (*Counting All*), physical objects or fingers are used to represent each of the addends, and then the union of the two sets is counted, starting with one. Theoretically, there are two ways in which this basic strategy might be carried out. Once the two sets have been constructed, they could be physically joined by moving them together or by adding one set to the other, or the total could be counted without physically joining the sets. This distinction could be important. The first case would best represent the action of the Change–Join problems whereas the second would best represent the static relationship implied by the Combine problems. However, children generally do not distinguish between the two strategies in solving either Change–Join or Combine problems. Thus, it appears that there is a single *Counting All With Models* strategy. The strategy may be accompanied by different ways of organizing the physical objects, but the arrangements do not represent distinct strategies or different interpretations of addition.

A third alternative is also possible. A child could construct a set representing one addend and then increment this set by the number of elements given by the other addend without ever constructing a second set. Such a strategy would seem to represent best a unary conception of addition (Weaver, 1982). This strategy is seldom used.

There are three distinct strategies involving counting sequences. In the most elementary strategy, the counting sequence begins with one and continues until the answer is reached. This strategy, which is also a Counting All strategy, is the SUM strategy identified by Suppes and Groen (1967) and Groen and Parkman (1972). It is similar to the Counting All With Models strategy except that physical objects or fingers are not used to represent the addends. However, this strategy and the two following counting strategies require some method of keeping track of the number of counting steps that represent the second addend in order to know when to stop counting. Keeping track procedures are discussed in detail in Fuson (1982). Most children use their fingers to keep track of the number of counts, but a substantial number give no evidence of any physical action accompanying their counting. When counting is carried out mentally, it is difficult to determine how a child knows when to stop counting. Some children appear to use some sort of rhythmic or cadence counting such that counting words are clustered into groups of two or three. Others explicitly describe a double count (e.g., 6 is 1, 7 is 2, 8 is 3), but children generally have difficulty explaining this process.

When fingers are used, they appear to play a very different role than in the direct modeling strategy. In this case, the fingers do not represent the second addend per se, but are used to keep track of the number of steps incremented in the counting sequence. When using fingers, children often do not appear to have to count their fingers, but can immediately tell when they have put up a given amount of fingers.

The other two counting strategies are more efficient and imply a less mechanical application of counting. In applying these strategies, a child recognizes that it is not necessary to reconstruct the entire counting sequence. In the *Counting On From First Strategy*, a child begins counting forward with the first addend in the problem. The *Counting On From Larger* strategy is identical except that the child begins counting forward with the larger of the two addends. This strategy is the MIN strategy of Groen and Parkman (1972).

Although learning of basic number facts appears to occur over a protracted span of time, most children ultimately solve simple addition and subtraction problems by recall of number combinations rather than by using counting or modeling strategies. Certain number combinations are learned earlier than others, and before they have completely mastered their addition tables many children use a small set of memorized facts to derive solutions for addition and subtraction problems involving other number combinations. These solutions usually are based on doubles or numbers whose sum is 10. For example, to solve a problem representing $6 + 8 = ?$ a child might respond that $6 + 6 = 12$ and $6 + 8$ is just two more than 12. In an example involving the operation $4 + 7 = ?$ the solution may involve the following analysis: $4 + 6 = 10$ and $4 + 7$ is just 1 more then 10.

SUBTRACTION STRATEGIES

The three levels of abstraction described for addition strategies also exist for the solution of subtraction problems. However, whereas a single basic interpretation of addition has been the rule, a number of distinct classes of subtraction strategies have been observed at the direct modeling and counting levels.

One of the basic strategies involves a subtractive action. In this case, the larger quantity in the subtraction is initially represented and the smaller quantity is subsequently removed from it. When concrete objects are used, the strategy is called *Separating From.* The child constructs the larger given set and then takes away or separates, one at a time, a number of objects equal to the number given in the problem. Counting the set of remaining objects yields the answer. There is also a parallel strategy based on counting called *Counting Down From.* A child initiates a backward counting sequence beginning with the given larger number. The backward counting sequence contains as many counting number words as the given smaller number. The last number uttered in the counting sequence is the answer.

The *Separating To* strategy is similar to the Separating From strategy except that elements are removed from the larger set until the number of objects remaining is equal to the smaller number given in the problem. Counting the number of objects removed provides the answer. Similarly, the backward counting sequence in the *Counting Down To* strategy continues until the smaller number is reached and the number of words in the counting sequence is the solution of the problem.

The third pair of strategies involves an additive action. In an additive solution, the child starts with the smaller quantity and constructs the larger. With concrete objects (*Adding On*), the child sets out a number of objects equal to the smaller given number (an addend). The child then adds objects to that set one at a time until the new collection is equal to the larger given number. Counting the number of objects added on gives the answer. In the parallel counting strategy (*Counting Up From Given*), a child initiates a forward counting strategy beginning with the smaller given number. The sequence ends with the larger given number. Again, by keeping track of the number of counting words uttered in the sequence, the child determines the answer.

The fourth basic strategy is called *Matching*. Matching is only feasible when concrete objects are available. The child puts out two sets of cubes, each set standing for one of the given numbers. The sets are then matched one-to-one. Counting the unmatched cubes gives the answer.

A fifth strategy (*Choice*), involves a combination of Counting Down From and Counting Up From Given, depending on which is the most efficient. In this case, a child decides which strategy requires the fewest number of counts and solves the problem accordingly. For example, to find $8 - 2$, it would be more efficient to Count Down From whereas the Counting Up From Given strategy would be more efficient for $8 - 6$. Although most strategies can be observed on a single problem, the Choice strategy can only be inferred from performance on several problems with appropriate number combinations.

As with addition, modeling and counting strategies eventually give way to the use of recalled number facts or derived facts. Children's explanations of their solutions suggest that the number combinations they are calling upon are often addition combinations. To explain how they found the answer $13 - 7$, many children respond that they just knew that $7 + 6 = 13$. Many of the derived subtraction facts are also based on addition. For example, to explain the solution to find $14 - 8$, one child reported, "Seven and 7 is 14; 8 is 1 more than 7; so the answer is 6" (Carpenter, 1980b, p. 319).

Problem Structure and Solution Processes

SUBTRACTION

As can be seen from the descriptions of problem structure and children's processes, certain of the strategies naturally model the action described in specific subtraction problems. The Separate–Result Unknown problems (Table 2.2, Problem 2) are most clearly modeled by the Separating and Counting Down From Given strategies, whereas the Separate–Change Unknown problems (Problem 4) are best modeled by the Separating To and Counting Down To strategies. On the other hand, the implied joining action of the Join–Change Unknown problem (Problem 3) is most closely modeled by the Adding On and Counting

Up strategies. Compare–Difference Unknown problems (Problems 9 and 10) deal with static relationships between sets rather than action. In this case, the Matching strategy appears to provide the best model.

For the Combine subtraction problem, the situation is more ambiguous. Because Combine problems have no implied action, neither the Separating nor the Adding On strategy, exactly models the given relationship between entities. Because one of the given entities is a subset of the other, there are not distinct sets that can be matched.

For Equalize problems the situation is reversed. Because Equalize problems involve both a comparison and some implied action, two different strategies would be consistent with the problem structure. The Equalize–Join problem (Problem 15) involves a comparison of two entities and a decision about how much should be joined to the smaller to make them equivalent. Either the Matching or the Adding On (Counting Up From Given) strategy might be appropriate. For the Equalize–Separate problems (Problem 16), the implied action involves removing elements from the larger set until the two sets are equivalent. This action seems to be best modeled by the Separating To strategy, although the Matching strategy is again appropriate for the comparison aspect of the problem.

The results of a number of studies consistently show that young children have a variety of strategies available to solve different subtraction problems and that the strategies used generally tend to be consistent with the action or relationships described in the problem. This tendency is especially pronounced for children below the second grade, but for some children the structure of the problem influences the choice of strategy at least through third grade.

The results summarized in Table 2.3 are from a 3-year longitudinal study of the processes that children use to solve basic addition and subtraction word problems. Details of the design and procedures used in the study can be found in Carpenter and Moser (1979). The study involved approximately 100 children who were individually interviewed, three times a year in the first and second grades and twice in the third grade. The study included problems similar to Problems 1, 2, 3, 7, 8, and 9 in Table 2.2. Problems were administered under a variety of conditions. In some instances, manipulative objects were available; in others, they were not. Four separate number domains were included: basic facts with sums less than 10, basic facts with sums between 11 and 16, two-digit numbers that did not require regrouping to use an addition or subtraction algorithm, and two-digit numbers that required regrouping. The results reported in Table 2.3 are for problems involving subtraction facts with the larger number between 11 and 16. Manipulative objects were available to aid in the solution, but children were not required to use them. To simplify the table, only results for the January interviews are included. At the time of the Grade 1 interview, the children in the study had received no formal instruction in addition and subtraction. By the second-grade interview, they had received about 8 months of in-

TABLE 2.3

Relation of Strategy to Problem Structure: Longitudinal Study Results[a]

			Strategy[b]						
			Subtractive		Additive			Numerical	
Problem	Grade	Percentage correct	Separate from	Count down from	Add on	Count up from given	Match	Recalled fact	Derived fact
Separate–Result Unknown	1	61	68	1	1	3	0	1	2
	2	83	34	8	1	10	0	20	9
	3	95	9	3	1	12	0	54	13
Join–Change Unknown	1	57	2	0	42	12	1	2	4
	2	93	1	2	18	31	0	25	16
	3	95	0	1	6	27	1	48	14
Compare–Difference Unknown	1	41	8	0	3	9	30	1	1
	2	70	11	6	2	17	14	19	7
	3	89	3	3	2	14	2	52	17
Combine–Part Unknown	1	45	45	0	4	3	0	2	2
	2	78	36	5	0	11	0	20	14
	3	91	6	1	0	13	0	53	18

[a]Results from the January interview. Number combinations involved seems between 11 and 16. Manipulative objects were provided. Source: Carpenter and Moser (1983).

[b]Percentage responding.

struction in addition and subtraction. Progress toward mastery of number facts was expected, but there had been no instruction on the two-digit subtraction algorithm. By the third-grade interview, students were expected to have learned their number facts and to have learned the addition and subtraction algorithms.

In Grade 1, the vast majority of responses were based on problem structure. Almost all of the first graders who solved the problems correctly used the Separating From or the Counting Down From strategy for the Separate problem and the Adding On or Counting Up From Given strategy for the Join–Change Unknown problem. The results were not quite so overwhelming for the Compare problem, but the Matching strategy was used by the majority of children who solved the problem correctly. Furthermore, it was the only problem for which more than two children used a Matching strategy. The results for the Combine problem, for which there is no clear action to represent, generally tended to parallel those of the Separate problem.

By the second grade, about a third of the responses were based on number facts and the effect of problem structure was not quite so dominant; however, the structure of the problem continued to influence the responses of a large number of second graders. Forty-two percent of the second graders used a subtractive strategy to solve the Separate problem, whereas only 11% used an additive strategy. For the Join problem, 49% used an additive strategy and only 3% used a subtractive strategy. Thus, for these two problems, most of the children who used a counting or modeling strategy continued to represent the action described in the problem. The structure of the Compare problem did not continue to exert as strong an influence, and many second graders abandoned the Matching strategy for the more efficient Separating From or Counting Up From Given strategies.

By the third grade, about two-thirds of the responses were based on number facts, and there was more flexibility in the use of counting strategies. For the Separate problem, the most popular strategy apart from recalled number facts was Counting Up From Given. The Matching strategy was seldom used for the Compare problem. For the Join problem, however, almost all children who did not use number facts used an additive strategy. This represented about a third of the third-grade children in the study.

There are two plausible explanations for the continued reliance on additive strategies for the Join problem. The Counting Up From Given strategy may simply be the most efficient strategy available to some third graders to solve subtraction problems of any kind. In other words, for some children, the choice of a Counting Up From Given strategy for the Join problem may not have been dictated by the additive structure of the Join problem. The Counting Up From Given strategy may simply be the strategy they use to solve subtraction problems. The fact that 12–14% of the third graders used the Counting Up From Given strategy for the other three subtraction problems supports this hypothesis.

On the other hand, almost twice as many third graders used the additive strategy for the Join problem as for the other three problems.

The influence of the additive structure of the Join problem is apparent in the solution of two-digit problems, which are most efficiently solved using the subtraction algorithm. By the time of the January interview, the third-grade students in the study were expected to have mastered the subtraction algorithm. For the other five types of problems, between 80 and 95% of the children used the standard addition or subtraction algorithm; but for the Join–Change Unknown problem, only 53% used the subtraction algorithm. Almost a third used some form of additive strategy that paralleled the structure of the problem.

Results of other studies investigating the processes that children use to solve word problems are generally consistent with those of the longitudinal study reported above; (Anick, in preparation; Blume, 1981; Carpenter, Hiebert, & Moser, 1981; Hiebert, 1982). These studies also included problems not administered in the longitudinal study. Perhaps the most compelling evidence for the effect of problem structure is found on Separate–Change Unknown problems. The strategy that best represents the action described in these problems is the Separating To strategy. In general, this strategy appears somewhat inefficent and unnatural. The strategy involves removing elements from a set until the number remaining is equal to a given value. With the Separating From and Adding On strategies, the elements that are removed or added can be sequentially counted as they are removed or added. Within the Separating To strategy, however, the size of the remaining set must regularly be reevaluated. Furthermore, the Separating To strategy is virtually never taught explicitly. In spite of these limitations, Anick (in preparation) and Hiebert (1982) found that the Separating To strategy was used by approximately half of the children in their studies who were able to solve the Separate–Change Unknown problem.

Results for the Equalize problems also generally followed the predicted pattern (Carpenter, Hiebert, & Moser, 1981). The one case in which results were not consistent across studies involves the use of the Matching strategy. Matching was the primary strategy for Compare problems in both the longitudinal study and in Carpenter, Hiebert, and Moser (1981). It was almost never found, however, in the studies of Anick (in preparation) and Riley *et al.* (1982). In the studies in which Matching was used, the mathematics program was Developing Mathematical Processes (DMP) (Romberg *et al.,* 1974). This program provides early experience comparing the relative size of two sets by matching. It appears that children who have used matching to compare sets can extend this process to find the magnitude of the difference in Compare problems without explicit instruction. If children have no experience matching sets, they do not spontaneously apply the process to Compare problems. In this case, they have no way to represent the relationship described in the Compare problems. As a consequence,

Compare problems are very difficult for these children (Anick, in preparation; Riley *et al.*, 1982).

These results clearly illustrate the importance of examining children's solution processes. Although there were significant differences between the two sets of studies in the level of success for Compare problems, both sets of results are generally consistent with the conclusion that children's earliest solution processes for word problems are based on modeling the action or relationships described in the problem. In one instance, children had available a process to model the relationship described in Compare problems; in the other, they did not.

The results for the Compare problem demonstrate that if children do not have a process available to model the action or relationships in a given problem, then the problem is much more difficult than if it can be directly modeled. Certain types of problems are difficult to model. For example, in the Change–Start Unknown problems (Table 2.2, Problems 5 and 6), the unknown is the initial entity that is operated on to yield a given result. To model directly the action in a Change problem requires that there be an initial set to be either increased or decreased. To model the action in the Start Unkown problems would require some sort of trial and error in which one guessed at the size of the initial set and performed the specified transformation to check whether it produced the given result. Rosenthal and Resnick (1974) hypothesized that children may use such a process. Hiebert (1982) found that some children do use a trial and error strategy with Start Unknown problems, but most children are unable to model these problems at all. Recent data from Anick (in preparation) support this conclusion.

Because of the difficulty of modeling Start Unknown problems, they should be significantly more difficult to solve than are problems that can be modeled directly. The analysis of the relative difficulty of different types of word problems indicates that problems that cannot be easily modeled are significantly more difficult that those that can. The results of a study reported by Riley *et al.* (1982) are summarized in Table 2.4. The Start Unknown problems were significantly more difficult than the other four Change problems. Because subjects in the study had not been exposed to the Matching strategy, all of the Compare problems were difficult. The other problem that proved to be especially difficult was the Combine–Part Unknown problem. Because there is no clear action to be modeled in the Combine problem, these results are also consistent with the hypothesized effect of problem structure.

Differences in wording of problems, experimental protocols and procedures, and students' backgrounds complicated the comparison of different studies examining problem difficulty. However, although most studies have not found the clear differences reported by Riley *et al.* (1982), other studies support the conclusions that Start Unknown problems are more difficult than are other Change problems (Anick, in preparation; Ibarra & Lindvall, 1979) and that Compare and

TABLE 2.4

Relative Difficulty of Word Problems[a]

Problem type	Problem number[b]	Grade K	Grade 1	Grade 2	Grade 3
Joint–Result Unknown	1	.87[c]	1.00	1.00	1.00
Separate–Change Unknown	2	1.00	1.00	1.00	1.00
Join–Change Unknown	3	.61	.56	1.00	1.00
Separate–Change Unknown	4	.91	.78	1.00	1.00
Join–Start Unknown	5	.09	.28	.80	.95
Separate–Start Unknown	6	.22	.39	.70	.80
Combine Addition	7	1.00	1.00	1.00	1.00
Combine Subtraction	8	.22	.39	.70	1.00
Compare Difference Unknown	9	.17	.28	.85	1.00
Compare Difference Unknown	10	.04	.22	.75	1.00
Compare Compared Quantity Unknown	11	.13	.17	.80	1.00
Compare Compared Quantity Unknown	14	.17	.28	.90	.95
Compare Referent Unknown	13	.17	.11	.65	.75
Compare Referent Unknown	12	.00	.06	.35	.75

[a]Number combinations with sums less than 10. Manipulative objects were provided. Source: Riley (1981).
[b]Problem number in Table 2.2.
[c]Proportion correct.

Combine subtraction problems are relatively difficult (Anick, in preparation; Gibb, 1956; Nesher, 1982; Schnell & Burns, 1962; Shores & Underhill, 1976).

ADDITION

Whereas children have multiple conceptions of subtraction, reflected in different solution processes for different problems, they appear to have a reasonably unified concept of addition. As noted above, subtle differences in the processes used could reflect differences in the structure of Join and Combine addition problems, but children fail to make these distinctions and appear to treat Join and Combine addition problems as though they were equivalent. Not only are the same basic processes used for both problems, but the same pattern of responses appears for both. The similarity of responses for the two types of problems is illustrated by the results from the longitudinal study of Carpenter and Moser (1982) summarized in Table 2.5. The results are from the September interviews for problems with sums between 11 and 16 in which manipulative objects were not available.

The fact that there is no difference in children's solutions to Join and Combine addition problems does not mean that there is no difference in performance on

TABLE 2.5

Combine and Join Addition Problems: Longitudinal Study Results[a]

			Strategy[b]				
Grade	Problem	Percentage correct	Count all	Count on from first	Count on from larger	Derived fact	Recalled fact
1	Combine	50	52	3	3	1	1
	Join	47	46	3	8	2	1
2	Combine	72	39	6	29	4	7
	Join	84	41	14	26	6	6
3	Combine	91	13	7	33	11	30
	Join	90	11	15	32	9	32

[a]Results from September interview. Number combinations involved sums between 11 and 16. Manipulative objects were not provided. Source: Carpenter and Moser (1983).

[b]Percentage responding.

any of the addition word problems. The Separate–Start Unknown and the Compare addition problems are significantly more difficult than are the Join or Combine addition problems (see Table 2.4). The action and relationships described in the Separate and Compare addition problems are not directly modeled by the addition strategies children have available. This lack of congruence between problem structure and available solution strategies makes these problems more difficult than the Join and Combine addition problems for which the Counting All and Counting On strategies provide a reasonable model.

Thus, children appear to use the same basic processes to solve all types of addition problems (Carpenter, *et al.*, 1981). However, the problem structure also influences children's solutions of addition problems. The problems that can be reasonably modeled by the available processes are relatively easy, whereas those that cannot are significantly more difficult.

Development of Addition and Subtraction Processes

Much of the early research on addition and subtraction focused on factors affecting problem difficulty. This approach provided a relatively static view of children's performance. Certain problems were identified as being more difficult than others, but generally there was no attempt to describe these differences in difficulty in terms of a developmental hierarchy in which the acquisition of ability to solve more difficult problems built upon the abilities used to solve the easier problems.

A primary focus of current research is to describe how addition and subtraction concepts are acquired over time. This emphasis reflects the influence of Piaget; however, whereas Piaget's description of development focuses on children's limitations and misconceptions at early stages of development, much of the current research describes a sequence of development of addition and subtraction concepts whereby perfectly valid solution processes are replaced by increasingly efficient and abstract processes.

There is reasonably clear evidence that there is a level at which children can solve addition and subtraction problems only by directly modeling with physical objects or fingers the action or relationships described in the problem. At this level, children cannot solve problems that cannot be directly modeled. The results for the lowest grade levels reported in Tables 2.3, 2.4, and 2.5 provide reasonably compelling evidence for the existence of this level in children's acquisition of addition and subtraction concepts and skills. The great majority of responses of the youngest children were limited to direct modeling strategies (Tables 2.3 and 2.5), and few of the younger children could solve any problem that could not be directly modeled (Table 2.4).

The development of more advanced levels of children's solution processes proceeds along two dimensions: an increase in the levels of abstraction and an increase in the flexibility of choice in strategy.

LEVEL OF ABSTRACTION

At the most primitive level at which children can solve addition and subtraction problems, they completely model the action or relationships in the problem using physical objects or fingers. In this case, they actually construct sets to represent all the quantities described in the problems. The Counting All strategy is the addition strategy used at this level. The parallel subtraction strategies are Separating From, Separating To, Adding On, and Matching.

At the next level, the external direct modeling actions of this initial concrete level are internalized and abstracted, allowing greater flexibility and efficiency. At this level of abstraction, children no longer have to represent physically each of the quantities described in the problem. Instead they are able to focus on the counting sequence itself. They realize that in order to find the number of elements in the union of two sets they do not actually have to construct the sets or even go through the complete counting sequence. They can start at a number representing one quantity and count on the number of counts representing the other quantity. Although children do use finger patterns in conjunction with the Counting On strategy, fingers are used in a very different sense than at the direct modeling level. Children's explanations of their use of Counting On suggest that the finger patterns do not represent the second set per se but are simply a tally of the number of steps counted on. The abstraction of the subtraction strategies essentially involves the same basic pattern of development except that in the case of the Separating strategies a backward counting sequence is required.

The shift from complete modeling to counting strategies depends upon the development of certain basic number concepts and counting skills (Fuson & Hall, 1983; Steffe, Thompson, & Richards, 1982). Fuson (1982) has argued that counting-on depends on understanding basic principles involving the relation between cardinality and counting, and on recognizing that each addend plays a double role in that it is both an addend and a part of the sum. The counting skills required include the ability to begin a counting sequence at any number; the ability to maintain a double count; and in the case of subtraction, the ability to count backwards.

The role of these concepts and skills in the development of the Counting On strategy is discussed in detail in Fuson (1982) and Secada, Fuson, and Hall (1983). The available evidence suggests that counting strategies are not simply mechanical techniques that children have learned in order to solve addition and subtraction problems. They are conceptually based strategies that directly build upon the direct modeling strategies of the previous level. The abstraction and flexibility demonstrated by their application imply a deeper understanding of number and addition and subtraction than was found at the direct modeling level, but this understanding is based upon the conceptualizations of the initial concrete actions. In other words, the levels are not independent; one builds on the other.

There is a clear parallel between the development of counting strategies and the stages of development described by Piaget. In both cases, the operations implied by external concrete actions are internalized, providing for greater flexibility in their application.

Children continue to use counting strategies for an extended period of time. Lankford (1972) found that as many as 36% of the seventh graders he interviewed continued to use counting strategies to find some basic addition and subtraction facts. Furthermore, children become so proficient and quick in the use of the counting strategies and so covert in the use of their fingers as tallying devices that it is often difficult to distinguish between the use of a counting strategy and recall of a basic number fact. However, to rely on counting strategies to solve more complex problems requiring algorithms is inefficient and provides too much opportunity for miscalculation. To learn number facts at a recall level remains a viable goal of the mathematics curriculum, and most students eventually attain this level.

There is not a clearly distinct broad shift from counting strategies to use of number facts. As the research on difficulty of different number facts indicated, some facts are learned and used earlier than others; there is a long period during which children use a combination of number facts in conjunction with direct modeling or counting strategies.

Not a great deal is known about how the use of counting strategies evolves into or affects the learning of basic number facts at the recall level. Leutzinger (1979) investigated the effect of the Counting On strategy on the learning of basic addition facts, and other studies have provided instruction in counting strategies along with a broad range of other types of strategies to provide some structure to facilitate recall of basic facts (Swenson, 1949; Theile, 1938; Thornton, 1978). Aside from the finding that such instruction has generally proved effective, it is still not clear exactly how the recall of number facts is related to children's counting strategies.

The relation between counting strategies is one issue in the learning of basic facts. A second is the relation between different facts. As noted earlier, children occasionally use known facts to derive facts that they do not know at the recall level. Results from the longitudinal study (Carpenter, 1980b) indicate that the use of derived strategies is not limited to a select group of superior students. By the end of first grade, over half the students had used a derived strategy at least once, and by the middle of the second grade, over three-fourths of the students had done so. Children who derived strategies, however, did not use them consistently. For the smaller number, only one child ever used a derived strategy more than three times on the 12 problems administered. For the larger number, only one first-grade child used more than three derived strategies. Four second graders used more than three derived strategies at the September interview and 12 did so in January.

It is tempting to assume that derived strategies are used during a transitional level between the use of counting strategies and the routinization of number facts. The data from the longitudinal study suggest, however, that things are not that simple. So far we have been unable to establish a clear connection between the use of derived strategies and the levels of development of modeling and counting strategies. Derived strategies are occasionally used by children relying on the most primitive modeling and counting strategies.

Some studies have claimed success for explicit instruction in strategies that can be used to derive unknown facts from known facts (Swenson, 1949; Theile, 1938; Thornton, 1978). But the role that derived strategies play in the learning of basic facts at the recall level is far from clear (Rathmell, 1979; Steffe, 1979).

GENERALIZABILITY OF LEVELS OF ABSTRACTION

One of the questions regarding the development of different levels of abstraction is whether the use of more advanced strategies is broadly based across all problems or whether children may use more advanced strategies with some problems that with others. Results from the longitudinal study (Carpenter & Moser, 1982) suggest that there are different patterns of increasing abstraction for different strategies and problems. The Counting Up From Given strategy is used much earlier and more frequently than are the Counting Down strategies. For the Separate subtraction problem, no more than 15% of the children use the Counting Down From strategy for any interview. The Counting Up From Given strategy, on the other hand, accounted for as much as 50% of the responses to the Join–Change Unknown problems. In fact, it appears that some children may never use a Counting Down strategy.

There does seem to be a relation between the use of Counting On strategies for addition and the Counting Up From Given strategy for subtraction. Many children appear to see them as essentially the same strategy. A number of children, when asked to explain their use of the Counting Up From Given strategy, said that they did the same thing that they had done on the previous problem, which was an addition problem that had been solved by Counting On.

CHOICE OF STRATEGY

The second dimension along which development occurs is the flexibility of choice of strategy. At first, the only problems that young children can solve are those for which the action or relationships described in the problem can be directly modeled. By the second or third grade, however, many children are able to use strategies that are not entirely consistent with the structure of the problem (Tables 2.3 and 2.4).

As with the case of the shift to more abstract counting strategies, children's flexibility in the choice of strategy is not consistent over problems. Children soon abandon the somewhat complicated Matching strategy for the Compare problem but are much less flexible in the choice of strategy for the Separate and Join problems. The results in Table 2.3 indicate that even though children were much more successful in applying Counting Up From Given than they were with Counting Down strategies, fewer than 15% ever used Counting Up From Given to solve Separate problems.

The flexibility in choice of strategy also affects children's ability to solve numerical open sentence problems. Woods *et al.* (1975) and Groen and Poll (1973) suggest that the increasing flexibility to choose between strategies is also reflected in the choice of strategy for solving numerical subtraction problems. They present response latency data that they argue is best explained by the Choice strategy, which involves choosing either a Counting Up From Given or a Counting Down From strategy, depending on which requires fewer steps.

The support for the widespread use of such a strategy has not been uniformly consistent. Although it does seem to present the best fit with the data from response latency studies, Blume (1981) found little evidence of such a strategy in clinical interviews. He found that most young children consistently use a Separating From strategy to solve problems of the form $a - b = ?$ and the Counting Up From Given strategy to solve problems of the form $a + ? = c$. Intuitively, it would seem that children would be more likely to use such a strategy with problems in which the chosen strategy would involve few steps (e.g., 11 − 3, or 11 − 9), but not in situations in which the choice was unclear (e.g., 12 − 5). One also has to be cautious in interpreting the data in support of the Choice strategy. Some of the children to whom the Choice strategy has been attributed were as old as the fourth grade. Most children of this age have memorized the basic facts used in these studies. Consequently, the latency data may reflect something other than the overt use of counting strategies.

To have a completely developed concept of subtraction, children should recognize the equivalence of the different strategies. This knowledge should make the Choice strategy possible. But the evidence suggests that many children avoid Counting Down strategies. Consequently, some caution should be exercised in assuming that there is widespread use of the Choice strategy over a broad class of problems.

Relation between Dimensions of Development

The previous section of this chapter described the increase in flexibility and abstraction that occurs over time in children's processes for solving addition and subtraction problems. An important question is whether certain levels of abstraction require a more flexible choice of strategy, or vice versa. Attempting to characterize the relation between these two dimensions over which children's strategies evolve is complicated by the fact that children do not consistently use their optimal strategies. For example, throughout first grade, children in the longitudinal study solved Join–Change Unknown problems by using counting strategies, rather than direct modeling strategies, almost twice as often when cubes were not available as when they were. In fact, direct modeling strategies were used more frequently for problems with smaller numbers than for problems with larger numbers.

The available evidence suggests, however, that there is a shift in the level of abstraction of children's solution processes before they begin to recognize the equivalence of different subtraction strategies. In other words, the first evidence of growth is that children begin to use counting strategies that parallel the action in a problem rather than to model completely the action with objects. The results summarized in Table 2.6 indicate that almost half of the first-grade children could use a counting strategy rather than completely model the problem. On the other hand, most children used either a modeling or a counting strategy that directly represented the action in the problem.

The relationship between children's understanding of the equivalence of different addition and subtraction strategies and their ability to use either derived or recalled number facts is a bit more different to establish. Many derived number

TABLE 2.6

Results for Selected Separate and Join Subtraction Problems[a]

Problem	Percentage correct	Strategy[b]				
		Separating	Count down from	Add on	Count up from given	Derived–recalled number fact
Separate–Result Unknown	45	23	14	1	8	15
Join–Change Unknown	61	1	0	14	43	13

[a]Results from Grade 1 June interview. Number combinations involved the difference of a two-digit and a one-digit subtraction fact. Manipulative objects were provided. Source: Carpenter and Moser (1983).
[b]Percentage responding.

facts are based on understanding relationships between addition and subtraction that would suggest an understanding of the equivalence of different subtraction strategies. Similarly, for many children recall of subtraction number facts is based on their knowledge of addition facts. On the other hand, many children learn some number facts and generate derived facts before they give any evidence of being able to use modeling or counting strategies that are not consistent with the structure of the problem.

In fact, some children's ability to solve problems that cannot be directly modeled may be based on their ability to relate the problem to known number facts rather than on an understanding of the relationships in the problem that would allow them to choose different counting or modeling strategies. Many children do not solve problems that cannot be readily modeled until they would be expected to have learned the related number facts. However, although Anick (in preparation) found a high incidence of recall of number facts in children's solutions to problems that could not be directly modeled, she also found many responses that were based on counting strategies. This suggests that even the problems that cannot be readily modeled are not just solved at an abstract numerical level. Children can solve these problems by understanding how the action or relationships in the problems are related to the counting processes that represent different actions or relationships. How this may occur is the topic in the next section of this chapter.

A Model of Verbal Problem-Solving

Several models have been proposed to describe stages that children go through in learning to solve addition and subtraction problems and to hypothesize knowledge structures sufficient to account for the behavior in these stages. Briars and Larkin (1982) and Riley *et al.* (1982) have developed computer simulation models that solve addition and subtraction word problems; the models are based on similar analyses of performance, but somewhat different characterizations of knowledge at each stage of problem solving.

Riley identifies three basic kinds of knowledge involved in problem solving: (*a* Problem Schemata, which are used to represent the problem situation; (*b*) Action Schemata which, at the most global level, essentially correspond to the solution processes described earlier, and (*c*) Strategic Knowledge for planning solution to problems.

Based on the results summarized in Table 2.4, Riley identified the levels of skill for solving Change, Combine, and Compare problems, and a computer simulation model was constructed for each level.

For Change problems, children in Level 1 are limited to external representations of problem situations using physical objects. They rely on Counting All and Separating strategies. They cannot use the Adding On strategy because they have no way to keep track of set–subset relationships.

The major advance of Level 2 over Level 1 is that it includes a schema that makes it possible to keep a mental record of the role of each piece of data in the problem. This allows children in Level 2 to solve Change Unknown problems (Table 2.2, Problems 3 and 4). Level 2 children are also able to use Counting On procedures. Level 2 children are limited to direct representation of problem action and are unable to solve Start Unknown problems (Table 2.2, Problems 5 and 6) because the initial set cannot be represented.

Both Level 1 and Level 2 children are limited to direct representation of problem structure. Level 3 includes a schema for representing part–whole relations that allows children to proceed in a top-down direction in order to construct a representation of the relationships among all the pieces of information in the problem before solving it. This frees children from relying on solutions that directly represent the action of the problem. Level 3 children can solve all six Change problems.

The model proposed by Briars and Larkin hypothesizes the same three levels of performance. Although the details differ, both models predict essentially the same set of cognitive mechanisms governing the first two levels. At Level 3, however, there are some fundamental differences between the models. Although both models predict the same responses for all problems, they attribute the responses to different knowledge structures. Briars and Larkin hypothesize that two schema are required to represent problems in Level 3. There is a subset equivalence schema, which is similar to Riley's part–whole schema, but it is more limited in its application. It allows children to interchange subsets, but it does not provide a basis for analyzing subset–superset relations. This schema provides the basis for Counting-On From Larger and for solving Join start-unknown problems (Table 2.2, Problem 5) but it does not provide a means for solving Separate Start-Unknown problems (Table 2.2, Problem 6). These solutions require a time reversal schema, which allows joining and separating actions to be reversed in time.

At present, there is no empirical basis for deciding which model is most appropriate. Neither model is entirely consistent with the available data regarding children's solutions to word problems. For example, both models hypothesize that children will solve simple Join and Separate problems (Table 2.2, problems 1 and 2) before they will be able to solve Join missing-addend problems (Table 2.2, Problem 3). The results of the longitudinal study by Carpenter and Moser (1983) suggest that this is not the case.

The simulation models are only first approximations for representing children's behavior. There is a great deal that they either oversimplify or do not explain. For example, the models are limited to operations on sets and fail to take into account children's knowledge of number facts. There is also a great deal less uniformity in children's behavior than is implied by the models. Children are not consistent in their use of strategies (Blume, 1981; Carpenter, Hiebert, & Moser, 1981). Siegler and Robinson (1981) have argued that it is not sufficient to build a model of how children may apply a particular strategy; it is also necessary to account for how they choose between alternative strategies.

Discussion

Research on how children solve basic addition and subtraction problems has made considerable progress in the past few years. A framework for characterizing problems has evolved that facilitates understanding how children solve different problems and why certain problems are more difficult than others. The strategies that children use to solve addition and subtraction problems have been clearly documented in such a way as to make it possible to identify major stages in the acquisition of addition and subtraction skills, especially at the early levels. Very recently, models have been constructed that go a long way toward characterizing internal cognitive processes that may account for children's behavior.

There is, however, a great deal that is yet unknown about how addition and subtraction concepts and skills are acquired. One of the basic assumptions of much of the research and theory building in the area is that the processes that children use to solve an addition or subtraction problem are intrinsically related to the structure of the problem. There is support for this assumption in that generally consistent results have been reported over a variety of instructional programs and within a number of different population groups.

The effects of instruction, however, are still unclear, especially at the more advanced levels of children's acquisition of addition and subtraction. The specific strategies that children use may be influenced by instruction, as children's use of the Matching strategy clearly shows. Hatano (1982) suggests that Japanese children may not rely predominantly on the counting strategies that have been so prevalent in American research.

There is also very little known about the transition from the informal modeling and counting strategies that children appear to invent themselves to the formal algorithms and memorized number facts that children learn as part of the mathematics curriculum. Some evidence suggests that, at first, children do not see a connection between their informal modeling and counting strategies and many of the formal skills they learn in their mathematics classes (Carpenter, Moser, & Hiebert, 1981). How or whether this connection is made is an important issue that so far has received relatively little attention in the growing body of research on addition and subtraction.

There is clear evidence that young children's responses to addition and subtraction problems are based on the semantic structure of the problem, but little is known regarding how children extract the meaning from the particular wording or different problems. Children's solutions clearly are not based exclusively on semantic structure.

Several recent investigations demonstrate the effect on performance of differences in wording of problems with the same semantic structure. In a study of kindergarten and first-grade children, Hudson (1980) found significant differences in children's performance on a basic subtraction problem that asked children to compare the number of birds in a picture to the number of worms. In one case, children were asked how many more birds there were than worms. The problem was significantly easier, however, when the children were asked the following question: "Each of these birds wants to eat a worm. How many of them will not get a worm?" Similar differences have been found for different wording of Join–Change Unknown problems (Carpenter & Moser, 1982; Riley *et al.*, 1982).

These studies indicate that the semantic structure of addition and subtraction problems does not completely determine children's performance, and it is necessary to be cautious in drawing conclusions about children's processing from specific studies, especially when difficulty level is used as the criterion measure. The fact that alternative versions of problems with the same semantic structure produce significant differences in performance does not threaten the general conclusions regarding young children's attention to problem structure. Although the difficulty levels are affected by changes in wording or syntax, the processes that children use remain relatively consistent. It appears that some wordings make the semantic structure of problems more transparent than others, but beyond that the processes used to extract meaning from the verbal statements of the problems remain something of a black box in most of the current research and theory.

Another limitation of most current research on addition and subtraction is that it does not deal with the question of individual differences. Most of the theory at least tacitly assumes that children go through essentially the same stages in acquiring addition and subtraction concepts and skills. There clearly are differences between children of a given grade, but generally these differences have been attributed to individual children's being at different levels in acquiring basic addition and subtraction concepts and skills. So far this assumption has not been seriously examined. Little is known about whether there are fundamentally different ways that individual children acquire addition and subtraction concepts and skills.

One of the reasons for the recent progress in understanding how children learn to add and subtract is the clear focus of much of the current research on a well-defined domain. This research could be criticized for being too narrowly focused. Siegler and Robinson (1981) argue for the importance of building large-scale, integrative models that specify how performance in addition and subtraction is related to performance in other basic content domains. They provide one example of such a model. Fuson (1982) and Steffe *et al.* (1982) have also been attempting to examine addition and subtraction within a larger context. They have focused on the relation between addition and subtraction and the development of counting skills.

It has also been proposed that the development of specific addition and subtraction processes may depend on the development of more basic cognitive capacities. Case (1982) has proposed that the acquisition of addition and subtraction processes is constrained by the growth of central information-processing capacity. Research by Romberg and Collis (1980) provides some support for this conclusion. But a great deal is yet to be done to really understand how information processing capacity affects the addition and subtraction processes children are capable of using.

There is certainly a great deal left to be explained about how children learn to add and subtract. However, these details are insignificant compared to the disparity between what is known about how children solve addition and subtraction problems and current programs of instruction (Carpenter, 1981). There is a compelling need for research that attempts to establish how the knowledge that has accumulated about children's addition and subtraction processes can be applied to design instruction.

Acknowledgments

We gratefully acknowledge the assistance of Diane Briars, Karen Fuson, James Hiebert, and Leslie Steffe, who provided valuable critical comments on earlier drafts of this paper.

References

Anick, C. M. *Strategies used by urban children to solve addition and subtraction word problems.* (Doctoral dissertation, University of Wisconsin, in preparation).

Arnett, L. D. Counting and adding. *American Journal of Psychology,* 1905, *16,* 327–336.

Beattle, I. D., & Deichmann, J. W. *Error trends in solving number sentences in relation to workbook format across 1st and 2nd graders.* Paper presented at the American Educational Research Association Annual Meeting, Chicago, April 1972. (ERIC Document Reproduction Service No. ED 064 170)

Blume, G. *Kindergarten and first-grade children's strategies for solving addition and subtraction and missing addend problems in symbolic and verbal problem contexts.* (Doctoral dissertation, University of Wisconsin, 1981).

Bolduc, E. J., Jr. A factorial study of the effects of three variables on the ability of first grade children to solve arithmetic problems (Doctoral dissertation, University of Tennessee, 1969). *Dissertation Abstracts International,* 1970, *30,* 3358A. (University Microfilm No. 70–2094).

Briars, D. J., & Larkin, J. H. *An integrative model of skill in solving elementary word problems.* (Technical Report ACP #2) Pittsburg: Carnegie Mellon University, 1982.

Brown, J. S., & Van Lehn, K. Toward a generative theory of "bugs." In T. P. Carpenter, J. M. Moser, & T. A. Romberg (Eds.), *Addition and subtraction: A cognitive perspective.* Hillsdale, New Jersey: Erlbaum, 1982.

Browne, C. E. The psychology of simple arithmetic processes: A study of certain habits of attention and association. *American Journal of Psychology,* 1906, *17,* 1–37.

Brownell, W. A. *The development of children's number ideas in the primary grades.* Supplementary Educational Monographs, No. 35. Chicago: University of Chicago Press, 1928.

Brownell, W. A. *Arithmetic in grades I and II: A critical summary of new and previously reported research.* Duke University Research Studies in Education, No. 5. Durham: Duke University Press, 1941.

Brownell, W. A., & Carper, D. V. *Learning and multiplication combinations.* Duke University Research Studies in Education, No. 7. Durham: Duke University Press, 1943.

Buswell, G. T., & Judd, H. Summary of educational investigations relating to arithmetic. *Supplementary Educational Monographs,* 1925, *27,* 65.

Carpenter, T. P. Cognitive development and mathematics learning. In R. Shumway (Ed.), *Research in mathematics education.* Reston, Virginia: National Council of Teachers of Mathematics, 1980. (a)

Carpenter, T. P. Heuristic strategies used to solve addition and subtraction problems. In *Proceedings of the Fourth International Congress for the Psychology of Mathematics Education,* Berkeley, 1980. (b)

Carpenter, T. P. Initial instruction in addition and subtraction: A target of opportunity for curriculum development. In *Proceedings of the National Science Foundation Directors Meeting,* Washington, 1981.

Carpenter, T. P., Blume, G., Hiebert, J., Anick, C. M., & Pimm, D. *A review of research on addition and subtraction* (Technical Report). Madison: Wisconsin Research and Development Center for Individualized Schooling, 1981.

Carpenter, T. P., Hiebert, J., & Moser, J. M. Problem structure and first grade children's initial solution processes for simple addition and subtraction problems. *Journal for Research in Mathematics Education,* 1981, *12,* 27–39.

Carpenter, T. P., & Moser, J. M. *An investigation of the learning of addition and subtraction* (Theoretical Paper No. 79). Madison: Wisconsin Research and Development Center for Individualized Schooling, 1979.

Carpenter, T. P., & Moser, J. M. The development of addition and subtraction problem-solving skills. In T. P. Carpenter, J. M. Moser, & T. A. Romberg (Eds.), *Addition and subtraction: A cognitive perspective*. Hillsdale, New Jersey: Erlbaum, 1982.

Carptenter, T. P. & Moser, J. M. *Addition and subtraction operations: How they develop* (Project paper). Madison: Wisconsin Center for Education Research, 1983.

Carpenter, T. P., Moser, J. M., & Hiebert, J. *The effect of instruction on first-grade children's solutions of basic addition and subtraction problems* (Working Paper No. 304). Madison: Wisconsin Research and Development Center for Individualized Schooling, 1981.

Case, R. General developmental influences on the acquisition of elementary concepts and algorithms in arithmetic. In T. P. Carpenter, J. M. Moser, & T. A. Romberg (Eds.), *Addition and subtraction: A cognitive perspective*. Hillsdale, New Jersey: Erlbaum, 1982.

Cronbach, L. J., & Suppes, P. (Eds.). *Research for tommorow's schools: Disciplined inquiry for education*. National Academy of Education, 1969.

Davidov, V. V. The psychological characteristics of the foundation of elementary mathematical operations in children. In T. P. Carpenter, J. M. Moser, & T. A. Romberg (Eds.), *Addition and subtraction: A cognitive perspective*. Hillsdale, New Jersey: Erlbaum, 1982.

Flavell, J. H. Concept development. In P. H. Mussen (Ed.), *Charmichael's manual on child psychology* (Vol. 1). New York: Wiley, 1970.

Fuson, K. *Counting solution procedures in addition and subtraction*. Paper presented at the Wingspread Conference on Addition and Subtraction, Racine, Wisconsin, 1979.

Fuson, K. An analysis of the counting-on solution procedure in addition. In T. P. Carpenter, J. M. Moser, & T. A. Romberg (Eds.), *Addition and subtraction: A cognitive perpective*. Hillsdale, New Jersey: Erlbaum, 1982.

Fuson, K. C. & Hall, J. W. The acquisition of early number word meanings: A conceptual analysis and review. In H. Ginsburg (Ed.), *The development of mathematical thinking*. New York: Academic Press, 1982.

Gelman, R., & Gallistel, C. R. *The child's understanding of number*. Cambridge, Massachusetts: Harvard University Press, 1978.

Gibb, E. Children's thinking in the process of subtraction. *Journal of Experimental Education*, 1956, *25*, 71–80.

Ginsburg, H. *Children's arithmetic: The learning process*. New York: Van Nostrand, 1977.

Ginbayashi, H. *Theory of quantity*. Paper presented at the Fourth International Congress on Mathematics Education, Berkeley, California, 1980.

Groen, G. J., & Parkman, J. M. A chronometric analysis of simple addition. *Psychological Review*, 1972, *79*, 329–343.

Groen, G. J., & Poll, M. Subtraction and the solution of open sentence problems. *Journal of Experimental Child Psychology*, 1973, *16*, 292–302.

Groen, G., & Resnick, L. B. Can preschool children invent addition algorithms? *Journal of Educational Psychology*, 1977, *69*, 645–652.

Grouws, D. A. Open sentences: Some instructional considerations from research. *Arithmetic Teacher*, 1972, *19*, 595–599.

Hatano, G. Learning to add and subtract: A Japanese perspective. In T. P. Carpenter, J. M. Moser, & T. A. Romberg (Eds.), *Addition and subtraction: A cognitive perspective*. Hillsdale, New Jersey: Erlbaum, 1982.

Hebbeler, K. Young children's addition. *The Journal of Children's Mathematical Behavior*, 1977, 1(4), 108–121.

Hiebert, J. Young children's solution processes for verbal addition and subtraction problems: The effect of the position of the unknown set. *Journal For Research in Mathematics Education*, 1982, *13*, 341–349.

Hiebert, J., & Carpenter, T. P. Piagetian tasks as readiness measures in mathematics instruction: A critical review. *Educational Studies in Mathematics,* 1982, *13,* 329–345.

Hiebert, J., Carpenter, T. P., & Moser, J. Cognitive development and children's performance on addition and subtraction problems. *Journal for Research in Mathematics Education,* 1982, *13,* 83–98.

Hirstein, J. J. Children's counting in addition, subtraction, and numeration contexts (Doctoral dissertation, University of Georgia, 1978). *Dissertation Abstracts International,* 1979, *39,* 7203A. (University Microfilms No. 7914032).

Hudson, T. Young children's difficulty with "How many more . . . than . . . are there?" questions. (Doctoral dissertation, Indiana University, 1980). *Dissertation Abstracts International,* July 1980, *41,* No. 1.

Ibarra, C. G., & Lindvall, C. M. *An investigation of factors associated with children's comprehension of simple story problems involving addition and subtraction prior to formal instruction on these operations.* Paper presented at the National Council of Teachers of Mathematics Annual Meeting, Boston, Massachusetts, April 1979.

Ilg, F., & Ames, L. B. Developmental trends in arithmetic. *Journal of Genetic Psychology,* 1951, *79,* 3–28.

Jerman, M. Problem length as a structural variable in verbal arithmetic problems. *Educational Studies in Mathematics,* 1973, *5,* 109–123.

Jerman, M., & Mirman, S. Linguistic and computational variables in problem solving in elementary mathematics. *Education Studies in Mathematics,* 1974, *5,* 3–28.

Jerman, M., & Rees, R. Predicting the relative difficulty of verbal arithmetic problems. *Educational Studies in Mathematics,* 1972, *4,* 306–323.

Klahr, D., & Wallace, J. G. *Cognitive development: An information processing view.* Hillsdale, New Jersey: Erlbaum, 1976.

Knight, F. B., & Behrens, M. S. *The learning of the 100 addition combinations and the 100 subtraction combinations.* New York: Longmans, Green, and Co., 1928.

Kouba, V., & Moser, J. *Development and validation of curriculum units related to initial sentence writing* (Technical Report No. 522). Madison: Wisconsin Research and Development Center for Individualized Schooling, 1979.

Lankford, F. G. Some computational strategies of seventh grade pupils. U.S. O.E. project No. 2-C-013, 1972.

LeBlanc, J. F. *The performance of first grade children in four levels of conservation of numerousness and three IQ groups when solving subtraction problems* (Technical Report No. 171). Madison: Wisconsin Research and Development Center for Cognitive Learning, 1971.

Leutzinger, L. D. *The effects of counting on the acquisition of addition facts in first grade.* (Doctoral dissertation, University of Iowa), 1979.

Lindvall, C. M., & Ibarra, C. G. *The development of problem-solving capabilities in kindergarten and first grade children.* Paper presented at the annual meeting of the National Council of Teachers of Mathematics, Seattle, April 1980.

MacLatchy, J. H. Another measure of the difficulty of addition combinations. *Educational Research Bulletin,* 1933, *12*(3), 57–61.

Marshall, G. G. A study of training and transfer effects of comparison subtraction and one-to-one correspondence (Doctoral dissertation, University of Houston, 1976). *Dissertation Abstracts International,* 1977, *37,* 4936A. (University Microfilms No. 77–1516).

Murray, J. *Studies in arithmetic* (Vol. 1). Scottish Council for Research in Education. London: University of London Press, 1939.

Nesher, P. Levels of description in the analysis of addition and subtraction word problems. In T. P. Carpetner, J. M. Moser, & T. A. Romberg (Eds.), *Addition and subtraction: A cognitive perspective.* Hillsdale, New Jersey: Erlbaum, 1982.

Nesher, P., & Katriel, T. Two cognitive modes in arithmetic word problem solving. In E. Cohors-Fresenborg & I. Wacksmith (Eds.), *Asnabruckes schriften zur mathematik* (Proccedings of the Second International Conference of the Psychology of Mathematics Education.) Osnabruck, Germany: University of Osnabruck, 1978.

Piaget, J. *The child's conception of number*. New York: Norton, 1952.

Pottle, H. L. *An analysis of errors made in arithmetic addition*. Unpublished doctoral dissertation, University of Toronto, 1937.

Rathmell, E. C. A reply to formal thinking strategies: A prerequisite for learning basic facts? *Journal for Research in Mathematics Education*, 1979, *10*, 374–377.

Resnick, L. B. Syntax and semantics in learning to subtract. In T. P. Carpenter, J. M. Moser, & T. A. Romberg (Eds.), *Addition and subtraction: A cognitive perspective*. Hillsdale, New Jersey: Erlbaum, 1982.

Riley, M. S. *Conceptual and procedural knowledge in development*. Unpublished Master's thesis, University of Pittsburg, 1981.

Riley, M. S., Greeno, J. G., & Heller, J. I. Development of children's problem-solving ability in arithmetic. In H. Ginsburg (Ed.), *The development of mathematical thinking*. New York: Academic Press, 1982.

Romberg, T. A., & Collis, K. A. Cognitive level and performance on addition and subtraction problems. In *Proceeding of the Fourth International Conference for the Psychology of Mathematics Education*, Berkeley, CA, 1980.

Romberg, T. A., Harvey, J. G., Moser, J. M., & Montgomery, M. E. *Developing mathematical processes*. Chicago: McNally, 1974.

Rosenthal, D. J. A., & Resnick, L. B. Children's solution processes in arithmetic word problems. *Journal of Educational Psychology*, 1974, *66* 817–825.

Schaeffer, B., Eggleston, V., & Scott, J. L. Number development in young children. *Cognitive Psychology*, 1974, *6*, 357–379.

Schell, L. M., & Burns, P. C. Pupil performance with three types of subtraction situations. *School Science and Mathematics*, 1962, *62*, 208–214.

Secada, W. G., Fuson, K. C., & Hall, J. W. The transition from counting all to counting on in addition. *Journal for Research in Mathematics Education*, 1983, *14*, 47–57.

Shores, J. H., & Underhill, R. G. *An analysis of kindergarten and first grade children's addition and subtraction problem-solving modeling and accuracy*. Paper presented at the Annual Meeting of the American Educational Research Association, San Francisco, April 1976. (ERIC Document Reproduction Service No. ED 121–626)

Siegler, R. S., & Robinson, M. The development of numerical understandings. In H. W. Reese & L. P. Lipsitt (Eds.), *Advances in Child Development and Behavior*, (Vol. 16) New York: Academic Press, 1981.

Smith, J. H. Arithmetical combinations. *Elementary School Journal*, 1921, *21*, 762–770.

Starkey, P., & Gelman, R. The development of addition and subtraction abilities prior to formal school in arithmetic. In T. P. Carpenter, J. M. Moser, & T. A. Romberg (Eds.), *Addition and subtraction: A cognitive perceptive*. Hillsdale, New Jersey: Erlbaum, 1982.

Steffe, L. P. Differential performance of first-grade children when solving arithmetic addition problems. *Journal for Research in Mathematics Education*, 1970, *1*, 144–161.

Steffe, L. P. A reply to "Formal thinking strategies: A prerequisite for learning basic facts?" *Journal for Research in Mathematics Education*, 1979, *10*, 370–374.

Steffe, L. P., & Johnson, D. C. Problem-solving performances of first-grade children. *Journal for Research in Mathematics Education*, 1971, *2*, 50–64.

Steffe, L. P., Spikes, W. C., & Hirstein, J. J. *Quantitive comparisons and class inclusion as readiness variables for learning first grade arithmetical content*. Athens, Georgia: The Georgia Center for the Study of Learning and Teaching Mathematics. University of Georgia, 1976.

Steffe, L. P., Thompson, D. W., & Richards, J. Children's counting in arithmetical problem solving. In T. P. Carpenter, J. M. Moser, & T. A. Romberg (Eds.), *Addition and subtraction: A cognitive perspective*. Hillsdale, New Jersey: Erlbaum, 1982.

Suppes, P., & Groen, G. Some counting models for first grade performance data on simple facts. In J. M. Scandura (Ed.), *Research in Mathematics Education*. Washington, D.C.: National Council of Teachers of Mathematics, 1967.

Suppes, P., Jerman, M., & Brian, D. *Computer-assisted instruction: The 1965–1966 Stanford arithmetic program*. New York: Academic Press, 1968.

Suppes, P., Loftus, E. F., & Jerman, M. Problem-solving on a computer-based teletype. *Education Studies in Mathematics*, 1969, *2*, 1–15.

Suppes, P., & Morningstar, M. *Computer-assisted instruction at Stanford, 1966–68: Data, models, and evaluation of the arithmetic programs*. New York: Academic Press, 1972.

Swenson, E. J. Organization and generalization as factors in learning, transfer, and retroactive inhibition. *Learning theories in school situations*. University of Minnesota Studies in Education (Vol. 2). Minneapolis: University of Minnesota Press, 1949.

Theile, C. L. *The contribution of generalization to the learning of addition facts* (Contributions to Education, No. 763). New York: Teacher's College, 1938.

Thornton, C. A. Emphasizing thinking strategies in basic fact instruction. *Journal for Research in Mathematics Education*, 1978, *9*, 214–227.

Vergnaud, G. A classification of cognitive tasks and operations of thought involved in addition and subtraction problems. In T. P. Carpenter, J. M. Moser, & T. A. Romberg (Eds.), *Addition and subtraction: A cognitive perspective*. Hillsdale, New Jersey: Erlbaum, 1982.

Weaver, J. F. Some factors associated with pupils' performance levels on simple open addition and subtraction sentences. *Arithmetic Teacher*, 1971, *18*, 513–519.

Weaver, J. F. Interpretations of number operations and symbolic representations of addition and subtraction. In T. P. Carpenter, J. M. Moser, & T. A. Romberg (Eds.), *Addition and subtraction: A cognitive perspective*. Hillsdale, New Jersey: Erlbaum, 1982.

Woods, S. S., Resnick, L. B., & Groen, G. J. An experimental test of five process models for subtraction. *Journal of Educational Psychology*, 1975, *67*, 17–21.

CHAPTER 3

Proportional Reasoning of Early Adolescents*

Robert Karplus, Steven Pulos,
and Elizabeth K. Stage

You have probably heard the brain teaser about the poultry farm, where a chicken and a half lays an egg and a half in a day and a half. How many eggs do five chickens lay in six days? To answer the question, you have to interpret it as dealing with averages or assume that all chickens are alike and produce eggs at a uniform rate. Yet the proportional reasoning applicable here is also vital in other ratio or rate relationships, such as earnings at a fixed wage for various lengths of working time, the exchange of dollars for a foreign currency, distances traveled by a car at constant speed, and the size of the projected image as a screen is moved away from the projector.

Background

The concepts of *ratio* and *proportion* as applied by young people have been widely studied, with the earliest research reports coming to our attention dated almost seventy years ago (Winch, 1913–1914). Though Winch was primarily concerned with elementary school students' ability to solve proportion problems correctly, he also reported the results of interviews in which students described their solution methods retrospectively—usually the unit measures approach, but also additive and qualitative reasoning. One hypothesis, that the explicit mention

*This research was supported in part by the National Science Foundation under Grant No. SED 79-18962. Any opinions, findings, and conclusions expressed in this report are those of the authors and do not necessarily reflect the views of the National Science Foundation.

ISBN No. 0-12-444220-X

of unit amounts in a problem would make it easier than problems in which no such reference was made, was supported by his data.

Piagetian Studies

Piaget and his collaborators (Inhelder & Piaget, 1958; Piaget, 1970; Piaget, Grize, Szeminska & Bang, 1977; Piaget & Inhelder, 1975) have called attention to qualitative and quantitative features of childrens' and adolescents' approaches to tasks that require proportional reasoning. They used situations in which linear functional relationships governed the interdependence of two or more variables, as in the distance moved by a toy car when the wheels make one, two, or more revolutions. The principal aims of this research were the clarification of young people's development with respect to the concepts of functions, probability, speed, and compensating effects of variables that describe a *physical* system (balance beam) or *logico-mathematical* concept (correlation). Understandably, therefore, the tasks chosen for this research were not selected primarily to illustrate proportional reasoning.

To characterize the subjects' reasoning, Piaget described (*a*) early stages in which thinking made use of qualitative correspondences and seriations, (*b*) intermediate stages of additive compensations or the employment of 2:1 ratios, and (*c*) advanced stages in which proportional reasoning was applied regardless of the numerical values of the data and their ratios. For Piaget, proportionality was the prime exemplar of function. Yet Piaget's work concentrated on children's attainment of proportional reasoning and left unanswered many questions that are aimed at the ways in which proportional reasoning is applied in problems with differing contexts and numerical relationships.

A few years later, Lovell (1961) and Lovell and Butterworth (1966) replicated and extended the Genevan studies. They found that fewer than half of the 15-year-olds in their sample solved proportion problems. Lunzer and Pumfrey (1966) used Cuisenaire rods, a pantograph, and a balance beam to investigate proportionality. They identified iterative strategies in which children established a basic correspondence (e.g., 2:3) and then generated equivalent correspondences by adding (e.g., 2 + 2 + 2:3 + 3 + 3, or 6:9). The authors stated that "when a child does try to establish a relation between relations, he prefers to add or subtract, and perhaps multiply, but refuses to divide" (Lunzer & Pumfrey, 1966, p. 11).

Some recent research using Piagetian formal reasoning tasks (Inhelder & Piaget, 1958) applied a rule-assessment approach to investigate the development of children's understanding of phenomena governed by two variables (Siegler, 1981). Even though the tasks were consistently identified as proportionality tasks, proportional reasoning played a very minor role in the study in that it was represented in only one of the four rules, and was required for only one out of six tasks. We interpret the very low performance of young adults on these tasks as evidence that they assessed much more complex reasoning than proportionality.

Surveys of Proportional Reasoning

In a series of papers published during the past 10 years (Karplus, 1979a, 1981; Karplus, Adi, & Lawson, 1980; Karplus & Karplus, 1972; Karplus, Karplus, Formisano & Paulsen, 1979; Karplus, Karplus, & Wollman, 1974; Karplus & Peterson, 1970; Kurtz & Karplus, 1979; Wollman & Karplus, 1974), the research group at the Lawrence Hall of Science, Berkeley, described the proportional reasoning displayed by secondary school students on several tasks that were designed to minimize the need for knowledge of physical principles. Most widely used was the group-administered Paper Clips Task (Karplus, 1981), in which the subjects were asked to explain how they found the height of a Mr. Tall figure measured in paper clips, using the heights of the Mr. Tall and a Mr. Short figure given in buttons (6 and 4 buttons tall, respectively) and an individually made measurement of Mr. Short's height in paper clips (about 6). The diverse reasoning approaches used by the subjects were grouped into four major categories: incomplete or no use of the data given, additive or constant difference relations, transitional (iterative, graphic, or only partial use of proportions), and explicit use of equal ratios. Though academically upper-track or upper middle-class students used proportional reasoning increasingly after about age 12 years, only a small fraction of urban low-income and academically lower-track students used proportions at age 14 or even 17 years.

Similar results were obtained with the Paper Clips Task and similar problems by Suarez (1977; Suarez & Rhonheimer, 1974) in Switzerland and Hart (1978; 1981) in Britain. It was also found that successful application of a 1:2 ratio appeared to be much easier than application of a 3:2 ratio (Hart, 1981; Wollman & Karplus, 1974), a result in keeping with the observations of Kieren and Nelson (1978), among others. The latter authors asked their subjects to apply fractions as operators and reported that many children under 11 years of age mastered the "½" tasks but did not solve tasks involving other unit fractions or composite fractions.

The inconsistency with which students apply proportional reasoning to problems embodying a 2:1 as compared to a 3:2 or other nonintegral ratio (Hart, 1981; Wollman & Karplus, 1974) has raised questions about the extent to which proportional reasoning is part of a larger operational structure as suggested by Piaget (Inhelder & Piaget, 1958) or is itself a unitary structure. Variations in difficulty have been investigated more deeply by researchers who assigned their subjects problems that differed in their numerical content (Abramowitz, 1976; Hart, 1978; Noelting, 1978, 1980a; Rupley, 1981).

In his extensive study, Rupley (1981) found that lack of integer ratios, an unknown smaller than the given data, and inclusion of numerical values larger than about 30 increased item difficulty. A very successful solution method used by many of Rupley's subjects made use of two steps: (*a*) finding a unit ratio, and (*b*) applying the unit ratio to compute the answer. Rupley's findings confirmed and extended results obtained by Abramowitz (1976), who had used a much smaller student sample and fewer problems. Hart (1981) used a more varied set of problems, including recipe and geometrical scaling questions. She found addition strategies widely used by students when simple integral ratios and a plausible unit measure did not occur.

STAGES OF PROPORTIONAL REASONING

Noelting (1980a; 1980b) studied proportional reasoning with tasks that required subjects to compare two ratios (orange juice to water, to prepare two drinks of the same or differing orange taste) rather than requiring them to compute an answer that would produce a desired ratio. Noelting's 321 subjects between ages 6 and 16 were given 25 problems that ranged widely in difficulty. The problems were then arranged in order of success (correct answer justified by proportional reasoning) and grouped into eight stages characterized by the numerical relationships (e.g., equal ½ ratios—about 75% success; unequal ratios with corresponding terms that were integral multiples—about 35% success). These problems, which we call *comparison problems*, thus revealed difficulty factors similar to those on the *missing-value problems* that required calculation of an answer.

The reasoning used by Noelting's subjects also displayed the diversity observed by investigators of missing-value problems: nonuse of data, qualitative comparisons, additive and iterative approaches, and calculation of ratios. Noelting (1980b) proposed a theory of *adaptive restructuring* to account for the transition from one stage to the next.

TYPES OF COMPARISON

In his theory, Noelting (1980b) found it useful to distinguish the type of comparison made by a subject when solving a problem. He noted that subjects might proceed by comparing the amounts of orange juice and water within each recipe first, and then comparing these two relations; or they might compare the two amounts of orange juice and the two amounts of water first. The former approach he called a *Within* strategy, the latter a *Between* strategy.

The Within–Between distinction has also been made by others. Freudenthal (1978) noted that ancient scholars had stated ratios of "like" quantities only, such as orange juice to orange juice in the Between strategy in Noelting's tasks, or buttons to buttons in the Paper Clips Task (Karplus, 1981; Karplus & Peterson, 1970) described above. With the development of modern science, ratios of unlike quantities, such as distance–time or mass–volume, came to be used also. These correspond to the orange juice to water ratio as in the Within strategy in Noelting's tasks, or paper clips to buttons in the Paper Clips Task.

Kurtz (1976), who studied the reasoning of beginning secondary school students, classified the additive reasoning used by most of them according to the Within–Between distinction. He found the Within pattern to be preferred on a missing value task using the turns of two different-sized pulleys linked by a belt (i.e., small pulley and large pulley), but the Between pattern to be preferred by the same subjects on an adjustment task using distances in a given drawing and a hypothesized enlarged version of it. Because the numerical values were closely similar in these two tasks, it appears that the context influenced the type of comparison.

Lybeck (1978) and Vergnaud (1980) have described the preferred type of comparison in tasks that involved more diverse variables, such as mass, length, time, and fuel consumption. In his report of the reasoning of 26 high school students, Lybeck (1978) described their preference for the Within approach on the Cylinder Task (Suarez & Rhonheimer, 1974), which is very similar to the Paper Clips Task (Kurtz, 1976). The same students had a moderate preference for the Between strategy on a task concerning weights stretching a spring that satisfied Hooke's law. Vergnaud's (1980) tasks involved the amount of fuel consumed in various periods of time in a particular house's central heating system. More than half of the 80 subjects of ages 11 to 15 years succeeded in solving the problems whose numerical values were integral multiples of one another, with a substantial preference for the Between comparison (i.e., time–time and fuel–fuel rather than fuel–time).

A close reading of more recent publications of Noelting (1981) and Vergnaud (Chapter 5, this volume) suggests that the types of comparison are more complicated than may appear at first glance. Rather than combining pairs of data only as we have described, we found that some subjects took into account all four data in the two recipes in a single more complex operation. The analysis of such a more complex operation has not been described clearly by the other authors, with the result that our classification according to types of comparison may differ from those of Noelting and Vergnaud.

THEORIES OF PROPORTIONAL REASONING

In Piaget's theory, the proportionality schema arises in conjunction with the INRC group and proportional reasoning (Inhelder & Piaget, 1958). Weaknesses of this theory have been pointed out by Suarez (1977), who stressed the shortcomings of the INRC group and Piaget's de-emphasis of the context in which proportional reasoning was to be applied. We also are not convinced by Piaget's presentation because it appears to us that additive compensations are as compatible with the INRC group as multiplicative compensations, though they have a different logical structure. As we have already stated, we consider the balance beam used by Inhelder and Piaget (1958) to be too complicated a system to allow a fruitful analysis of proportional reasoning in its context.

In the reasoning pattern theory of Karplus (1979b, 1981), proportional reasoning is an independent entity, not integrally linked to other reasoning patterns, that may be applied under simple or complex conditions and correspondingly may be used at various levels. Just what these conditions are has not been specified—hence the theory is only global at the present time.

It is clear, however, that the essential nature of the proportionality of linear interdependence of two variables x and y may be described as follows: whatever correspondence exists between x and y, the same correspondence exists between all equal multiples of x and y. In other words, proportionality cannot be divorced from the understanding of linear functions (Suarez, 1977) or Vergnaud's (Chapter 5, this volume) more general notion of *isomorphism of measures*. In our opinion, theories that deal only with arithmetic relations among particular sets of numbers and not with the relation between the variables themselves cannot satisfactorily account for proportional reasoning.

Noelting's (1980b) theory of the development of proportional reasoning fits within the reasoning pattern approach in that it deals only with children's proportional reasoning and no other cognitive operations. It goes beyond Karplus's (1979b; 1981) global theory in that it seeks to identify stages and mechanisms whereby proportional reasoning is progressively applied to more and more demanding tasks. Noelting proposed two periods of development, the second of which occurs during adolescence. In this second period, the elements of the Within ratio are gradually differentiated, first only for equal ratios whose elements are integral multiples, later for all ratios and comparisons whether equal or not.

By appealing to the gradual growth in the effectiveness with which working memory is used, Case (1978; 1979) attempted to explain development. He postulated distinct concrete and formal-level structures in working memory to account for additive and proportional reasoning (Case, 1979). He did not, however, define the differences operationally. His theory is therefore incomplete. Similar efforts have been made by de Ribaupierre and Pascual-Leone (1979) and by Furman (1980).

Research Questions

When planning the present research, we were interested in unifying the results described above regarding the development of proportional reasoning; factors causing difficulty in proportional reasoning; alternative strategies used by students instead of proportional reasoning; and the relation of comparison tasks to missing-value problems. To complement Noelting's studies (1980a), we intended to concentrate on early adolescent students, who were expected to be in the second period of development he described (Noelting, 1980b). Accordingly, our tasks were chosen to be comparable to the more difficult ones used by Noelting.

Our research plan aimed at finding answers to the following questions regarding such students and tasks:

1. What reasoning patterns are used?
2. What are the distributions of reasoning patterns on tasks differing systematically in
   a. the occurrence of Within and Between integral ratios?
   b. the occurrence of equal and unequal ratios?
   c. the requirement to compare ratios or adjust the quantities (i.e., find a missing value) in originally unequal ratios so they will be equal?
3. Are there differences in these distributions according to the grade level and/or sex of the students?
4. How consistently does a particular student use proportional reasoning?
5. How do individual differences among students with respect to information processing capacity, intelligence, spatial ability, and other factors influence their responses to proportional reasoning tasks?
6. How do a student's attitudes toward school and mathematics correlate with success on proportional reasoning tasks?

The next section of this chapter presents our findings from the use of the lemonade puzzles, a proportional reasoning task designed for this research. The following section describes the relationships of proportional reasoning to the cognitive and attitudinal factors that were included in our study. These results are compared with expectations based on the previous research and are used to suggest a theory of proportional reasoning. We conclude this chapter with our views on the implications of proportional reasoning research for teaching mathematics during the middle school years.

The Lemonade Puzzle Interviews

The lemonade puzzles were designed as tasks that require proportional reasoning for successful solution, but that allow subjects to use other forms of reasoning. In the puzzles, John and Mary each prepare lemonade concentrate by mixing a specified number of spoonfuls of sugar with a different given number of spoonfuls of lemon juice until the sugar dissolves. For instance, John might mix 4 spoonfuls of sugar with 6 of lemon juice, whereas Mary uses 10 and 15 spoonfuls, respectively. Subjects were then asked to compare the flavors or sweetnesses of the two mixtures. Because the flavor depended on the ratio of the two ingredients and not on their absolute amounts, comparison of sweetness required proportional reasoning. In this example, the two concentrates tasted the same.

We chose the lemonade puzzles because (*a*) lemonade is familiar to early adolescents, (*b*) the concentrate might plausibly be made with various amounts of the two ingredients, (*c*) both lemon juice and sugar are materials with distinct properties, (*d*) spoonfuls are relatively small so that the amounts can be considered as continuous quantities, (*e*) the amounts can be illustrated easily, and (*f*) we wanted a task that was similiar to Noelting's (1980a) to facilitate a comparison with his results. The specific instruments, the research subjects, the task administration, and the scoring procedures are now described.

Method

THE INSTRUMENTS

Each lemonade puzzle was administered by means of an illustrated card, such as that shown in Figure 3.1. Each lemonade puzzle interview consisted of eight puzzles that used the amounts of sugar and lemon juice given in Table 3.1. Two practice puzzles (Table 3.1) were used to introduce the task and to explain the procedure to the subjects. Note that the puzzles in other than the practice items all required the formation of ratios between quantities that were unequal and neither was equal to one. In four of the puzzles—WB, W, B, and W—the ratios of ingredients, and therefore the sweetnesses of the concentrates, were equal. In the four remaining puzzles—WBX, WX, BX, and NX—the ratios and therefore the sweetnesses were unequal. Furthermore, the numerical values were chosen so that integral ratios of ingredients occurred in carefully chosen ways, as designated by the letters W, B, and N in Table 3.1.

THE SUBJECTS

Our subjects were 60 sixth and 60 eighth graders from a middle-class school with an ethnically mixed population performing near the California state average on standardized mathematics achievement tests. Half of each group were boys and half girls, selected randomly from an enrollment of about 150 at each grade level.

PROCEDURE

The lemonade puzzles were administered during the last 15 minutes of a 30-minute interview. The subjects' responses were tape recorded and were also recorded by interviewers, who noted the occasional use of paper and pencil.

Figure 3.1 Card for presentation of Lemonade Puzzle W.

Each puzzle was read aloud by the interviewer, who pointed to the various amounts of sugar and lemon juice on the puzzle page as they were mentioned. Two or four questions were asked:

1. (Comparison) Whose lemonade concentrate is sweeter, John's or Mary's, or do they taste the same?
2. (After the response to Question Number 1) How did you come up with this answer?
3. (Adjustment, a form of missing value problem: after the response to Question Number 2, if response #1 indicated unequal taste) How much lemon

TABLE 3.1

Data Used in the Lemonade Puzzle Tasks[a]

	Equal ratios					Unequal ratios			
	John		Mary			John		Mary	
Item[b]	Sugar	Lemon	Sugar	Lemon	Item[b]	Sugar	Lemon	Sugar	Lemon
Practice A	3	3	5	5	Practice B	4	6	4	8
WB	2	6	8	24	WBX	3	6	9	15
	2	8	6	24		3	9	6	15
W	3	12	5	20	WX	4	12	6	16
	3	6	7	14		6	12	8	15
B	3	7	6	14	BX	6	8	12	15
	3	5	12	20		4	6	12	16
N	4	6	10	15	NX	3	5	7	11
	4	10	6	15		3	7	5	11

[a]Two versions of data were used, each with 60 subjects. This chapter always refers to the combined results.
[b]W, Within recipe ratio integral; B, between recipe ratio integral; N, No ratio integral; X, unequal ratios.

juice would Mary need with her ___ spoons of sugar to make her concentrate taste just like John's?

4. (After the response to Question Number 3) Please explain your answer.

RESPONSE CATEGORIES

The subjects' explanations in answer to Questions 2 and 4 were classified into about 40 subcategories that were collapsed into the following four major categories:

Category I (incomplete, illogical)—don't know, guess, inappropriate quantitive operations.

Category Q (qualitative)—qualitative comparison of four given amounts, using *more, less,* or equivalent terms.

Category A (additive)—using the data to compute and compare differences or remainders.

Category P (proportional)—using the data to compute and compare proportional relationships, possibly with arithmetic errors.

Listings of sample responses in Categories P and A, which were the most frequent, are presented in Tables 3.2 and 3.3, respectively.

In addition to classifying the subjects' explanations according to the categories described above, we considered the type of comparison by which the subject related the data to one another. Thus, in some explanations the first relation

TABLE 3.2

Examples of Proportional Explanations

Example	Puzzle[a]	Explanation	Type of comparison[b]
A	WB (2/6, 8/24)	Same taste: 8 goes into 24 three times, 2 goes into 6 three times.	w
B	B (3/7, 6/14)	Same taste: 3 times 2 is 6, 7 times 2 is 14.	b
C	BX (4/6, 12/16)	Mary sweeter: She used 3 times as much sugar, should have used 3 times as much lemon juice, but didn't use enough.	b
		Adjustment: 18 lemon juice.	b
D	W (3/12, 5/20)	Same taste: 3 goes into 12 four times, 5 times 4 is 20.	w
E	BX (6/8, 12/15)	Mary sweeter: She's got ⅘ and John's only got ¾.	w
		Adjustment: 16, because 12 is ¾ of 16.	w
F	N (4/10, 6/15)	Same taste: 4 and 4 and half of 4 is 10, 6 and 6 and half of 6 is 15.	w
G	N (4/6, 10/15)	Same taste: If I make this (John's) smaller, it becomes 2 to 3, and if I reduce this (Mary's) it becomes 2 to 3.	w
H	NX (3/5, 7/11)	Mary sweeter: 5 times 7 is 35, 11 times 3 is 33, so John has more lemon juice, and it is sour.	u

[a]W, within recipe ratio integral; B, between recipe ratio integral; N, no ratio integral; X, unequal ratios.
[b]w, within recipe comparison; b, between recipe comparison; u, comparison unclassifiable.

TABLE 3.3

Examples of Additive Explanations

Example	Puzzle[a]	Explanation	Type of comparison[b]
A	NX (3/7, 5/11)	John sweeter: Mary has 2 more sugars and she has 4 more lemon juices.	b
		Adjustment: 9 lemon juices, 5 and 4 is 9.	w
B	WX (6/12, 8/15)	John sweeter: I subtracted this time, 8 minus 6 is 2, 15 minus 12 is 3, so there's more lemon juice than sugar.	b
		Adjustment: 14 lemon juices, 14 minus 12 is 2.	b
C	NX (3/7, 5/11)	Same taste: 3 times 2 is 6 plus 1 is 7, 5 times 2 is 10 plus 1 is 11.	w
D	WBX (3/6, 9/15)	John sweeter: 3 into 6 goes once with 3 left over, 9 into 15 goes once with 6 left over.	w
		Adjustment: 12 lemon juice, 9 into 12 goes once, with 3 left over.	w

[a]W, within recipe ratio integral; B, between recipe ratio integral; N, no ratio integral; X, unequal ratios.
[b]w, within recipe comparison; b, between recipe comparison; u, comparison unclassifiable.

involved the two ingredients of sugar and lemon of the same recipe, that is, a *Within recipe* or *Within* comparison. In other explanations the relation(s) involved corresponding ingredients of the two recipes, that is, John's and Mary's sugar or John's and Mary's lemon juice, a *Between recipe* or *Between* comparison. In Tables 3.2 and 3.3 we have identified the type of comparison used in each example by *w* or *b*. In some explanations, one of which is identified in Table 3.2 by *u*, the type of relation could not be inferred from the subject's statements, which made reference to a complex operation involving all four data in the two recipes.

Results and Discussion

Our presentation of the results consists of three parts. First we describe the distributions of responses to the comparison and adjustment questions. The second part of this section is concerned with the consistency of any one subject using the same reasoning pattern on several puzzles. In the last part, we concentrate on cognitive elements of proportional reasoning as they are related to puzzle structure.

COMPARISON QUESTIONS

The frequencies with which the four reasoning patterns I, Q, A, and P were used on the comparison questions of the eight puzzles are shown in Table 3.4. For the additive and proportional reasoning patterns, the Within, Between, and unclassifiable comparisons are listed separately in the lower section of the table.

Several striking results can be observed immediately. First, the relatively high rates of proportional reasoning of 54 and 63% on puzzles WB and W, respectively, decrease progressively for the other puzzle structures to about 18 and 10% for puzzles N and NX, respectively. In fact, the eight items can be divided into three subsets: *easy* puzzles WB and W, with proportional reasoning near 60%; *medium* difficulty puzzles B, WBX, and WX, with proportional reasoning frequencies near 35%; and *difficult* puzzles BX, N, and NX with proportional reasoning frequencies below 20%.

Note that the presence of Within integer ratios or equal ratios for the two recipes facilitated the application of proportional reasoning. Conversely, noninteger ratios were factors that inhibited proportional reasoning to some extent separately and even more in combination.

Second, illogical and qualitative reasoning were used at approximately the same frequencies of 20–30% and 5–10% of the subjects, respectively, independently of puzzle structure. Because these two patterns did not employ any specific relations among the numerical data that were provided, it is plausible that their use would not be appreciably affected by the inhibiting factors mentioned above.

TABLE 3.4

Frequencies of the Four Reasoning Patterns on the Comparison Questions[a]

Reasoning category[c]	Puzzle structure[b]							
	WB	W	B	N	WBX	WX	BX	NX
I	32.5	21.7	27.5	33.3	24.2	25.8	26.7	27.5
Q	7.5	7.5	5.0	5.0	10.8	8.3	10.0	7.5
(A-w)[d]	(4.2)	(7.5)	(30.0)	(37.5)	(32.5)	(25.0)	(40.8)	(45.0)
(A-b)	(1.7)	(0.8)	(0.8)	(6.7)	(0.8)	(6.7)	(1.7)	(9.2)
(A-u)	(0.0)	(0.0)	(0.0)	(0.0)	(0.0)	(0.0)	(0.8)	(0.8)
A	5.9	8.3	30.8	44.2	33.3	31.7	43.3	55.0
(P-w)	(41.7)	(62.5)	(7.5)	(16.7)	(22.5)	(33.3)	(9.2)	(5.8)
(P-b)	(12.5)	(0.0)	(29.2)	(0.8)	(7.5)	(0.0)	(10.8)	(1.7)
(P-u)	(0.0)	(0.0)	(0.0)	(0.0)	(1.7)	(0.8)	(0.0)	(1.5)
P	54.2	62.5	36.7	17.5	31.7	34.1	20.0	10.0

[a]Percentage; $N = 120$.

[b]W, within recipe ratio integral; B, between recipe ratio integral; N, no ratio integral; X, unequal ratios.

[c]I, illogical; Q, qualitative, A, additive; P, proportional; w, within recipe comparison; b, between recipe comparison; u, comparison unclassifiable.

[d]Figures in parentheses refer to subclassifications of additive and proportional reasoning, which total to the A and P categories, respectively.

On the other hand, it is curious that their use did not increase in conjunction with the decrease in proportional reasoning on the more difficult puzzles.

Third, it is quite clear that additive reasoning replaced proportional reasoning to a very large degree on the more difficult puzzles. Note that the sum of proportional and additive reasoning varied only between 60 and 70% on all the puzzles.

Though the data in Table 3.4 do not reveal the consistency with which a particular student reasoned on the eight puzzles, they give the impression that there were three distinct groups of subjects. About 30% of the subjects appear to have used primarily illogical or qualitative reasoning, about 65% appear to have used primarily proportional and additive reasoning, and about 5% appear to have used a less consistent combination of approaches. This impression is supported by data we present in a subsequent section.

Fourth, the Within-recipe comparisons were made considerably more frequently than the Between comparisons, except for the subjects who used proportional reasoning on the puzzles with integral ratios between recipes. On the puzzles where the subjects could choose either of two simple ratios, about 75% preferred the Within ratio. On the puzzles with no integer ratio, a majority of subjects also chose the Within type of explanation. The exception on Puzzles B and BX merely reflects the students' response to the integral ratio between recipes in those puzzles. Even the students who used additive reasoning on any puzzle greatly preferred—by more than 75%—the Within comparison on that puzzle.

The presentation of the lemonade puzzles clearly emphasized the Within relationships because each recipe was described as an entity and formed a conceptual role. Our subjects' preference for Within comparisons, therefore, does not imply that the same preference will be displayed on all proportional reasoning tasks.

ADJUSTMENT QUESTIONS

The frequencies of illogical, qualitative, additive, and proportional reasoning on the adjustment questions are presented in Table 3.5. The adjustment question was, of course, omitted for all subjects who used proportional reasoning correctly on the comparison questions for "Equal" items, because they concluded that the two recipes had the same taste. The question was also omitted for a few other subjects who concluded that the tastes were equal by using illogical or qualitative reasoning.

The other differences between Tables 3.4 and 3.5 show reduced illogical reasoning, greatly reduced qualitative reasoning, and somewhat enhanced additive reasoning on adjustment. Because the adjustment question required the determination of a missing value of lemon juice, the question itself discouraged the use of qualitative reasoning.

TABLE 3.5

Frequencies of the Four Reasoning Patterns on the Adjustment Questions[a]

Reasoning category[c]	Puzzle structure[b]							
	WB	W	B	N	WBX	WX	BX	NX
I	22.5	20.0	20.0	25.0	21.7	22.5	24.2	27.5
Q	4.2	0.8	0.8	0.0	0.8	0.0	0.0	0.0
(A-w)[d]	(12.5)	(10.0)	(33.3)	(43.3)	(30.8)	(26.7)	(44.2)	(39.2)
(A-b)	(1.7)	(1.7)	(1.7)	(5.8)	(2.5)	(6.7)	(2.5)	(8.3)
(A-u)	(0.8)	(1.7)	(0.8)	(0.0)	(0.8)	(1.7)	(0.0)	(0.8)
A	15.0	13.4	35.8	49.2	34.1	35.0	46.7	48.3
(P-w)	(0.8)	(0.0)	(0.0)	(0.0)	(25.8)	(30.8)	(6.7)	(1.7)
(P-b)	(0.0)	(0.0)	(0.0)	(0.0)	(8.3)	(0.8)	(15.8)	(0.8)
(P-u)	(0.0)	(0.0)	(0.0)	(0.0)	(0.8)	(0.8)	(0.8)	(1.7)
P	0.8	0.0	0.0	0.0	35.0	32.5	23.3	4.2
omitted	57.5	65.8	43.3	25.8	8.3	10.0	5.8	20.0

[a]Percentage; $N = 120$.

[b]W, within recipe ratio integral; B, between recipe ratio integral; N, no ratio integral; X, unequal ratios.

[c]I, illogical; Q, qualitative, A, additive; P, proportional; w, within recipe comparison; b, between recipe comparison; u, comparison unclassifiable.

[d]Figures in parentheses refer to subclassifications of additive and proportional reasoning, which total to the A and P categories, respectively.

When we compared the reasoning on comparison and adjustment questions of the "Unequal" puzzles, we found that about 75% of the students used the same pattern. Between 5 and 20% incorrectly concluded that the tastes were equal and therefore were not asked the adjustment question. This error was particularly frequent on Item NX, and often took the form of additive reasoning with incorrect treatment of a remainder, as illustrated by Example D, Table 3.3.

Because the adjustment and comparison questions produced similar results, one might incorrectly infer that a missing value problem is also similar to a comparison question. We believe that this inference is incorrect because of the selection and the psychological set that arise from answering an Unequal lemonade puzzle comparison before the adjustment question. Because adjustment requires the subject to conceptualize equal ratios, we believe that an adjustment question is similar to an Equal lemonade puzzle.

EFFECTS OF GRADE AND SEX

Half of our 120 subjects were sixth graders and half were eighth graders. For the two easiest puzzles (WB and W) and for the most difficult puzzle (NX), the differences in explanations between these two groups were not significant. On the remaining five puzzles, the eighth graders used proportional reasoning more frequently than did the sixth graders, and they used additive and "illogical" reasoning less frequently. The differences were especially large for Puzzles N and BX, on each of which only 3 sixth graders used proportional reasoning, compared to about 20 eighth graders.

A comparison of the frequency distributions of reasoning on the lemonade puzzles of the 60 boys and girls in our sample showed smaller differences than those associated with school grade. Most striking was a tendency for boys to succeed on Puzzles B and BX that could be solved by proportional reasoning using integer ratios between the two recipes. Girls more than boys used additive and illogical reasoning more extensively than proportional reasoning on these two puzzles. A smaller but generally similar difference in reasoning appeared on Puzzle N, which required the comparison of equal noninteger ratios.

CONSISTENCY OF PROPORTIONAL REASONING

The frequency distributions of reasoning presented in Tables 3.4 and 3.5 show differing success in proportional reasoning on the eight puzzles with various structures, but they do not show how consistently any one subject used this reasoning pattern. We provide this information in two stages. First, in Table 3.6, we present the number of times a subject applied proportional reasoning on the eight comparison questions, according to subsamples of grade and sex. Second,

TABLE 3.6

Distributions of the Numbers of Proportional Reasoning Responses by Format, Grade, and Sex[a]

Sample	Number of proportional reasoning responses									
	0	1	2	3	4	5	6	7	8	χ^2
Total sample	30.0	10.8	12.5	15.0	6.7	6.7	5.0	4.2	8.3	
Grade 6	40.0	15.0	10.0	21.7	3.3	6.7	0.0	0.0	3.3	25.9 ($p < .001$)
Grade 8	21.7	6.7	15.0	8.3	10.0	6.7	10.0	8.3	13.3	
Male	23.3	13.3	8.3	15.0	5.0	8.3	5.0	6.7	15.0	13.7 ($p < .1$)
Female	38.3	8.3	16.7	15.0	8.3	5.0	5.0	1.7	1.7	

[a]Percentage.

we consider certain cognitive elements (defined below) that were used to solve the puzzles of progressively increasing difficulty.

One can see in Table 3.6 that about 30% of the subjects never used proportional reasoning, with sixth graders and girls somewhat more likely than eighth graders and boys never to use proportional reasoning. Success in proportional reasoning on many puzzles was achieved by far fewer students than on small numbers of puzzles. The overall distributions for sixth and eighth graders differed at the statistically significant level of $p < .001$ (chi-square test), but the difference between boys and girls was not significant. It is clear that a small number of students used proportional reasoning consistently, but that much larger numbers solved a few of the easier puzzles only.

COGNITIVE ELEMENTS

A comparison of the proportional reasoning procedures used by our subjects on the eight puzzles has led us to identify what we call *cognitive elements*. These elements are individual steps in proportional reasoning that are required on some of the eight lemonade puzzles but not on others. Here are the cognitive elements we have found to be important:

1. Comparing or constructing equal integer ratios (integer–integer–equal); termed the *integral strategy* by Rupley (1981).
2. Comparing an integer ratio with a noninteger ratio (integer–noninteger–unequal).
3. Comparing or constructing equal noninteger ratios (noninteger–noninteger–equal).
4. Comparing unequal noninteger ratios (noninteger–noninteger–unequal).
5. Using only one type of comparison, (Within or Between) on the puzzles solved by proportional reasoning.

6. Using both Within and Between types of comparison or combinations of all four data on the puzzles solved by proportional reasoning.

To find evidence of the use of these cognitive elements, we examined the students' answers to the comparison and adjustment questions. Clearly each cognitive element could be applied on several puzzles. In the analysis we ascribed competence with respect to a cognitive element if the subject used it appropriately on at least one puzzle. To identify a hierarchical relation among the cognitive elements, we tested the assignments of competence with respect to the first four and determined that they formed a Guttman scale (Torgerson, 1958). The results are shown in Table 3.7. Thirty-seven subjects identified in Table 3.7 did not use proportional reasoning at all and therefore lacked competence on any of these cognitive elements. Of the remaining 83 subjects, all but 3 fit the scale of difficulty suggested by our ordering of the elements, with a coefficient of scalability in excess of .90.

Progress from one to two types of comparison could not be ordered into the sequence formed by the other four cognitive elements without a substantial reduction in scalability. We are therefore left with the conclusion that advances in proportional reasoning take place along two partly independent dimensions.

To investigate this two-dimensional progress in greater detail, we have tabulated the sixth- and eighth-grade students in our sample according to the most advanced cognitive element they used (Table 3.8). One can see that most sixth graders either used no cognitive element or were limited to one type of comparison and equal integer ratios. Many eighth graders, however, used both types of comparison and unequal noninteger ratios. Though only a longitudinal study could reveal the order of acquisition of cognitive elements by any one student,

TABLE 3.7

Guttman Scale for Cognitive Elements[a]

Element	Types of comparison		Total	Errors
	One	Two		
None			37	—
Integer–integer, equal (IIE)	38	44	82	—
Integer–noninteger, unequal (INU)	16	34	50	1[b]
Noninteger–noninteger, equal (NNE)	4	19	23	1[c]
Noninteger–noninteger, unequal (NNU)	1	17	18	1[c,d]

[a]$N = 120$.
[b]Omitted IIE.
[c]Omitted INU.
[d]Omitted NNE.

TABLE 3.8

Development of Cognitive Elements[a]

Element	Grade 6 Types of comparison		Grade 8 Types of comparison		Combined total
	One	Two	One	Two	
None	(24)		(13)		(37)
Integer–integer, equal	11	7	11	2	31
Integer–noninteger, unequal	10	5	3	10	28
Noninteger–noninteger, equal	1	0	2	3	6
Noninteger–noninteger, unequal	0	2	1	15	18
	22	14	17	30	120

[a]$N = 120$.

the distribution of numbers in Table 3.8 implies that making both types of comparison is acquired after integer–noninteger–unequal, and before noninteger–noninteger by many students, but not by all.

Our analysis of the tasks and student reasoning according to cognitive elements has identified the difficulty factors that reduced proportional reasoning on certain of the lemonade puzzles compared to others, as shown in Tables 3.4 and 3.5. The least severe of these factors was comparing an integer and noninteger ratio, next was dealing with both types of comparison, and most severe was applying equal or unequal noninteger ratios. Furthermore, students using only one type of comparison preferred the Within type.

Cognitive and Attitudinal Correlates

The wide variations in reasoning on the lemonade puzzles raise questions about the possibility of explaining them in terms of individual differences in cognitive and attitudinal variables. Such an explanation has value because (*a*) the composition of a population sample can be characterized in a generalizable way; (*b*) observed differences in achievement can be related to theories of learning and development; (*c*) apparent effects of sex and grade can be expressed partly and more informatively in terms of these variables; and (*d*) teachers can respond more appropriately to their students' individual needs by stating them with the help of these variables. The variables we used in the present research were information-processing capacity, two components of field dependence–independence, two components of general ability, formal reasoning, divergent thinking, and general attitudes toward mathematics.

Previous Research

Some of the cognitive variables have been studied in conjunction with Piagetian proportional reasoning tasks during the past 10 years. Investigations have employed field dependence (Adi & Pulos, 1980; de Ribaupierre & Pascual-Leone, 1979; Lawson, 1976; Lawson & Wollman, 1977; Nummedal & Collea, 1981), processing capacity (de Ribaupierre & Pascaul-Leone, 1979; Furman, 1980), general ability (Adi & Pulos, 1980; Bart, 1972; Cloutier & Goldschmid, 1976; de Ribaupierre, 1975; Keating & Schaefer, 1975), divergent thinking (Cloutier & Goldschmid, 1976), and formal reasoning (Bart, 1971; Bentley, 1977; Docherty, 1977; Lawson, 1976; 1977; 1979a; 1979b; Martorano, 1977).

Several problems exist in applying these findings to proportional reasoning, however. First, because most of the studies employed Piagetian tasks with their complex apparatus (Inhelder & Piaget, 1958), the correlations could reflect a relationship with demands of the task other than proportional reasoning. Second, few studies employed enough cognitive tasks to determine whether an observed relationship was direct or was moderated by a third variable. Third, the studies treated proportional reasoning as a unitary construct and did not provide for individual cognitive elements. Fourth, suggested components of general ability and field dependence were assessed separately. We intended the present study to identify correlates of proportional reasoning while minimizing these problems.

No research relating attitudes toward mathematics, proportional reasoning, or cognitive variables has come to our attention. Yet there have been reports of gender-related differences on mathematical tasks, and these were observed in conjunction with more negative attitudes by females and toward females' achievement in mathematics (Fennema and Sherman, 1978). Because gender differences in proportional reasoning have been found occasionally (e.g., Karplus *et al.*, 1979), we anticipated that we might find similar effects. Hence we included attitudes toward mathematics among the variables investigated in the present study.

Variables and Instruments

Because the variables are defined operationally by means of their assessment instruments, we shall give an overview of this information before continuing with the specific method employed in our investigation.

INFORMATION-PROCESSING CAPACITY

The information-processing capacity (M *capacity*) construct refers to the number of schemes or psychological units a person simultaneously coordinates while solving a problem. *M* capacity was assessed in the current study with the Figural

Intersection Test (FIT) (deAvila, Havassy, & Pascual-Leone, 1976; Pascual-Leone & Burtis, 1974). The test consists of a series of pages on which there are sets of overlapping geometrical figures. The student's task is to put a dot on the intersection of all but one of the overlapping figures. The items include from three to nine figures and consequently vary in the amount of processing required to solve them. The score was the largest number of intersecting figures solved correctly, up to seven.

FIELD DEPENDENCE–INDEPENDENCE

Recent analyses of the field dependence construct suggest that it contains at least two major components (Linn & Kyllonen, 1981; Pascual-Leone, 1974; Witkin & Goodenough, 1977), (*a*) cognitive restructuring and (*b*) susceptibility to irrelevant salient cues. Cognitive restructuring was assessed with the Find-a-Shape puzzle (Pulos & Linn, 1979), which is an embedded figure test having a simple shape and, on the same page, five complex shapes in which it may be imbedded. It is similar to the Group Embedded Figures Test (Oltman, Raskin, & Witkin, 1971) except that the memory component of the latter is eliminated. The Find-a-Shape puzzle consists of two pages on which the subject has to color the embedded figures, with 1 minute allowed for each page. The score was determined by the number of complex shapes colored correctly.

The second component of field dependence was assessed with the water level task (deAvila, Havassy, & Pascual-Leone, 1976; Pascual-Leone, 1980), in which the student's task is to draw a line representing the water in each of four tilted bottles. Students susceptible to irrelevant clues tend to be influenced by the sides of the bottle and draw the lines more nearly parallel or perpendicular to the sides than is correct (Pascual-Leone, 1980). The score was determined by the closeness of a subject's lines to the horizontal.

GENERAL ABILITY

Many recent analyses of general ability have recognized two distinct components (Horn, 1978): (*a*) crystallized intelligence describing an individual's general accumulation of knowledge and (*b*) fluid intelligence describing an individual's cognitive adaptability to a novel problem situation. Our measure of crystallized intelligence was a 50-item modification of the vocabulary test from the Primary Mental Ability test for Grades 6–9 (Thurstone, 1963). Our measure of fluid intelligence was a 30-item alphabetic series completion task adapted from the same battery. The modifications consisted of additional items to increase the reliability and range of difficulty. On both tasks the scores were the numbers of items correct.

FORMAL REASONING

Formal reasoning was determined by a volume displacement task (Piaget & Inhelder, 1974) in which students had to identify which of two square blocks shown pictorially and labeled according to their volume and weight would raise the level of water higher if dropped into partly filled glasses (Pulos, de Benedictis, Linn, Sullivan, & Clement, 1982). Equality and inequality of volume and weight were combined in various ways on the eight items. The score was the number of items answered correctly.

DIVERGENT THINKING

We assessed divergent thinking by an alternative-uses task (Wallach & Kogan, 1965). In this task, the subjects were asked to generate as many possible uses for a brick as they could think of. The score was the number of different uses listed.

ATTITUDES TOWARD MATHEMATICS

For the investigation of attitudes, we used a 24-item written Mathematics Attitude Survey. The survey items, which were presented as 4-point Likert scales, were 18 items adapted from the Fennema and Sherman (1976) scales and 6 items devised locally to reflect mathematics learning procedures. The attitude score was the sum of a subject's ratings, with negative items reversed so that a score of 4 always indicated a positive attitude.

Method

PROCEDURE

The group tasks described above were administered to most of the sixth and eighth graders in the cooperating school in classroom groups, with the absentees on the day of assessment not participating. Two periods were required, one for the Mathematics Attitude Survey, vocabulary test, series completion test, Find-a-Shape puzzle, water level tasks, and alternative uses test, and the second for the Figural Intersection Test. Test booklets were distributed at the beginning of the period, instructions for each part were given orally, and the students were asked to wait after completing a page or pages until further instructions were given. Work on timed tests was concluded at a signal of the experimenter, who then instructed all students to proceed to the next test.

THE SUBJECTS

The student body of the school where this research was conducted has been briefly characterized in the section describing the lemonade puzzles. Because of absentees and students who moved from the school's attendence area, only 95 students who were interviewed with the lemonade puzzles also contributed to the group tasks.

Results and Discussion

In this section we describe and interpret the understanding of proportional reasoning afforded by our investigation of cognitive and attitudinal variables. Our primary procedure is multiple regression analyses in which we attempt to account for the variances in competence on the cognitive elements at least partly in terms of correlation with information-processing capacity, fluid intelligence, attitudes toward mathematics, and so on. In this procedure, one has to keep two limitations in mind. First, both the interview and the group task results were subject to possible misunderstandings, ambiguous or incomplete responses, and other experimental uncertainties that introduced unpredictable variations. Second, because the variables were related to one another, the multiple regression procedure did not lead to unique results.

CORRELATIONS AMONG THE VARIABLES

The correlations among the group tests for the total sample are presented in the appendix, along with the reliabilities for the tests, average scores, and standard deviations. All tasks had acceptable reliabilities ranging from the high .70s to the low .90s. These variables were not highly related to either grade level or gender, but considerable overlap was evident amont them, with the exception of the alternative-uses task. Such a finding is not surprising because many of the tasks share common requirements. For example, success on the series completion task probably requires a certain amount of M capacity, so that a theoretical overlap between the series completion task and the Figural Intersection test exists.

MULTIPLE REGRESSION ANALYSES

The regressions were conducted with competence on the four cognitive elements of proportional reasoning as dependent variables. We wished to examine these elements separately, in the sense that we looked for prediction of noninteger–noninteger–equal in addition to integer–noninteger–unequal, integer–integer–equal, and the two types of comparison. The noninteger–noninteger–equal analysis included only the 40 subjects having competence on integer–noninteger–unequal, and the integer–noninteger–unequal and two types of comparison analyses included the 67 subjects who had competence on integer–integer–equal. The integer–integer–equal analysis, of course, made use of all 95 subjects for whom group and interview data were available, because integer–integer–equal was the lowest cognitive element on the Guttman scale. No analysis was conducted for the noninteger–noninteger–unequal element separately because of the small number of subjects involved. Correlations and means of the independent variables for the reduced samples are included for reference in the appendix.

The starting point for the multiple regression analyses was Table 3.9, with the simple correlations between the independent variables and the cognitive elements for each of the three different samples. The Figural Intersection Test, series completion, Find-a-Shape puzzle and water level task scores had highly significant correlations with integer–integer–equal. Only series completion was correlated with integer–noninteger–unequal. All the scores except for series completion, volume displacement, and alternative uses were correlated with noninteger–noninteger–equal; only grade level was correlated with use of both types of comparison.

TABLE 3.9

Correlations of Variables with Cognitive Elements

	IIE[a]	INU[b]	NNE[c]	w & b[d]
Vocabulary	.17	−.06	.41*	.02
Figural intersection	.58***	.06	.31*	.05
Series completion	.39***	.25*	.22	.23
Find-a-Shape	.42***	.17	.38*	.20
Water level	.39***	.01	.37*	−.08
Volume	.24*	.07	.25	.14
Alternative uses	.10	−.02	.02	−.04
Attitude survey	.07	.05	.45**	.14
Grade	.21	.20	.41*	.28*
Sex[e]	−.14	−.05	−.43**	−.08*

[a]IIE, integer–integer–equal; $N = 95$.
[b]INU, integer–noninteger–unequal; $N = 67$.
[c]NNE, noninteger–noninteger–equal; $N = 40$.
[d]w & b, both types of comparison; $N = 67$.
[e]Male = 1, female = 2.
*$p < .05$.
**$p < .01$.
***$p < .001$.

Because the alternative uses score had no significant correlations with any of the other variables (Appendix), it was dropped from further analysis. Formal reasoning, which had significant correlations with the other variables, was kept for the multiple regression analyses to determine any indirect effects it might have, in addition to minor direct effects.

To reduce the problem of ambiguity in the multiple regressions due to correlation among the variables, we conducted hierarchical multiple regressions (Cohen & Cohen, 1975) with the order of entry of independent variables determined by their uniqueness and theoretical overlap. The order, from general to progressively more specific capabilities, was vocabulary score, Figural Intersection Test, series completion, Find-a-Shape puzzle, water level task, volume displacement, and mathematical attitude survey. Grade and gender were entered last to find whether they were related to proportional reasoning after the variance attributable to the other variables had been removed. To reduce spurious effects of order, we investigated other orders also and only report robust effects.

A summary of the multiple regressions is presented in Table 3.10. Each of the table's four columns deals with the cognitive element given at the top. The variables are listed in the order of entry in the left column. The three columns of data consist of the squares of the simple correlations coefficients (r^2), of the changes in the squares of the multiple correlation coefficients (ΔR^2), and of the F to enter. The value r^2 is equal to a variable's contribution to the variance if it were entered first in the multiple regression. It is instructive to compare r^2 and ΔR^2 to observe how much of the contribution to the variance has been accounted for by a variable entered earlier.

COGNITIVE ELEMENT INTEGER–INTEGER–EQUAL

The most prominent effect revealed in Table 3.10 is a large contribution of the Figural Intersection Test score to the variance in element integer–integer–equal. In a more detailed examination of this relationship, we found that only 24% of the 26 subjects with M capacity three or four showed competence in integer–integer–equal, but that 83% of the 24 subjects with M capacity of five did so, as did 91% of 45 subjects with M capacity of six or more.

To see how this result may apply to a lemonade puzzle, consider Example D in Table 3.2, an explanation to Puzzle W (3/12, 5/20): "Same taste, 3 goes into 12 four times, 5 times 4 is twenty." The first step appears to have an M demand of four—John's 3 spoons sugar, divide into amount of lemon, John's 12 spoons of lemon juice, quotient 4. The second step may have a demand of 5—direction of operation from sugar to lemon juice, Mary's 5 spoons sugar, multiply, quotient 4 (from first step), match product to Mary's 20 spoons lemon juice.

Also of statistical significance was a small contribution of the Find-a-Shape puzzle score to the variance in element integer–integer–equal. In other words,

TABLE 3.10

Summary of Hierarchical Multiple Regressions for Elements of Proportional Reasoning

	IIE[a]			INU[b]			NNE[c]			w & b[d]		
	ΔR^2	r^2	F	ΔR^2	r^2	F	ΔR^2	r^2	F	ΔR^2	r^2	F
Vocabulary	.03	.03	2.75	.00	.00	.23	.17	.17	7.61**	.00	.00	.02
Figural intersection	.31	.34	42.84***	.01	.00	.41	.02	.10	.75	.00	.00	.14
Series	.00	.15	.54	.08	.07	5.80*	.00	.05	.01	.06	.05	3.89
Find-a-Shape	.05	.18	6.90**	.04	.03	2.97	.03	.14	1.23	.04	.04	3.10
Water level	.01	.15	.80	.00	.01	.11	.06	.14	2.58	.03	.01	1.85
Volume	.00	.06	.63	.01	.00	.46	.00	.07	.05	.00	.02	.01
Attitude	.01	.00	2.04	.01	.00	.32	.12	.21	6.22*	.02	.02	1.64
Grade	.02	.04	2.48	.01	.00	.92	.16	.17	11.22**	.06	.08	4.37*
Sex	.00	.02	.06	.02	.00	1.32	.07	.18	5.56*	.02	.01	1.25
Unexplained variance	.57			.82			.38			.77		

[a]IIE, integer–integer–equal; $N = 95$.
[b]INU, integer–noninteger–unequal; $N = 67$.
[c]NNE, noninteger–noninteger–equal; $N = 40$.
[d]w & b, both types of comparison; $N = 67$.
*$p < .05$.
**$p < .01$.
***$p < .001$.

some cognitive restructuring was related to proportional reasoning with equal integer comparisons.

Series completion and water level task scores, which also had highly significant correlations with integer–integer–equal, did not contribute significantly in the multiple regression. Because they were highly correlated with the Figural Intersection test score, their effects were presumably included in the effect of M capacity.

COGNITIVE ELEMENT INTEGER–NONINTEGER–UNEQUAL

Among our variables, only the series completion score, a measure of fluid intelligence, was significantly related to the integer–noninteger–unequal element, accounting for about 8% of the variance. Once an adolescent has the competence to solve proportionality problems requiring the integer–integer–equal element, therefore, fluid intelligence helps him or her to carry out the additional steps to determine which of two unequal ratios is greater. This effect and its statistical significance are approximately the same, regardless of the order in which the six cognitive variables are entered into the multiple regression.

The very large unexplained variance of .82 suggests that our subjects were less systematic in their use of integer–noninteger–unequal versus integer–integer–equal or that specific knowledge, not related to the variables, was involved in the use of integer–noninteger–unequal.

COGNITIVE ELEMENT NONINTEGER–NONINTEGER–EQUAL

Four variables contributed significantly to the variance in noninteger–noninteger–equal: crystallized intelligence, as measured by the vocabulary test, attitude, as measured by the mathematics attitude survey, and grade level each accounted for about 15% of the variance, whereas gender accounted for about half that amount. Thus, the cognitive obstacle facing students who had mastered integer–noninteger–unequal but had difficulty with noninteger–noninteger–equal appeared to be a deficiency in general knowledge. For instance, these students may not have had a method for representing noninteger ratios.

In addition, a positive attitude toward mathematics was associated with successful application of the element noninteger–noninteger–equal. Also, unspecified factors associated with grade level, such as instruction in ratios and fractions, were related to the use of the cognitive element, as was gender. The large simple correlation of gender with noninteger–noninteger–equal was reduced when the vocabulary and mathematical attitude survey scores were entered in the regression.

THE TYPES OF COMPARISON

In addition to the regression analyses of the three cognitive elements—integer–integer–equal, integer–noninteger–unequal, and noninteger–noninteger–equal—an analysis was conducted of the use of Within and Between types of comparison. The sample consisted of the 67 subjects who exhibited proportional reasoning on at least one puzzle. They were given a score of 2 if they used both types. The only statistically significant contribution to the variance came from the grade level, and this was too small to be meaningful. Furthermore, the multiple regression for the type of comparison score was nonsignificant. This result suggests that the type of comparison may involve specific cognitive skills some adolescents acquire and others do not.

OVERVIEW OF THE REGRESSION ANALYSES

The hierarchical approach to factors that facilitate proportional reasoning has led us to identify information-processing capacity and cognitive restructuring as contributing to the comparison of equal integers, fluid intelligence as contributing to the comparison of an integer with·a noninteger ratio, and crystallized intelligence, favorable attitude, and grade level as contributing to the comparison of noninteger ratios. This result is, of course, not definitive because variables not included in our study may contribute, and the proportional reasoning of other populations—especially younger or older subjects—may relate to the variables in different ways.

It is interesting to note that three of the variables, (*a*) formal reasoning as measured by the volume task, (*b*) susceptibility to irrevelant cues as measured by the water level task, and (*c*) divergent thinking made no significant contributions to any of the cognitive elements in the regression analysis as we conducted it. As a matter of fact, a two-way contingency table in which proportional reasoning was identified with the achievement of integer–noninteger–unequal and formal reasoning with the fifth level on the volume displacement task, only yielded chi square = 6.48, $p < .05$. These results indicate that proportionality is a reasoning pattern separate from reasoning with variables (weight and volume) and physical causality (water is displaced according to volume and not weight of an immersed object) that are considered part of formal reasoning (Inhelder & Piaget, 1958; Piaget & Inhelder, 1974).

The score on the water level task had significant simple correlations with the integer–integer–equal and noninteger–noninteger–equal scores (Table 3.9). The earlier entry into the multiple regression of the Figural Intersection test and Find-a-Shape puzzle scores, however, reduced the water level task contribution to the variances in integer–integer–equal and noninteger–noninteger–equal scores to statistical nonsignificance. Thus the water level task did not measure a unique trait that was important for the two cognitive elements apart from what was measured by the Figural Intersection test and Find-a-Shape puzzle.

Our results lend support to the distinction between competence and performance (Flavell & Wohlwill 1969; Stone & Day, 1978). Competence in proportional reasoning, exemplified by the use of cognitive element integer–integer–equal, required a minimum *M* capacity, as suggested by de Ribaupierre and Pascual-Leone (1979) and Case (1979). Yet competence of a student did not assure good performance as exemplified by use of the other cognitive elements. Performance was aided by fluid intelligence, general knowledge, and a positive attitude toward mathematics.

Conclusions

In this section we consolidate the results of our study in relation to earlier research on proportional reasoning. We also propose extensions of cognitive theories of proportional reasoning, leading to the discussion of implications for teaching in the final section of this chapter.

Relation to Previous Research

How do our findings relate to the previous research cited earlier in this chapter? We believe that our results complement and extend what was known before. They generally substantiate results concerning the diversity of reasoning approaches, provide for the first time data on the consistency of proportional reasoning, provide an understanding of difficulty factors in terms of cognitive elements of proportional reasoning, and identify cognitive correlates of the elements.

DIVERSITY AND FREQUENCY OF PROPORTIONAL REASONING

Our results reveal again the high frequency of additive reasoning that has been commonly reported on missing-value proportionality problems. At the same time, however, the comparison questions elicited a major group of qualitative explanations that did not occur on the adjustment questions and had not been specifically recognized in research using missing-value problems. In fact, about 30% of the subjects rarely used more than qualitative reasoning on any item and did not appear to grasp the quantitative aspects of the lemonade puzzles. This group of students may be expected to have quite different educational needs compared to students who use systematic quantitative reasoning.

The actual diversity of reasoning we reported in Tables 3.2 and 3.3 is similar to that described in detail, but without relative frequencies, by Noelting (1980a) for the similar orange juice experiment. Important differences between our subjects and Noelting's were that his often computed the total amounts of beverage in a recipe, whereas very few of our subjects did so. Rupley (1981) reported that most successful seventh graders used the integral multiple strategy on missing value problems (Examples B and C in Table 3.2), but that older students solved the same problems more frequently by a unit ratio approach (Example A in Table 3.2).

To our knowledge, there has been no previous study extensively describing the consistency of proportional reasoning by a student sample on a set of tasks. Hence, many of the results we described in the previous section, such as the scalability of cognitive elements, are new and cannot be compared with other research. A partial study of consistency was carried out by Noelting (1980a) and Noelting and Gagne (1980) when they performed a scalogram analysis of the success frequencies. Unfortunately, their analysis was based only on their subjects' success frequencies on the various items, and not on the procedures their subjects actually used to solve problems. These authors' stages relate to our cognitive elements and are discussed below.

LEVELS OF PROPORTIONAL REASONING

The neo-Piagetian theory of Case (1978, 1979), as well as the more phenomenological proposals of Noelting (1980b) and Noelting and Gagne (1980), can be compared with our observations regarding the cognitive elements. We have support for their views of levels or substages in that the four cognitive elements formed a Guttman scale that resembles Noelting's (1980b) proposals. Yet, contrary to Noelting's (1980b) suggestions, we found that the restructuring involved in using both Within and Between comparisons could not be included in this scale. The acquisition of proportional reasoning, whether developmental or through instruction (Kurtz and Karplus, 1979), is therefore more complicated than has been proposed.

Nevertheless, we can compare our analysis with the sequences proposed by Noelting (1980b) and Case (1979). The summary in Table 3.11 reveals many similarities among the three proposed sequences. The third and fifth rows of the table are virtually alike in their criteria. Differences in the first row are minor in that the *a:a* ratio used by Noelting in his tasks was simpler than any we used.

The principal difference between our results and those of earlier studies is that we require a fourth level, noninteger–noninteger–equal, which Noelting and

TABLE 3.11

Noelting's (1980b), Case's (1979), and This Study's Progression of Proportional Reasoning

Piagetian substage	Noelting	Case	This Study
II A	Equivalence of *a:a* ratios	1. No comparison of ratios	Noncompetence on the set of eight lemonade puzzles
II B	Equivalence of any equal ratios	2. Equivalence of any equal ratios	Equivalence of equal integer ratios only
III A	Nonequivalence of integer and noninteger ratios	3. Nonequivalence of unequal ratios with an integral ''between'' ratio	Nonequivalence of integer and noninteger ratios Equivalence of equal noninteger ratios
III B	Nonequivalence of unequal noninteger ratios	4. Nonequivalence of unequal noninteger ratios by finding common denominator	Nonequivalence of unequal noninteger ratios

Case included in their second substage. Thus there are differences between our results and earlier findings that may be ascribed to differences in samples or method, or to errors.

TYPES OF COMPARISON

Our subjects used Within comparisons more frequently than Between comparisons when both led to integer or to noninteger ratios. Our results differ from Vergnaud's (1980) findings on the fuel consumption problems and Lybeck's (1978) findings with stretching springs, but they agree with Lybeck's (1978) finding for the Cylinder Task. We therefore have further evidence that the context as well as the numerical relationships of a proportions problem significantly influences the type of comparison preferred by many students.

COGNITIVE AND ATTITUDINAL CORRELATES

The multiple regression analyses of the cognitive elements of proportional reasoning have largely overcome the four problems of previous research of cognitive variables related to proportional reasoning. Our most clear-cut result, the importance of five units of information processing capacity, is in accord with earlier findings (de Ribaupierre & Pascual-Leone, 1979; Furman, 1980) of an *M* demand of five when proportional reasoning was assessed by balance beam or projection of shadows tasks.

Because our study was the first one concerned with student attitudes and proportional reasoning, our finding of a relationship for the cognitive element noninteger–noninteger–equal cannot be compared with earlier results. The reduction in the variance associated with gender by the inclusion of the attitude score is consistent with studies of mathematics achievement more generally (Fennema & Sherman, 1978). With respect to divergent thinking, the lack of relationship we observed suggests that Cloutier and Goldschmid's (1976) report of a connection with proportional reasoning may depend on the population studied. The lack of relation between formal reasoning and proportional reasoning is consistent with the analysis of Lawson, Karplus, and Adi (1978), who used the Cylinder Puzzle (Suarez, 1977) to assess proportional reasoning, but differs from the results of other researchers who assessed proportional reasoning with a more complex task derived from Inhelder and Piaget (1958).

Theories of Proportional Reasoning

Because our findings are generally consistent with earlier empirical and theoretical views but add detail, it is appropriate for us to propose interpretations that consolidate these results and permit them to be used to suggest further research and implications for teaching.

DEVELOPMENT OF PROPORTIONAL REASONING

Even though our subjects were limited to a 2-year range, our data led to some implications for developmental theories of proportional reasoning, if advances in proportional reasoning beyond the integer–integer–equal level are developmental. We found that the cognitive elements could not be ordered along a single scale, but that they formed a two-dimensional grid, as illustrated in Table 3.8. In other words, development has to account for students' progress both in their ability to reason with equal or unequal integer and noninteger ratios, as well as for their progress in applying both Within and Between types of comparison. The result shows that a reasoning pattern such as proportional reasoning may be a more complicated mental process than expected. Given this two-dimensional character of advances in proportional reasoning, it is clear that the single-scale theories of Noelting (1980b) and Case (1979) are not adequate.

HIERARCHY OF STRATEGIES

We conclude this section on a speculative note by proposing a new cognitive theory of proportional reasoning. We suggest that one has to consider three hierarchical levels of strategies in proportional reasoning: (*a*) the top level involves the decision of whether to apply direct proportion, inverse proportions, additive reasoning, or another numerical relationship; (*b*) the middle level concerns the type of numerical comparisons or combinations that will be made; (*c*) the bottom level deals with the execution of the numerical operations that are to be carried out. For a person who understands the relationship of the variables in whatever context the task is posed, the top-level decision is made firmly whereas the middle-level decision is left open, pending an exploration of the operations required at the bottom level. After some searching, a middle-level decision aimed at economy of effort is made and the bottom-level computations are carried out.

In our theory, the reduced use of proportional reasoning on the Unequal compared to Equal lemonade puzzles results from a search for economy that erroneously involves the top strategy level. Rather than deciding definitively on the numerical relationship from the problem context, some students reconsider their top-level decision to use proportional reasoning when they encounter unequal or noninteger ratios and apply a constant difference procedure instead.

A HYPOTHETICAL ILLUSTRATION

To make this theory more tangible, we have constructed a decision tree (Figure 3.2) that illustrates a sequence of steps if, indeed, the steps are carried out sequentially. The diagram represents the reasoning of a hypothetical subject who uses both types of comparison and one or more of the other cognitive elements we have discussed earlier.

Suppose our hypothetical subject begins at START and tests for an integer ratio of lemon to sugar for John. If the ratio is an integer the subject then determines whether Mary's lemon:sugar ratio is the same. If it is, the subject concludes that the recipes have the same taste. If it is not (Decision Point 1), then (*a*) the subject using integer–integer–equal switches to an additive approach, whereas (*b*) the subject using integer–noninteger–unequal computes the amount of lemon juice Mary would need for an equal taste (marked by *) and concludes from the inequality test that Mary's concentrate is sweeter.

Now return to START and consider the steps when the subject does not find an integer ratio originally. The subject then tests the sugar:sugar ratio (between comparison) and proceeds along the central column of the diagram if this is an integer. Decision Point 2, analogous to Decision Point 1, leads to different outcomes for subjects using elements integer–integer–equal or integer–noninteger–unequal.

If the sugar ratio is not an integer, the subject proceeds along the line to the right side of the diagram. At Decision Point 3, the integer branch leads to the difference comparison, whereas the noninteger branch leads to computation of the noninteger sugar-to-lemon juice ratios as fractions that can be compared. If

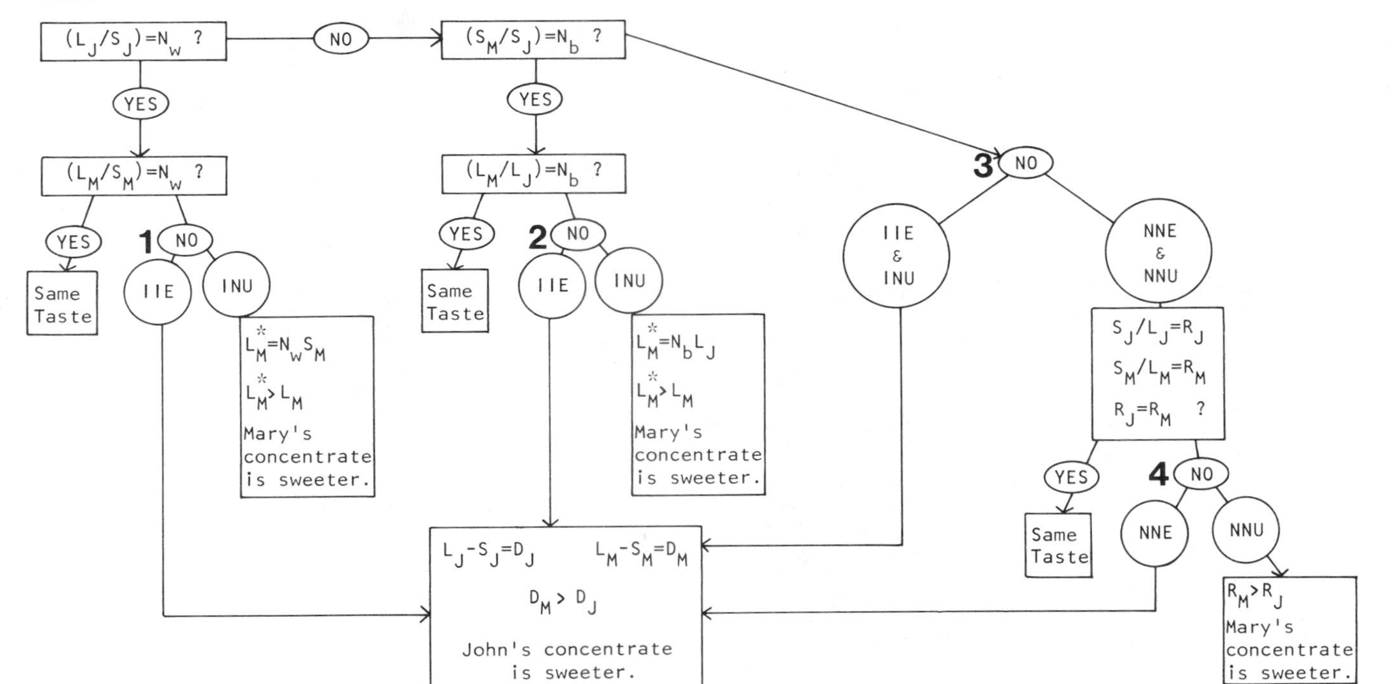

Figure 3.2 Decision tree showing the use of cognitive elements of proportional reasoning. L_J, S_J = John's amounts of lemon juice, sugar; L_M, S_M = Mary's amounts of lemon juice, sugar; L_M^* = amount of lemon juice Mary needs so her concentrate tastes like John's; N_w, N_b, D_J, D_M = counting numbers; R_J, R_M = rational numbers.

they are found to be equal, then the tastes are declared equal. If the two fractions are found unequal (Decision Point 4), then (*a*) the subject using noninteger–noninteger–equal switches to the additive approach, whereas (*b*) the subject using noninteger–noninteger–unequal concludes that Mary's ratio is greater and therefore her concentrate tastes sweeter.

PROPERTIES OF THE MODEL

It is clear from Tables 3.2 and 3.3 that we have used certain arithmetic steps in Figure 3.2 to represent whole classes of equivalent steps that may be used. Thus, the figure is intended to illustrate a possible sequence of reasoning only, and should not be taken too literally in its computational details.

It is also clear that comparable decision trees can be generated for individuals who prefer the Between comparison to the Within comparison or those who use cognitive element w/b instead of w & b. For the latter, Decision Point 3 would be reached immediately after they fail to find an integer Within-recipe ratio and they would not test for an integer Between-recipe ratio—they would skip over the central column of the diagram.

The key decision points of our theory, where the subject encounters a negative result, have been marked in Figure 3.2 with the numbers 1–4. The student who commands all cognitive elements makes the correct choice at each circle. The person who is less skilled switches sooner or later to a difference comparison process. We referred to this switch when we stated above that the unskilled subject reconsiders the top level decision by which the numerical relationship for testing the recipes was chosen. In our theory, therefore, use of only some cognitive elements is interpreted as evidence that the three decision-making levels are not ordered in the proper hierarchy.

The reader may wonder to what extent the decision tree in Figure 3.2 is based on more than inferences from the lack of consistency of proportional reasoning. As a matter of fact, several subjects changed their answers and explanations during the interviews, sometimes from an additive to proportional strategy, sometimes in the other direction. A few subjects were questioned about their procedures after the conclusion of the interview and indicated that their criteria for choosing a strategy lay in whether the numbers could be "divided evenly." For example, one subject explained, "Well, it's well . . . it's like if you can divide it, that's a good method. But subtracting is good, if you can't divide it this way. If it won't divide evenly, I would then go to subtracting."

THE CONTRIBUTIONS OF CORRELATES

How do our findings about the independent variables fit into this theory? Because we are dealing with subjects who have competence, they share the minimal information processing capacity and cognitive restructuring ability required. To apply cognitive element integer–noninteger–unequal at Decision Points 1 or 2 in Figure 3.2, they are aided by cognitive adaptability to accept the unexpected negative result of the ratio test. To apply the cognitive element noninteger–noninteger–equal at the upper Decision Point 3 in Figure 3.2, they are aided by the general knowledge of representing ratios as fractions and by a positive attitude toward mathematics that allows them to persist on the task on which they have so far experienced little success. In other words, the correlates of the individual cognitive elements make the theory more plausible.

What about the subjects who lack competence and therefore never display any operations represented in the diagram? As a matter of fact, about 40% of the 37 subjects used additive reasoning consistently and thus appeared to encode the data completely and correctly, but failed to use the correct relationship. Perhaps they began at START and immediately switched to the additive approach. The remaining 60% used illogical or qualitative strategies and thus appear to encode the problem incompletely and/or incorrectly. The limited information processing capacity of virtually all these students may have prevented them from proceeding more effectively.

Implications for Teaching

This chapter concludes with suggestions for teaching. After reviewing recent research into the teaching of proportions, we describe teaching approaches that promise to advance the level of proportional reasoning of early adolescents.

Previous Research

We have not been able to trace the research origins of the present instructional approach to proportional reasoning in the schools in the United States. Examination of the widely used texts reveals that students are shown how to represent the information in proportions word-problems as an equivalent fraction equation, and to solve by cross-multiplying and then dividing. Cross-multiplication is traditionally used to test for equality of ratios, whereas converting into equivalent fractions with a common denominator is used to compare unequal ratios. Our results and those cited earlier show conclusively that these approaches are used meaningfully by only a very small fraction of early adolescent students. Rupley (1981) found that, even among eleventh-grade students, fewer than 15% used such strategies.

In other countries, where proportional reasoning has been found considerably more frequently than in the United States (Karplus *et al.*, 1979; Suarez, 1977), teaching makes use of the unit ratio strategy, in which the proportions problem is divided into two distinct steps. In the first step, the solver uses the information about, say, the number of spoonfuls of sugar and lemon juice in John's concentrate to find how many spoonfuls of lemon juice are needed for one spoonful of sugar. In the second step, the solver finds the amount of lemon juice needed for the actual amount of sugar.

When the proportionality problem was presented to these students in an unconventional format, however, the number of lower-ability students who succeeded dwindled drastically (Karplus, 1979a). Apparently, therefore, many students had learned the unit ratio strategy only well enough to apply it to routine problem statements.

In the United States, Herron and Wheatley (1978) and Goodstein and Boelke (1980) have recently studied the unit ratio approach and developed activities for introducing it in secondary school classes. A process of matching subsets, as in "for every 2 spoons of sugar there are 5 of lemon juice," was used by Wollman and Lawson (1978) to teach seventh graders to solve proportions problems iteratively. It also served as the starting point for Gold's (1978; 1980) neo-Piagetian approach that proceeded to a unit ratio strategy and was governed by a stepwise procedure that increased complexity gradually.

A somewhat broader-based approach developed by Kurtz (1976; Kurtz & Karplus, 1979) attempted to deal with the students' need to differentiate constant ratio and constant difference relationships by properly identifying the problem variables and connecting them with everyday experience. Kurtz (1976) also used a learning cycle combining student autonomy and teacher input (Karplus, 1979b) to establish a secure connection between the newly learned mathematical strategies and the students' prior knowledge.

Our conclusion from this past work is that the dominant equivalent fraction approach to teaching ratio and proportion in the United States fails to teach more than a small minority of early adolescents. The unit ratio approach is more effective but nevertheless is likewise limited in its impact. The procedures of Wollman and Lawson (1978), Kurtz (1976), and Gold (1978; 1980) are promising but have not really been developed to the point of large-scale implementation. Yet the students' need to deal successfully with linear functions in their high school algebra, chemistry, and physics courses makes more successful teaching of proportional reasoning a high priority for junior high-school mathematics programs.

New Directions

Because our research generally extended previous findings without serious conflict, its implications for teaching overlap the ideas that have been introduced recently and were described above. We therefore build on this work and refine it as appropriate in the following suggestions.

RELATIONSHIPS AMONG VARIABLES

First, and perhaps most important, is our observation that few students conceptualized the taste of the concentrate as a variable that depended on certain other variables—the amounts of sugar and lemon juice—in a well-defined way. Instead, some subjects applied numerical relationships as they felt convenient, whereas others did not apply numerical relationships at all. This outcome is not unique to proportionality problems but helps one to understand why word problems in general are difficult for students to solve.

How can one teach students about variables and their relationships? Kurtz's (1976) approach made use of the learning cycle of exploration, concept introduction, and concept application that has been applied successfully at the elementary, secondary, and college levels (ADAPT, 1977; Eakin & Karplus, 1976; Karplus, Lawson, Wollman, Appel, Bernoff, Howe, Rusch, & Sullivan, 1977). The learning cycle incorporates the active learning approach advocated by a diverse group of learning specialists, including Gagne, Piaget, Lawson, Case, Ausubel, and Karplus (Lawson, 1979c). It helps students gain experience through their own investigations and then to adapt their thoughts to the requirements of the subject matter as presented by the teacher, textbook, or other authority.

During exploration, the students learn through their own actions and reactions to a new situation, new ideas, or novel problems, with a minimum of teacher guidance or expectation of specific accomplishments. Students might be shown a pencil held upright against a desk top and asked what affects the length of a stick's shadow when it is in the sunlight. They may suggest the stick's length, the time of day, the angle of the sun, the angle of the ground, the weather, and other factors.

The second phase, concept introduction, builds on exploration and focuses on the introduction of a new concept or principle (such as *variable*) that leads students to reason about their exploration in a new way. In our example, the teacher would explain that the stick's length, the angle to the sun, and other suggestions are called *variables* that might affect the length of the shadow. The students would then be encouraged to consider which of these variables would have a greater or smaller effect, or perhaps no effect at all, and whether there might be other variables to consider.

In the last phase of the learning cycle, concept application, the students apply the new concept to additional examples. The comparison of shadow lengths from various sticks making the same angle with the ground would allow some students to recognize a linear relationship between shadow and stick length. The effects of the other variables could be investigated also. Other phenomena to be discussed in a similar way include the number of swings of a pendulum in 10 seconds, the rate of cars passing a certain street location, and the flavor of lemonade made from lemon juice, sugar, and water. A very important element of the discussion would be to identify sufficiently many variables in order to ensure that most students agree that the outcome will be the same (i.e., same shadow length, same number of cars per minute, and so on) if all the variables have the same values.

DEVELOPING THE INTEGRAL STRATEGY

Use of the integral strategy (Rupley, 1981) required a minimum processing capacity from our subjects. We also reported that most of them used only one type of comparison, either within or between recipes, and that few subjects used paper and pencil to aid them in setting up the problem.

Several teaching approaches suggest themselves for overcoming these difficulties, Gold (1978; 1980) considered the *M* capacity requirement and recommended breaking the problems into simpler steps. He devised an instructional sequence that included matching a given unit ratio to find a missing element, repeating a unit ratio to set up a more general integral ratio, and reducing a given integral ratio to a unit ratio. He also recommended using manipulatives, such as Cuisenaire rods, poker chips, or even red and white beans, to model the ratios, verify correct procedures, and challenge errors. Pascual-Leone (1976) has suggested a "graded learning loop" for such a sequence.

To diversify the types of comparisons used by students, Gold's or similar procedures should include examples that require within and between comparisons in the exploration and application phases of a learning cycle. During the concept introduction phase, the teacher should model both types of comparison.

Other techniques that will be helpful and that should therefore be introduced by the teacher include suggesting visual images to model the numerical relationships graphically (Kurtz, 1976) or even by dot and cross-tally marks that are grouped to record the key data in a pattern that suggests the relationships.

SELECTING A REASONING PATTERN

Because many students alternated between constant difference and constant ratio comparisons, their problem-solving success will be increased by a technique that allows them to select the appropriate procedure reliably. Kurtz (1976) developed a learning cycle around this objective. In the exploration phase, the teacher presents a two-variable table with entries $X = 4$, $Y = 6$; $X = 8$, $Y = ?$ The students are challenged to propose values for the missing entry and to justify their suggestions with reference to illustrative examples. Many possibilities exist, of course. Some of the most commonly used ones are $Y = 12$ (4 candies cost 6 cents, 8 candies cost 12 cents—an application of the between integral strategy), $Y = 10$ (when John is 4 years old, Mary is 6; when John is 8 years old, Mary will be 10), and $Y = 2$ (if John takes 4 cookies from a package, Mary gets 6; if John takes 8 cookies from a similar package, Mary gets only 2).

Concept introduction in this learning cycle presents the idea that numbers, by themselves, are insufficient to determine an answer uniquely; that the meaning of the variables and the situation they describe must be used to infer which numerical relationship is most appropriate. Note the application of the variables concept described earlier in this section.

Concept application includes a variety of examples in which students have to select the correct relationship, test their selection with measurements on a real situation, evaluate the relationship used in a possibly incorrect "solution" that is described to them, and make up problems from given data as in the exploration phase.

COMPARING UNEQUAL RATIOS

In the lemonade puzzles, the subjects were asked which concentrates taste sweeter, or whether the tastes were the same. Puzzles with unequal ratios were considerably more difficult than were the equal-ratio puzzles. A promising approach for students who use the integral strategy is to ignore one of the given amounts temporarily and to compute this "missing" value for equal tastes (Figure 3.2). In a second step, the given and newly found values are compared. If they are unequal, additional reasoning is required to determine which recipe tastes sweeter. Here again, manipulatives or visual representations can be of great value to verify procedures and to discover errors.

COMPARING NONINTEGER RATIOS

We found that the comparison of unequal and noninteger ratios were the most difficult cognitive elements. Analysis of the cognitive and attitudinal variables indicated that crystallized intelligence and positive, persistent attitudes toward mathematics contributed to the use of element noninteger–noninteger–equal. To improve proportional reasoning in this respect, we recommend several procedures: (*a*) overcoming students' avoidance of fractions by games and other motivating activities in which the use of fractions is rewarded (Thier & Willis, 1981); (*b*) encouraging self-directed learning activities in which students formulate their own goals and persist to achieve them, as in the exploration and application phases of a learning cycle; (*c*) using a unit ratio approach with decimal number representations and providing calculators for the fraction–decimal conversions; and (*d*) dealing explicitly with the erroneous "remainder" strategy (see Table 3.3, Examples C and D).

In view of the very small numbers of students who compared noninteger and unequal ratios successfully, we expect that major changes in teaching programs will be required to improve early adolescents' proportional reasoning in these repects. Of course, it is possible that the study of relationships among variables and more effective selection of reasoning patterns will lead many students to forego giving an answer to a difficult proportionality problem rather than to use an inappropriate reasoning pattern to produce a wrong answer.

It is clear that a great deal of progress has been made during the past 10 years in the understanding of early adolescents' proportional reasoning. Most significant, there appear to be specific directions for the teaching of ratio and proportion that offer substantial advantages over the presently used algebraic approach.

Acknowledgments

Our study has benefited from the assistance of Peter Birns, Anne Bois, Diane Downie, Linda Kakigi, Verna Norris, and Geraldine Patterson, who helped in the task design, conducted interviews, and contributed to the data reduction. We are indebted to Margaret Brown, Richard B. Davis, Constance Kamii, Jonas Langer, Richard Lesh, William H. Rupley, Michael Shayer, and Richard Skemp for helpful discussions of our observations and ideas. The cooperation and patience of the students and teachers of Albany Middle School, Albany, California are gratefully noted.

Appendix

To provide documentation for the multiple regression analysis in the section on cognitive and attitudinal correlates of proportional reasoning, we present here the simple correlation coefficients of the variables with one another. There are three tables (Table 3.A1, 3.A2, 3.A3) because there were three differing populations ($N = 95$, $N = 67$, and $N = 40$). Because almost 50 correlation coefficients were calculated for each table, we adopted a statistical criterion of $p < .01$, which was used to indicate significant correlations in Table 3.A1. Because of the smaller number of subjects in Tables 3.A2 and 3.A3 compared to 3.A1, the threshold for significance increased and there were more correlation coefficients in the range $.01 < p < .05$. These are also in Tables 3.A2 and 3.A3, but must be interpreted very cautiously.

TABLE 3.A1

Correlations among Moderator Variables for the Regression Analysis of Cognitive Element IIE (N = 95)

	Vocabulary	FIT[a]	Series	FASP[b]	WLT[c]	Volume	Alternative uses	MAS[d]
ocabulary		.24						
IT[a]	.24							
eries	.39***	.57***						
ASP[b]	.45***	.39***	.35***					
VLT[c]	.13	.53***	.30**	.28**				
'olume	.25	.38***	.37***	.43***	.23			
lternative uses	.12	−.17	−.08	−.19	−.11	.09		
IAS[d]	.10	.29**	.16	.16	.23	.14	.12	
irade	.18	.05	.27**	.19	.06	.04	−.30**	−.09
ex[e]	−.09	−.23	.01	−.10	−.30**	−.28**	−.06	−.12
	22.30	5.42	12.56	9.10	21.29	4.39	17.89	54.89
D	8.11	1.37	5.57	5.34	9.73	2.48	18.02	8.89
Iaximum	50.00	7.00	30.00	20.00	32.00	8.00	—	96.00
	.82	.93	.77	.77	.84	.81	—	.76

[a]Figural Intersection Test.
[b]Find-a-Shape Puzzle.
[c]Water Level Test.
[d]Mathematics Attitude Survey.
[e]Male = 1, female = 2.
**$p < .01$.
***$p < .001$.

TABLE 3.A2

Correlations among Moderator Variables for the Regression Analysis of Cognitive Elements INU and w & b (N = 67)

	Vocabulary	FIT[a]	Series	FASP[b]	WLT[c]	Volume	Alternative uses	MAS[d]
FIT[a]	.24*							
Series	.36**	.48***						
FASP[b]	.43***	.23	.21					
WLT[c]	.24*	.41***	.28*	.28*				
Volume	.15	.26*	.33**	.39**	.22			
Alternative uses	.11	−.07	−.05	−.17	.05	.12		
MAS[d]	.06	.35**	.25	.20	.27	.18	.09	
Grade	.19	−.06	.21	.19	.03	.03	−.32**	−.19
Sex[e]	−.17	.01	.03	−.15	−.22	−.34**	−.13	−.09
$\bar{X}$	23.23	5.78	14.06	10.62	23.83	4.79	16.72	55.30
SD	8.87	1.10	5.26	5.32	9.20	2.67	15.37	9.50
Maximum	50.00	7.00	30.00	20.00	32.00	8.00	—	96.00

[a]Figural Intersection Test.
[b]Find-a-Shape Puzzle.
[c]Water Level Test.
[d]Mathematics Attitude Survey.
[e]Male = 1, female = 2.
*$p < .05$.
**$p < .01$.
***$p < .001$.

TABLE 3.A3

Correlations among Moderator Variables for the Regression Analysis of Cognitive Element NNE (N = 40)

	Vocabulary	FIT[a]	Series	FASP[b]	WLT[c]	Volume	Alternative uses	MAS[d]
FIT[a]	.47**							
Series	.48**	.49**						
FASP[b]	.50**	.45**	.31					
WLT[c]	.17	.37*	.19	.14				
Volume	.39*	.27	.32*	.57***	.34*			
Alternative uses	−.15	−.04	.08	−.14	.21	.30		
MAS[d]	.13	.37*	.23	.16	.29	.08	.15	
Grade	.12	.11	.28	.19	.15	.10	−.37*	−.08
Sex[e]	−.24	.16	.15	−.31*	−.07	−.42*	−.31*	−.10
$\bar{X}$	22.74	5.80	15.11	11.26	24.14	4.89	16.44	55.14
SD	9.73	1.08	4.82	5.96	9.72	2.81	15.41	8.32
Maximum	50.00	7.00	30.00	20.00	32.00	8.00	—	96.00

[a]Figural Intersection Test.
[b]Find-a-Shape Puzzle.
[c]Water Level Test.
[d]Mathematics Attitude Survey.
[e]Male = 1, female = 2.
*$p < .05$.
**$p < .01$.
***$p < .001$.

References

Abramowitz, S. *Investigation of adolescent understanding of proportionality* (Doctoral dissertation, Stanford University, 1974). Ann Arbor, Michigan: Xerox University Microfilms, 1976.

Accent on Development of Abstract Processes of Thought. *Multidisciplinary Piagetian based programs for college freshmen.* Lincoln: University of Nebraska, 1977.

Adi, H., & Pulos, S. Individual differences and formal operational performance of college students. *Journal for Research in Mathematics Education,* 1980, *11,* 150–156.

Bart, W. M. The factor structure of formal operations. *British Journal of Educational Psychology,* 1971, *41,* 70–77.

Bart, W. M. Construct and validation of formal reasoning instruments. *Psychological Reports,* 1972, *30,* 663–670.

Bentley, W. *An exploration of structuring in the stage of formal operations.* Paper presented at the Annual Symposium of the Jean Piaget Society, Philadelphia, 1977.

Case, R. Intellectual development from birth to adulthood: a neo-Piagetian interpretation. In R. Siegler (Ed.), *Children's thinking: What develops?* Hillsdale, New Jersey: 1978.

Case, R. Intellectual development and instruction: A neo-Piagetian view. In A. E. Lawson (Ed.), *The 1980 AETS Yearbook: The psychology of teaching for thinking and creativity.* Columbus, Ohio: ERIC/SMEAC, 1979.

Cloutier, R., & Goldschmid, M. L. Individual differences in the development of formal reasoning. *Child Development*, 1976, *47*, 1097–1102.

Cohen, J., & Cohen, P. *Applied multiple regression/correlation analysis for the behavioral sciences*. Hillsdale, New Jersey: Erlbaum, 1975.

de Avila, E. A., Havassy, B., & Pascual-Leone, J. *Mexican American schoolchildren: A neo-Piagetian analysis*. Washington, D.C.: Georgetown University Press, 1976.

de Ribaupierre, A. *Mental space and formal operations*. Unpublished doctoral dissertation, University of Toronto, 1975.

de Ribaupierre, A., & Pascual-Leone, J. Formal operations and *M* power: A neo-Piagetian investigation. In D. Kuhn (Ed.), *Intellectual development beyond childhood*. San Francisco: Jossey–Bass, 1979.

Docherty, E. M. Qualitative differences in concrete and formal operational tasks. *Contemporary Educational Psychology*, 1977, *2*, 25–30.

Eakin, J., & Karplus, R. *SCIS final report*. Berkeley: University of California, 1976.

Fennema, E. H., & Sherman, J. A. Fennema-Sherman mathematics attitude scales. *Journal Supplement Abstract Services*, 1976, Ms. 1225.

Fennema, E. H., & Sherman, J. A. Sex-related differences in mathematics achievement and related factors: A further study. *Journal for Research in Mathematics Education* 1978, *9*(3), 189–203.

Flavell, J., & Wohlwill, J. Formal and functional aspects of cognitive development. In D. Elkind & J. H. Flavell (Eds.), *Studies in cognitive development*. New York: Oxford University Press, 1969.

Freudenthal, H. *Weeding and sowing: A preface to a science of mathematical education*. Dordrecht, Netherlands: Reidel, 1978.

Furman, I. *M demand of problem solving strategies in Piaget's balance-scale task*. Paper presented at the annual meeting of the American Psychological Association, Toronto, 1980.

Gold, A. P. *Cumulative learning versus cognitive development: A comparison of two different theoretical bases for planning remedial instruction in arithmetic*. Doctoral dissertation, University of California, Berkeley, 1978.

Gold, A. P. A developmentally based approach to the teaching of proportionality. In R. Karplus (Ed.), *Proceedings of the Fourth International Conference of the Psychology of Mathematics Education*. Berkeley, California: Lawrence Hall of Science, 1980.

Goodstein, M. P., & Boelke, W. W. *A prechemistry course on proportional calculation*. Columbus, Ohio: ERIC/SMEAC, 1980.

Hart, K. The understanding of ratio in the secondary school. *Mathematics in School*, 1978, *7*(1), 4–6.

Hart, K. *Children's understanding of mathematics* (Vol. 11–16). London: Murray, 1981.

Herron, J. D., & Wheatley, G. H. A unit factor method for solving proportion problems. *The Mathematics Teacher*, 1978, *71*, 18–21.

Horn, J. L. Human ability systems. In P. B. Baltes (Ed.), *Life-span development and behavior*. New York: Academic Press, 1978.

Inhelder, B., & Piaget, J. *The growth of logical thinking from childhood to adolescence*. New York: Basic Books, 1958.

Karplus, R. Proportional reasoning in the People's Republic of China. In J. Lochhead & J. Clement (Eds.), *Cognitive process instruction*. Philadelphia, Pennsylvania: The Franklin Institute Press, 1979. (a)

Karplus, R. Teaching for the development of reasoning. In A. E. Lawson (Ed.), *The 1980 AETS Yearbook: The psychology of teaching for thinking and creativity*. Columbus, Ohio: ERIC/SMEAC, 1979. (b)

Karplus, R. Education and formal thought: A modest proposal. In I. E. Siegel, (Ed.), *New directions in Piagetian theory and practice*. Hillsdale, New Jersey: Erlbaum, 1981.

Karplus, R., Adi, H., & Lawson, A. E. Intellectual development beyond elementary school: Proportional, probabilistic, and correlational reasoning (Vol. 8). *School Science and Mathematics*, 1980, *80*, 673–683.

Karplus, R., & Karplus, E. F. Intellectual development beyond elementary school: Ratio, a longitudinal study. (Vol. 3). *School Science and Mathematics*, 1972, *72*, 735–742.

Karplus, R., Karplus, E. F., Formisano, M., & Paulson, A. C. Proportional reasoning and control of variables in seven countries. In J. Lochhead & J. Clement (Eds.), *Cognitive process instruction*. Philadelphia, Pennsylvania: The Franklin Institute Press, 1979.

Karplus, R., Karplus, E. F., & Wollman, W. Intellectual development beyond elementary school: Ratio, the influence of cognitive style (Vol. 4). *School Science and Mathematics*, 1974, *74*, 476–482.

Karplus, R., Lawson, A. E., Wollman, W., Appel, M., Bernoff, R., Howe, A., Rusch, J. J., & Sullivan, F. *Workshop on science teaching and the development of reasoning*. Berkeley, California: Lawrence Hall of Science, 1977.

Karplus, R., & Peterson, R. W. Intellectual development beyond elementary school II: Ratio, a survey. *School Science and Mathematics*, 1970, *70*(9), 813–820.

Keating, D. P., & Schaefer, R. A. Ability and sex differences in the aquisition of formal operations. *Developmental Psychology*, 1975, *11*, 531–532.

Kieren, T. E., & Nelson, D. The operator construct of rational numbers in childhood and adolescence—an exploratory study. *The Alberta Journal of Educational Research*, 1978, *24*, 22–30.

Kurtz, B. *A study of teaching for proportional reasoning* (Doctoral dissertation, University of California, Berkeley, 1976). Ann Arbor, Michigan: Ann Arbor University, Microfilms No. TS277–15747.

Kurtz, B., & Karplus, R. Intellectual development beyond elementary school VII: Teaching for proportional reasoning. *School Science and Mathematics*, 1979, *79*(5), 387–389.

Lawson, A. E. Formal operations and field independence in a heterogeneous sample. *Perceptual and Motor Skills*, 1976, *42*, 981–982.

Lawson, A. E. Relationships among performances on three formal operations tasks. *Journal of Psychology, 1977, 96*, 235–241.

Lawson, A. E. Combining variables, controlling variables, and proportions: Is there a psychological link? *Science Education*, 1979, *63* (1), 67–72. (a)

Lawson, A. E. Relationships among performances on group administered items of formal reasoning. *Perceptual and Motor Skills*, 1979, *48*, 71–78. (b)

Lawson, A. E. (Ed.). *1980 AETS Yearbook: The psychology of teaching for thinking and creativity*. Columbus, Ohio: ERIC/SMEAC, 1979. (c)

Lawson, A. E., Karplus, R., & Adi, H. The acquisition of propositional logic and formal operational schemata during the secondary school years. *Journal of Research in Science Teaching*, 1978, *15*(6), 465–478.

Lawson, A. E., & Wollman, W. T. Cognitive development, cognitive style, and value judgement. *Science Education*, 1977, *61*(3), 397–407.

Linn, M. C., & Kyllonen, P. The field dependence-independence construct: Some, one, or none. *Journal of Educational Psychology*, 1981, *73*(2) 261–273.

Lovell, K. A follow-up study of Inhelder and Piaget's The growth of logical thinking. *British Journal of Psychology*, 1961, *52*, 143–153.

Lovell, K., & Butterworth, I. P. Abilities underlying the understanding of proportionality. *Mathematics Teaching*, 1966, *37*, 5–9.

Lunzer, A. E., & Pumfrey, P. D. Understanding proportionality. *Mathematics Teaching*, 1966, *34*, 7–12.

Lybeck, L. *Studies of mathematics in teaching of science in Goteborg*. Goteborg, Sweden: Institute of Education, University of Goteborg, No. 72, 1978.

Martorano, S. C. A developmental analysis of performance on Piaget's formal operations task. *Developmental Psychology*, 1977, *13*, 666–670.

Noelting, G. The development of proportional reasoning in the child and adolescent through combination of logic and arithmetic. In E. Cohors-Fresenborg & I. Wachsmuth (Eds.), *Proceedings of the Second International Conference for the Psychology of Mathematics Education.* Osnabruck, West Germany: University of Osnabruck, 1978.

Noelting, G. The development of proportional reasoning and the ratio concept: Part I—Differentiation of stages. *Educational Studies in Mathematics,* 1980, *11,* 217–253. (a)

Noelting, G. The development of proportional reasoning and the ratio concept: Part II—Problem structure at successive stages; problem solving strategies and the mechanism of adaptive restructuring. *Educational Studies in Mathematics,* 1980, *11,* 331–363. (b)

Noelting, G. Level and metalevel in development and the passage from a stage to another. In T. R. Post & M. P. Roberts (Eds.), *Proceedings of the Third Annual Meeting of the North American Chapter of the International Group for the Psychology of Mathematics Education.* Minneapolis, Minnesota: University of Minnesota, 1981.

Noelting, G., & Gagne, L. The development of proportional reasoning in four contexts. In R. Karplus (Ed.), *Proceedings of the Fourth International Conference for the Psychology of Mathematics Education.* Berkeley, California: Lawrence Hall of Science, 1980.

Nummedal, S. G., & Collea, F. P. Field independence, task ambiguity and performance on a proportional reasoning task. *Journal of Research in Science Teaching,* 1981, *18*(3), 255–260.

Oltman, P. K., Raskin, E., & Witkin, H. A. *Group embedded figures test (GEFT).* Palo Alto, California: Consulting Psychologists Press, 1971.

Pascual-Leone, J. *A neo-Piagetian process–structural model of Witkin's psychological differentiation.* Paper presented at the International Association for Cross-Cultural Psychology, Kingston, 1974.

Pascual-Leone, J. On learning and development, Piagetian style: A reply to Lefevre-Pinard. *Canadian Psychological Review,* 1976, *17,* 220.

Pascual-Leone, J. Constructive problems for constructive theories. In R. H. Kluwe & H. Spada (Eds.), *Developmental models of thinking.* New York: Academic Press, 1980.

Pascual-Leone, J., & Burtis, J. *The FIT: Figural-Intersection Test.* A group measure of *M* capacity. Unpublished manuscript, York University, Toronto, Canada, 1974.

Piaget, J. *The child's conception of movement and speed.* London: Routledge Kegan Paul, 1970.

Piaget, J., Grize, J. B., Szeminska, A., & Bang, V. *Epistemology and psychology of functions.* Dordrecht, Holland: Reidel, 1977.

Piaget, J., & Inhelder, B. *The child's construction of quantities.* London: Routledge Kegan Paul, 1974.

Piaget, J., & Inhelder, B. *The origin of the idea of chance in children.* New York: Norton, 1975.

Pulos, S., & Linn, M. C. *The Find a Shape Puzzle (FASP).* A group measure of cognitive restructuring. Unpublished research report. Adolescent Reasoning Project, Lawrence Hall of Science, University of California, Berkeley, 1979.

Pulos, S., de Benedictis, T., Linn, M. C., Sullivan, P. A. & Clement, C. A. Modification of gender differences in the understanding of displaced volume. *Journal of Early Adolescence,* 1982, *2*(1), 61–74.

Rupley, W. H. *The effects of numerical characteristics on the difficulty of proportional problems.* Doctoral dissertation, University of California, Berkeley, 1981.

Siegler, R. S. Developmental sequences within and between concepts. *Monograph of the Society for Research in Child Development,* 1981, *46*(2), Serial No. 189.

Stone, C. A., & Day, M. C. Levels of availibility of a formal operational strategy. *Child Development,* 1978, *49,* 1054–1065.

Suarez, A. *Formales Denken und Eunktionsbegriff bei Jugendlichen.* Bern: Hans Huber, 1977.

Suarez, A., & Rhonheimer, M. *Lineare Funktion.* Zurich: Limmat Stiftug, 1974.

Thier, H. D., & Willis, R. *Comparing ratios.* Berkeley, California: Lawrence Hall of Science, 1981.

Thurstone, L. L. *An analysis of mechanical aptitude.* Chicago: University of Chicago Press, 1963.

Torgerson, W. S. *Theory and methods of scaling*. New York: Wiley, 1958.

Vergnaud, G. Didactics and acquisition of multiplicative structures in secondary schools. In W. F. Archenhold, R. H. Driver, A. Orton, and C. Wood-Robison (Eds.), *Cognitive development research in science and mathematics*. Leeds, England: University of Leeds Press, 1980.

Wallach, M., & Kogan, N. *Modes of thinking in young children*. New York: Holt, Rhinehart & Winston, 1965.

Winch, W. H. Should young children be taught arithmetical proportion? Parts 1, 2, & 3. *Journal of Experimental Pedagogy*, 1913–1914, *2*, 79–88, 319–330, 406–420.

Witkin, H., & Goodenough, D. *Field dependence revisited* (ETS Research Bulletin 77–16). Princeton, New Jersey: Educational Testing Service, 1977.

Wollman, W., & Karplus, R. Intellectual development beyond elementary school V: Using ratio in differing tasks. *School Science and Mathematics*, 1974, *74*, 593–613.

Wollman, W., & Lawson, A. E. The influence of instruction on proportional reasoning in seventh graders. *Journal of Research in Science Teaching*, 1978, *15*(3), 227–232.

CHAPTER 4

Rational-Number Concepts*

Merlyn J. Behr, Richard Lesh, Thomas R. Post, and Edward A. Silver

Rational-number concepts are among the most complex and important mathematical ideas children encounter during their presecondary school years. Their importance may be seen from a variety of perspectives: (*a*) from a practical perspective, the ability to deal effectively with these concepts vastly improves one's ability to understand and handle situations and problems in the real world; (*b*) from a psychological perspective, rational numbers provide a rich arena within which children can develop and expand the mental structures necessary for continued intellectual development; and (*c*) from a mathematical perspective, rational-number understandings provide the foundation upon which elementary algebraic operations can later be based.

NAEP (National Assessment of Education Progress) results (Carpenter, Coburn, Reys, & Wilson, 1976; Carpenter, Corbitt, Kepner, Lindquist, & Reys, 1980) have shown that children experience significant difficulty learning and applying rational-number concepts. For example, results of both assessments indicate that most 13- and 17-year olds could successfully add fractions with like denominators, but only one-third of the 13-year olds and two-thirds of the 17-year olds could correctly add $\frac{1}{2} + \frac{1}{3}$. NAEP findings are consistent with those of other studies (Coburn, Beardsley, & Payne, 1975; Lankford, 1972; SMSG [School Mathematics Study Group] 1968), indicating generally low performance on rational-number computation and problem solving. The low level of performance may seem quite surprising in light of the fact that school programs tend to emphasize procedural skills and computational algorithms for rational numbers.

*This research was supported in part by the National Science Foundation under Grant No. SED 79-20591. Any opinions, findings, or conclusions expressed in this report are those of the authors and do not necessarily reflect the views of the National Science Foundation.

ISBN No. 0-12-444220-X

However, the generally poor performance may be a direct result of this curricular emphasis on procedures rather than the careful development of important functional understandings.

Many of the "trouble spots" in elementary school mathematics are related to rational-number ideas. Moreover, the development of rational-number ideas is viewed as an ideal context in which to investigate general mathematical concept acquisition processes because:

1. Much of the development occurs on the threshold of a significant period of cognitive reorganization (that is, the transition from concrete to formal operational thinking);
2. interesting qualitative transitions occur not only in the structure of the underlying concepts but also in the representational systems used to describe and model these structures;
3. the roles of representational systems are quite differentiated and interact in psychologically interesting ways because both figurative and operational task characteristics are critical;
4. the rational-number concept involves a rich set of integrated subconstructs and processes, related to a wide range of elementary but deep concepts (e.g., measurement, probability, coordinate systems, graphing, etc.).[1]

Piaget focused on the operational aspects of tasks and concepts, using the term *horizontal decalage* to refer to the fact that, whereas it may be useful to think of a person as being characterized by a given cognitive structure, he will not necessarily be able to perform within that structure for all tasks. It is common to encounter horizontal decalage with respect to rational number concepts, in that models embodying the same concept vary radically, often by several years, in the ease with which they are understood by children. Therefore, information about how task variables, such as figurative content, influence task difficulty is important for those who must select or devise appropriate models to illustrate rational-number concepts.

A Mathematical and Curricular Analysis of Rational-Number Concepts

Analyses of the components of the *concept* of rational number (Kieren, 1976; Novillis, 1976; Rappaport, 1962; Riess, 1964; Usiskin, 1979) suggest one obvious reason why complete comprehension of rational numbers is a formidable

[1]These are presupposed by a variety of problem-solving situations and are often taken to be "easy," when, in fact, many of these concepts developed rather late in the history of science and are exceedingly unobvious to those who have not assimilated them (Hawkins, 1979).

learning task. Rational numbers can be interpreted in at least these six ways (referred to as *subconstructs*): a part-to-whole comparison, a decimal, a ratio, an indicated division (quotient), an operator, and a measure of continuous or discrete quantities. Kieren (1976) contends that a complete understanding of rational numbers requires not only an understanding of each of these separate subconstructs but also of how they interrelate. Theoretical analyses and recent empirical evidence suggest that different cognitive structures may be necessary for dealing with the various rational number subconstructs.

A number of studies have identified stages in children's rational-number thinking by examining the gradual differentiation and progressive integration of separate subconstructs. One important aspect of these studies has been to observe whether or not subjects performing at a given stage on tasks involving one subconstruct perform at a comparable level on tasks involving a different subconstruct. Relationships between specific skills and certain basic rational number understandings have also been investigated.

Kieren (1981) has identified and discussed five *faces* of mathematical knowledge building. They relate to the mathematical, visual, developmental, constructive, and symbolic nature of mathematics, learning mechanisms, and learners. Four mathematical subconstructs of rational numbers—*measure, quotient, ratio,* and *operator*—each provide quantitative and relational rational-number experience. Equivalence and partitioning are constructive mechanisms operating across the four subconstructs to extend images and build mathematical ideas.

The Part–Whole and Measure Subconstructs

The *part–whole interpretation* of rational number depends directly on the ability to partition either a continuous quantity or a set of discrete objects into equal-sized subparts or sets. This subconstruct is fundamental to all later interpretations and is considered by Kieren (1981) to be an important language-generating construct.

The part–whole interpretation is usually introduced very early in the school curriculum. Children in first and second grade have primitive understandings of the meaning of *one-half* and the basic partitioning process (Polkinghorne, 1935). It is not until fourth grade, however, that the fraction concept is treated in a systematic fashion. Students normally explore and extend rational-number ideas through the eighth grade, after which these understandings are applied in elementary algebra. Many student difficulties in algebra can be traced back to an incomplete understanding of earlier fraction ideas.

Geometric regions, sets of discrete objects, and the number line are the models most commonly used to represent fractions in the elementary and junior high school. Interpretation of geometric regions apparently involves an understanding of the notion of area. Owens (1980) and Sambo (1980) each examine the relationship between a child's concept of area and her or his ability to learn fraction concepts. Owens finds a positive relationship between success on area tasks and success in an instructional unit based on geometric regions. Sambo reports that deliberate teaching for transfer from area tasks aids children's ability to learn fraction concepts when geometric regions and measurement interpretations are involved.

Ellerbruch and Payne (1978) claim that research as well as classroom practice suggest the introduction of fraction concepts using a single model, and they recommend the part–whole measurement model as the most natural for young children and the most useful for addition of like fractions. The Initial Fraction Sequence (IFS), an instructional sequence based on the research of Payne and his colleagues (Ellerbruch & Payne, 1978), emphasizes the importance of developing a firm foundation of fraction concepts before introducing children to operations or relations on rational numbers. IFS uses rectangular regions because of the ease of making models from strips of paper, and proceeds carefully from concrete or pictorial models to oral fraction names, to natural language written names (e.g., three-fourths), and finally to formal mathematical symbols.

The number-line model adds an attribute not present in region or set models, particularly when a number line of more than one unit long is used. Novillis-Larson (1980) presented seventh-grade children with tasks involving the location of fractions on number lines that were one or two units long and for which the number of segments in each unit segment equaled or was twice the denominator of the fraction. Results of the study indicated that, among seventh-graders, associating proper fractions with points was significantly easier on number lines of length one and when the number of segments equaled the denominator. Novillis-Larson's findings suggest an apparent difficulty in perception of the unit of reference: when a number line of length two units was involved, almost 25% of the sample used the whole line as the unit. Her data also indicate that children do not associate the rational number *one-third* with a point for which partitioning suggests *two-sixths*. Such results suggest an imprecise and inflexible notion of fraction among seventh graders.

Whether or not the type of embodiment (continous quantity versus discrete quantity) demands different types of cognitive structures was investigated by Hiebert and Tonnessen (1978), who asked whether the part–whole interpretation given by Piaget, Inhelder, and Szeminska (1960) was appropriate for both the discrete and the continuous cases of length and area. Their tasks required children to divide a quantity equally and completely among a number of stuffed animals. They found that children performed considerably better on tasks involving the discrete case (set–subset) than the continuous case. One possible explanation is that solutions of the continuous quantity tasks (Piaget *et al.*, 1960) require a well-developed anticipatory scheme, whereas discrete quantity tasks can be solved simply by partitioning. In particular, the discrete tasks can be solved without treating the set as a whole and without anticipating the final solution. Because the strategies employed by children for discrete quantity tasks are so markedly different from those employed for continuous quantity tasks, it is reasonable to assume that cognitive structures involved in solving rational-number problems referring to a discrete model are different from those involved in solving rational-number problems referring to a continuous model.

Rational Number as Ratio

Ratio is a relation that conveys the notion of relative magnitude; therefore, it is more correctly considered as a comparative index rather than as a number. When two ratios are equal they are said to be in proportion to one another. A proportion is simply a statement equating two ratios. The use of proportions is a very powerful problem-solving tool in a variety of physical situations and problem settings that require comparisons of magnitudes.

Noelting (1978, pp. 242–277) used an orange-juice test to investigate subjects' ability to compare ratios. Noelting's tasks asked children to specify which of two mixtures of orange juice and water would taste more "orangy." Three stages were observed among subjects' responses, ranging from making judgments based only on comparisons of terms, to comparing ordered pairs using multiplicative rules, to the final stage in which ordered pairs were seen as belonging to a class.

The use of glasses of water and orange juice suggests a discrete model. Another line of inquiry, using continuous quantities, is represented by Karplus, Karplus, Formisano, and Paulsen (1979, pp. 47–103), Karplus, Karplus, and Wollman (1974), and Kurtz and Karplus (1979). Subjects were asked to find an unknown component of a proportionality statement by equating two ratios involving length, distance, or volume. Like Noelting, Karplus and his colleagues have identified various levels of cognitive functioning, ranging from random guessing to additive (rather than multiplicative) reasoning, to the highest level at which the data are utilized at a formal level of multiplicative ratio thinking.

Rational Numbers as Indicated Division and as Elements of a Quotient Field

According to the part–whole interpretation of rational numbers, the symbol a/b usually refers to a fractional part of a single quantity. In the ratio interpretation of rational numbers, the symbol a/b refers to a relationship between two quantities. The symbol a/b may also be used to refer to an operation. That is, a/b is sometimes used as a way of writing $a \div b$. This is the *indicated division* (or *indicated quotient*) interpretation of rational numbers.

Consideration of rational numbers as quotients involves at least two levels of sophistication. On the one hand, $\frac{8}{4}$ or $\frac{2}{3}$ interpreted as an indicated division results in establishing the equivalence of $\frac{8}{4}$ and 2, or $\frac{2}{3}$ and $.\overline{666}$. But rational numbers can also be considered as elements of a quotient field, and, as such, can be used to define equivalence, addition, multiplication, and other properties from a purely deductive perspective; all algorithms are derivable from equations via the field properties (Kieren, 1976). This level of sophistication generally requires intellectual structures not available to middle school children because it relates rational numbers to abstract algebraic systems.

Rational Number as Operator

The subconstruct of rational number as *operator* imposes on a rational number p/q an algebraic interpretation; p/q is thought of as a function that transforms geometric figures to similar geometric figures p/q times as big, or as a function that transforms a set into another set with p/q times as many elements. When operating on a continuous object (length), we think of p/q as a stretcher–shrinker combination. Any line segment of length L operated on by p/q is stretched to p times its length and then shrunk by a factor of q. A multiplier–divider interpretation is given to p/q when it operates on a discrete set. The rational number p/q transforms a set with n elements to a set with np elements and then this number is reduced to np/q.

This rational-number concept can be embodied in a function machine in which p/q is thought of as a "p for q" machine. Thus, $\frac{3}{4}$ is thought of as a 3 for 4 machine: an input of length or cardinality 4 produces an output of length or cardinality 3.

The operator interpretation of rational number is particularly useful in studying equivalence of fractions and the operation of multiplication. The problem of finding fractions equivalent to a given fraction is that of finding function machines that accomplish the same input–output transformations. Multiplication of fractions involves composition of functions.

A number of studies conducted by Kieren and his colleagues (Ganson & Kieren, 1980; Kieren & Nelson, 1978; Kieren & Southwell, 1979) and Noelting (1978, 242–277) have investigated the stage development of the operator and ratio constructs and the relationship between them in children's thinking.

Analysis of children's descriptions of how a machine works indicated that students thought subtractively and not multiplicatively. This was particularly true for students under 12. A second important finding was the role one-half played in subjects' thinking; 91% of the subjects mastered the "one-half" tasks. Even students who knew that a machine was not a one-half machine would give a one-half response when confused. Apparently the students' higher rate of success on one-half tasks, and greater familiarity with the number itself, led to misapplications. This type of error was made by 47% of the subjects (Kieren & Nelson, 1978).

Kieren and Southwell (1979) examined differences between children's ability to perform operator tasks when the task was embedded in a function machine compared with a "simpler" approach consisting of patterns of symbolic input–output number pairs. An analysis of variance of correct responses indicated no significant differences due to representation mode. Three levels of rational-number operator development were observed in data from both types of tasks. The authors suggested that understanding of equivalence class and partitioning were the important mechanisms underlying this development. *Partitioning* refers to the division of a set into subsets. Applying equivalence class thinking to a one-third task, a subject who correctly pairs 2 with 6 explains, "divided by 3." A more sophisticated use of the mechanism is required for success on the nonunit fraction task of "two-thirds;" to pair 90 with 60, a student thinks, "divide by 3 and take 2 of them." The general fractional operator appears to require the coordination of the partitioning of two subsets of numbers with a multiplicative operation, in this case doubling. This covaried partitioning strategy was used most often by subjects in the machine representation condition. In the pattern representation condition, a pattern explanation frequently accompanied a correct response. Thus, in pairing 24 with 16 in the two-thirds task, the subject would say, "Well, I know 12 went to 8 so I just doubled to get 16." A higher level of performance was observed in the machine group at a younger age compared with the pattern group.

Ganson and Kieren (1980) gave a single group of subjects both operator tasks (Kieren & Nelson, 1978) and orange juice tasks (Noelting, 1978). They concluded that (*a*) there is an indication that students who are able to partition are also able to perform comparisons and to recognize equivalences, and (*b*) the level of cognitive thinking necessary for successful performance on general operator tasks is relatively the same level as that needed for successful performance on the multiplicative-equivalence comparisons in the ratio tasks.

Summary

Because of an emphasis on the part–whole subconstruct in school mathematics curricula and the concomitant rapid progression to symbolic computation procedures, a disproportionate amount of past research on rational numbers was concerned with questions relating to which of several algorithmic procedures would best facilitate children's computation performance. More recently, research has included data-based observations concerned not with simple comparisons between two instructional procedures but with attempts to identify and describe the mental processes employed by children engaged in these tasks.

The large majority of current efforts are status studies. That is, the researcher gathers data relating to children's knowledge of a particular area without regard for concurrent instruction or consideration of the quality or extent of the child's past instructional experiences. Because much of what children know about the more formal aspects of mathematics is influenced by instruction, these status studies, although very useful, are inherently limited in the extent to which children's cognitive structures can be linked directly to instruction and/or specific experiences. Furthermore, they do not provide any insights into how concepts develop over time under the influence of a well-defined instructional sequence. Such information is undoubtedly crucial if research is to provide guidance for the redefinition of school curricula to promote the more effective learning of mathematics by all children.

One effort presently underway is attempting to investigate the development of cognitive structures for rational-number thinking within a well-defined, theoretically-based instructional program. It is discussed in the next section.

The Rational-Number Project

The NSF-supported (National Science Foundation) Rational Number Project consists of three interacting components: (*a*) an instructional component in which 18 fourth and fifth graders were observed, interviewed, and tested frequently during 16 weeks of theory-based instruction, (*b*) an evaluation component in which more than 1600 second through eighth grade children were tested using a battery of written tests, instruction mediated tests, and clinical interviews, and (*c*) a diagnostic–remedial component in which young adults who were experiencing difficulties with fractions were identified; their misunderstandings were isolated and remediated using materials borrowed from the evaluation component and activities borrowed from the instructional component.

General goals of the Rational Number Project are (*a*) to describe the development of the progressively complex systems of relations and operations that children in Grades 2–8 use to make judgments involving rational numbers, and (*b*) to describe the role that various representational systems (e.g., pictures, manipulative materials, spoken language, written symbols) play in the acquisition and use of rational-number concepts. The project aims to develop a psychological "map" describing (*a*) how various rational-number subconstructs (e.g., fractions, ratios, indicated quotients) gradually become differentiated and integrated to form a more mature understanding of rational numbers, (*b*) how various representational systems interact during the gradual development of rational number ideas, and (*c*) how a variety of theory-based interventions can further this development.

The project is concerned not only with what children can do "naturally," but also with what they can do accompanied by minimal guidance or following theory-based instruction. The interest is in exploring the "zone of proximal development" of children's rational-number concepts (Vygotsky, 1976), not only to describe the *schemas* children typically use to process rational-number information and to interpret rational-number situations, but also to describe how these schemas change as a result of theory-based instruction. Rather than seeking only to accelerate rational number understandings along narrow conceptual paths, the project is interested in studying the results of broadening and strengthening deficient conceptual models (see Lesh, Landau, & Hamilton, Chapter 9, this volume).

Theoretical Foundations

Theoretical foundations for the project were derived from four separate but mutually supportive theoretical bases: (*a*) Kieren's (1976) mathematical analysis of rational number into subconstructs; (*b*) Post and Reys's (1979) interpretation of Dienes's perceptual and mathematical variability principles; (*c*) Lesh's (1979) analysis of modes of representation related to mathematical concept acquisition and use; (*d*) an analysis of memory structures developed by a learner (Behr, Lesh, & Post, Note 1). These theoretical bases are discussed in turn below.

THE ANALYSIS OF RATIONAL-NUMBER SUBCONSTRUCTS

Our work has resulted in a redefinition of some of Kieren's (1976) categories and a subdivision of others. The scheme includes the following seven subconstructs.

The *fractional measure* subconstruct of rational number represents a reconceptualization of the part–whole notion of fraction. It addresses the question of how much there is of a quantity relative to a specified unit of that quantity.

The *ratio* subconstruct of rational number expresses a relationship between two quantities, for example, a relationship between the number of boys and girls in a room.

The *rate* subconstruct of rational number defines a new quantity as a relationship between two other quantities. For example, speed is defined as a relationship between distance and time. We observe here that although one adds rates in such a context as computing average speed, one seldom adds ratios.

The *quotient* subconstruct of rational number interprets a rational number as an indicated quotient. That is, a/b is interpreted as a divided by b. In a curricular context this subconstruct is exemplified by the following problem situation:

There are 4 cookies and 3 children. If the cookies are shared equally by the three children, how much cookie does each child get?

The *linear coordinate* subconstruct of rational number is similar to Kieren's notion of a measure interpretation. It emphasizes properties associated with the metric topology of the rational number line such as betweenness, density, distance, and (non)completeness. Rational numbers are interpreted as points on a number line, emphasizing that the rational numbers are a subset of the real numbers.

The *decimal* subconstruct of rational number emphasizes properties associated with the base-ten numeration system.

The *operator* subconstruct of rational number imposes on rational number a function concept; a rational number is a transformation. The stretcher–shrinker notions developed by UICSM (University of Illinois Committee on School Mathematics) CSMP (Comprehensive School Mathematics Project), and Dienes (1967) represent physical embodiments of this construct.

Questions regarding which of these subconstructs might best serve to develop in children the basic fraction concept, relations on rational numbers, operations with rational numbers, and applications of rational numbers remain unanswered. It seems plausible that the part–whole subconstruct, based both on continuous and discrete quantities, represents a fundamental construct for rational-number concept development. It is, in addition, a point of departure for instruction involving other subconstructs. The preliminary conceptualization of the interrelationships among the various subconstructs is depicted in Figure 4.1. The solid and dashed arrows suggest established and hypothesized relationships, respectively, among rational-number constructs, relations, and operations. The diagram suggests that (*a*) partitioning and the part–whole subconstruct of rational numbers are basic to learning other subconstructs of rational number; (*b*) the ratio subconstruct is most "natural" to promote the concept of equivalence; (*c*) the operator and measure subconstructs are very useful in developing an understanding of multiplication and addition.

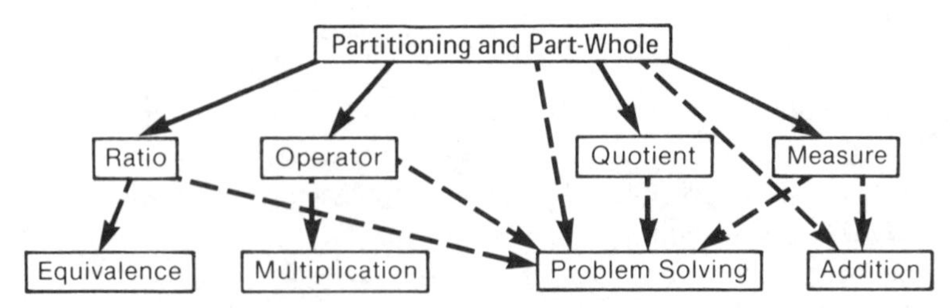

Figure 4.1 Conceptual scheme for instruction on rational numbers.

THE ANALYSIS OF CONCRETE MODELS RELATED TO RATIONAL-NUMBER CONCEPTS

Post and Reys (1979) interpret the mathematical and perceptual variability principles of Dienes (1967) by using a matrix framework. Kieren's (1976) rational-number subconstructs constitute the mathematical variability dimension of the matrix. The perceptual variability dimension includes discrete objects, length models, area models, and written symbolic models.

The Rational Number Project has refined Post and Reys's matrix to include the categories shown in Figure 4.2.

Each cell in this matrix implies both a type of physical or symbolic activity *and* a mathematical perspective. Discrete materials used in the project included counters, egg cartons, and other sets of objects. Continuous materials involved some quantity such as length or area, and included Cuisenaire rods, number lines, and sheets of paper. Countable-continuous materials involved a continuous quantity that had been partitioned into countable units of the same "size" but not necessarily the same shape. Examples of countable continuous materials included tiles and graph paper.

AN INTERACTIVE REPRESENTATIONAL SYSTEMS MODEL

Lesh (1979) reconceptualized Bruner's (1966) enactive mode, partitioned Bruner's iconic mode into manipulative materials and static figural models (i.e., pictures), and partitioned Bruner's symbolic mode into spoken language and written symbols. Furthermore, these systems of representation were interpreted as interactive rather than linear, and translations within and between modes were

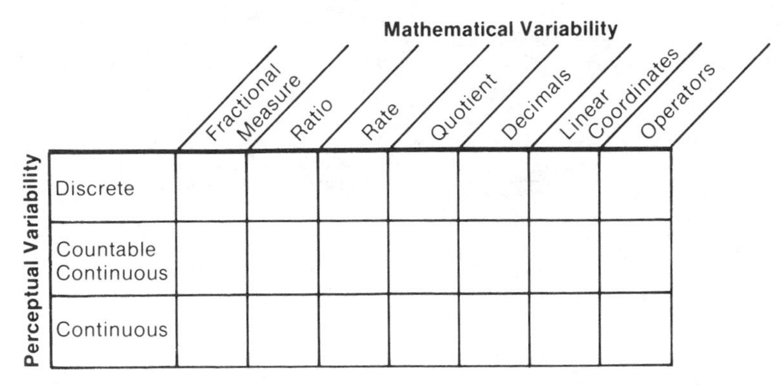

Figure 4.2. Sources of availability of rational-number concepts for instruction.

given as much emphasis as manipulation of single representational systems. Figure 4.3 shows the Rational Number Project's modified version of Lesh's model.

A major focus of the project is the role of manipulative materials in facilitating the acquisition and use of rational-number concepts as the child's understanding moves from concrete to abstract. Psychological analyses show that manipulatives are just one component in the development of representational systems and that other modes of representation also play a role in the acquisition and use of concepts (Lesh, Landau, & Hamilton, 1980). A major hypothesis of the project is that it is the ability to make translations among and within these several modes of representation that makes ideas meaningful to learners.

The Rational Number Project has shifted away from attempting to identify the "best" manipulative aid for illustrating (all) rational-number concepts toward the realization that different materials are useful for modeling different real-world situations or different rational-number subconstructs (i.e., part–whole fractions, ratios, operators, proportions), and different materials may be useful at different points in the development of rational-number concepts. For example, paper folding may be excellent for representing part–whole relationships or equivalent fractions, but may be misleading for representing addition of fractions. There is no single manipulative aid that is "best" for all children and for all rational-number situations. A concrete model that is meaningful for one child in one situation may not be meaningful to another child in the same situation nor

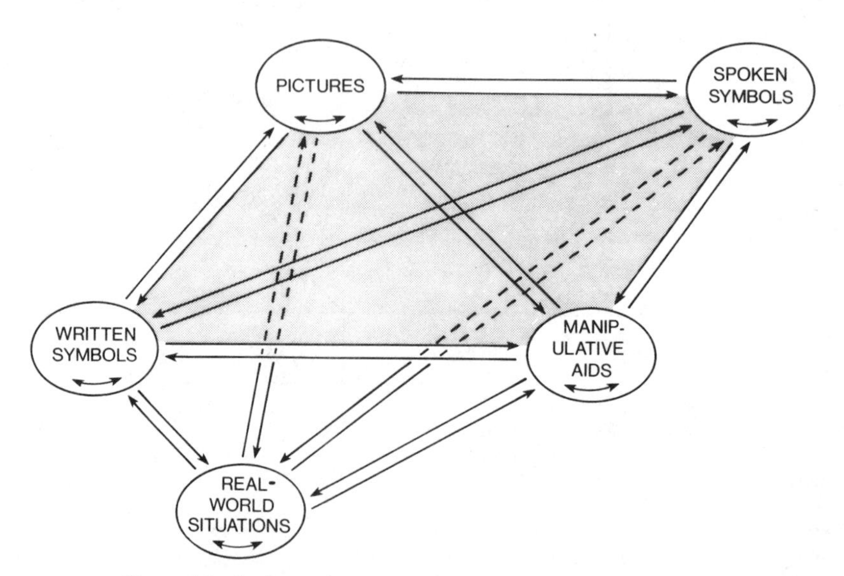

Figure 4.3. An interactive model for using representational systems.

to the same child in a different situation. The goal is to identify manipulative activities using concrete materials whose structure fits the structure of the particular rational-number concept being taught.

Our current research is focusing heavily on analyzing the cognitive structures children use to perform various rational-number tasks. Task analyses reveal that, even within the category of concrete materials, some are more concrete than others. One reasonable implication of this observation is that teachers attempting to concretize abstract rational-number concepts might be wise to begin instruction with those materials that are most concrete, least complex, and that draw upon useful intuitive understandings.

Because teachers can illustrate ideas such as "addition of fractions" using folded paper, Cuisenaire rods, or other manipulative materials, they may underestimate the level of sophistication that is required for performing these tasks. It is one thing for a child to know how to illustrate fractions such as $\frac{1}{2}$ or $\frac{1}{3}$ using Cuisenaire rods, and quite another to be able to illustrate $\frac{1}{2} + \frac{1}{3}$ using the rods. Concrete materials that are useful for illustrating fractions may not be useful for illustrating *addition* of fractions. That is, the addition of fractions may be more meaningful if it is built on a strong concrete understanding of individual fractions, but this does not imply that learning to add Cuisenaire rods or folded paper will facilitate the child's understanding of addition of fractions.

Young learners do not work in a single representational mode throughout the solution of a problem. They may think about one part of the problem (say, the number) in a concrete way, but may think about another part of the problem using other representational systems (i.e., actions, spoken language rules, or written symbol procedures).

Figure 4.3 is intended to suggest that realistic mathematical problems are frequently solved by: (*a*) translating from the real situation to some system of representation, (*b*) transforming or operating with the representational system to produce some result or prediction, and (*c*) translating the result back into the real situation.

Figure 4.3 is also intended to imply that many problems are solved using a sequence of partial mappings involving several representational systems. That is, pictures or concrete materials may be used as an intermediary between a real situation and written symbols, and spoken language may function as an intermediary between the real situation and the pictures, or the pictures and the written symbols.

Some problems inherently involve more than one mode at the start. For example, in real addition situations that involve fractions, the two items to be added may not always be two written symbols, two spoken symbols, or two pizzas; they may be one pizza and one written symbol, or one pizza and one spoken word. That is, the problem may involve showing the child half of a pizza and then asking how much the child would have if given another one-third of a pizza.

In such problems, which occur regularly in real situations, part of the difficulty is to represent both addends using a single representational system.

Although a goal of the Rational Number Project has been to trace the development of rational-number ideas, our data have made us sensitive to the need to explain concept stability and instability—as well as concept development. For example, even though our study used criteria for "mastery" that were considerably more stringent than those typically used in school instruction, it was common to observe significant regression in concept understanding over 2- or 3-week periods. Not only must "mastered" concepts be remembered, they must be integrated into progressively more complex systems of ideas; sometimes they must be reconceptualized when they are extended to new domains. Ideas that are true in restricted domains (e.g., "multiplication is like repeated addition" or "a fraction is part of a whole") are misleading, incorrect, or not useful when they are extended to new domains. Mathematical ideas usually exist at more than one level of sophistication. They do not simply go from "not understood" to "mastered." Therefore, as they develop they must be reconceptualized periodically, and they must be embedded in progressively more complex systems that may significantly alter their original interpretation.

The Rational Number Project has been especially interested in interactions between internal and external representations of problem situations. Frequently, when children solve problems, an internal interpretation of the problem influences the selection, generation, or modification of an external representation. The external representation may involve a picture, concrete materials, or written symbols. Often the external representation models only part of the problem. For example, a child's first picture for an "addition of fractions" problem might represent the fractions without any attempt to represent the addition process. An external representation typically allows children to refine their internal representations and interpretations—which may cycle back to the generation (or selection) of a refined external representation, or to a solution. External representations can, among other things, reduce memory load or increase storage capacity, code information in a form that is more manipulable, or simplify complex relationships.

MEMORY STRUCTURES

Gagne and White (1978) proposed a model for relating memory structures to learning outcomes. They considered four memory structures: (*a*) networks of propositions (which store verbal knowledge); (*b*) intellectual skills (which underlie the identification of concepts and the application of rules); (*c*) images (primarily visual, but also auditory or haptic representations corresponding more or less directly with concrete objects or events); and, (*d*) episodes (incorporating representations of personal experience in the form of "first I did this, then I did that''). Tulving (1972) distinguished episodic memory from semantic memory (the storage of organized linguistic knowledge), emphasizing the autobiographical nature of episodic memory.

The consideration of memory structures is relevant to the use of manipulative aids in mathematics teaching. Episodic experience that a child gains from concrete aids may not provide retrievable knowledge without semantic information about the episodes and about relationships among different episodic experiences. Verbal interaction by the learner with a teacher or peers to observe the similarities and differences among episodic experiences and the materials on which they are based is probably essential.

Major Project Components

THE INSTRUCTIONAL COMPONENT

Materials Development

Twenty weeks of student instructional materials (over 600 pages) have been developed by project staff, piloted, and used in small-group teaching experiments with fourth- and fifth-grade children. Extensive observation guides and interview protocols have been produced to collect data about children's cognitive behavior on a lesson-by-lesson basis.

Each 20-week teaching experiment (one in DeKalb, two in Minneapolis) involved groups of six students and utilized audio and video taping, extensive interviewing, and pre- and postinstruction achievement testing. Control students were identified for each group for comparison purposes.

The instructional materials reflect the project's underlying theoretical foundations. Part–whole, quotient, measure, and ratio interpretations of rational number, and translations within and among five representational modes are emphasized.

Data Collection and Analysis

Four major types of instruments have been used at the DeKalb and Minneapolis sites to identify and assess the development of children's rational-number concepts within the theory-based instruction.

1. The *Rational-Number Test* was used as a pre- and postmeasure with both experimental and randomly selected nonexperimental students. This test, mainly concerned with content mastery, identified levels of student achievement in three areas: rational-number concepts, relations, and operations. This instrument was also used with classroom groups in Grades 2–8 across five geographic locations ($N > 1600$).

2. *Class Observation Guides* were developed for each of the 12 lessons. Each lesson spanned 2–6 instructional days. These guides were designed to provide staff with information about the cognitive processes students may employ when dealing with situations involving rational-number concepts. Because the amount of information called for was extensive, the guides were often supplemented by audio or video tapes.

3. *Interview Protocols* involved audio or videotaped individual interviews, lasting from 15 to 50 minutes, and conducted with each student after each lesson. They provide extensive information on the inferred mental processes, memory structures, and understandings gained and utilized. These data will provide detailed longitudinal information on the development of rational-number concepts in individual students. Interview data were examined on a lesson-by-lesson basis to assess the impact of specific instructional "moves" on conceptual development.

The sequence of individual interviews that were conducted with the children over the 16–18-week teaching experiment were developed so that continuous information about the development of certain rational-number concepts would emerge. The interview protocols produced data in several separate but not mutually exclusive data strands that reflect children's ability to

- *a.* deal with visual perceptual distractors,
- *b.* deal with questions related to the equivalence and ordering of fractions,
- *c.* deal with the basic fraction concept,
- *d.* understand the concept of unit in rational-number situations,
- *e.* perform on tasks that require proportional reasoning,
- *f.* perform within and between mode translations and the relationship between this ability and performance on other rational number tasks, and
- *g.* accomplish problem-solving tasks involving rational numbers.

Data from the first strand are reported in a subsequent section of the chapter.

4. *Translation Coding System* was designed to provide specific information on students' translations within and between modes of representation, the relative frequency of each type, and the identification of those that proved particularly troublesome.

THE EVALUATION COMPONENT

There are three components to the testing program: paper and pencil tests, instruction mediated tests, and clinical interviews (Figure 4.4) (Lesh & Hamilton, 1981).

The paper and pencil portion consists of three tests: Concepts, Relations, and Operations. The first assesses basic fraction and ratio concepts. The second

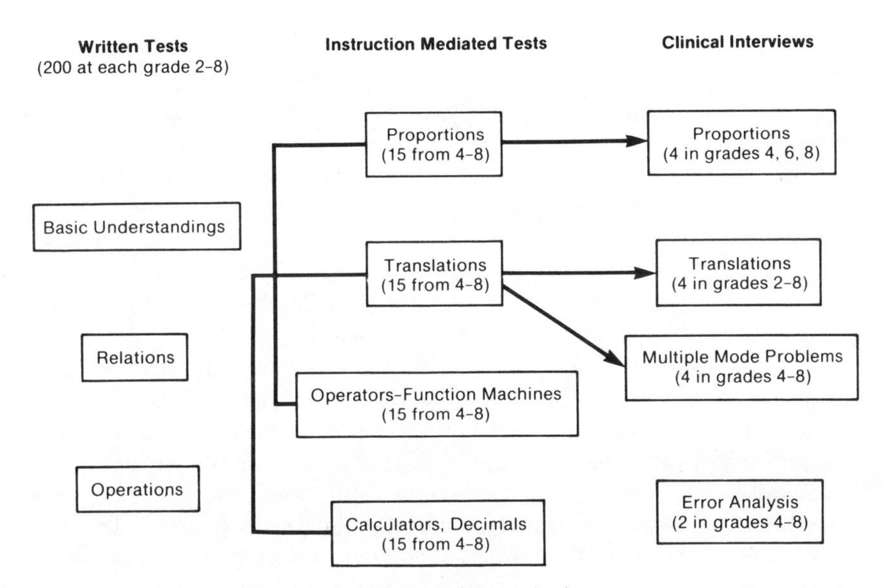

Figure 4.4. Components of the evaluation program.

assesses understanding of relationships between rational numbers, such as ordering, equivalent forms, and simple proportions. The third test assesses performance on addition and multiplication with fractions and various applications items. The tests are modularized to accommodate children in Grades 2–8. Several test items from previous studies were used intact or modified to make it possible to integrate results with past research (e.g., Carpenter, Coburn, Reys, & Wilson, 1978; Karplus, Pulos, & Stage, 1980; Kieren, 1976; Klahr & Siegler, 1978; Noelting, 1979) and to provide a basis for future research. A cross section of elementary and junior high school students were tested, producing baseline data on rational-number understandings for over 1600 children in Grades 2–8. Five geographic sites (Evanston, DeKalb, Minneapolis, San Diego, and Pittsburgh) were included in the data collection. The written tests were also used to identify students for follow-up interviews and for the diagnostic–remedial and error-analysis component of the project.

ERROR-ANALYSIS AND INTERVENTION COMPONENT

The point of view that led to the development of this component of the project was that a great deal could be learned from the careful study of rational-number knowledge possessed by young adults who had studied the topic for many years and had received typical school mathematics instruction. Two particular benefits to the Rational Number Project curriculum development efforts were planned:

(*a*) by identifying the understandings and misunderstandings of these individuals, useful insights could be gained that might guide instructional development, and (*b*) the instructional routines developed in the other components of the project could be used in this component to test their remedial utility.

Work on this component of the project was completed at the San Diego site and consisted of three parts: written testing, clinical interviews, and instructional intervention. A total of 161 community college students (enrolled in a remedial arithmetic class) and 59 college students (enrolled in a mathematics course for preservice elementary school teachers) completed the three multiple-choice, written tests developed in the evaluation component.

From the testing sample, 20 subjects were chosen for clinical interviews. Subjects were interviewed individually in sessions lasting from 45 to about 75 minutes. The clinical interview reviewed selected problems from the written tests, especially those that the subject had missed, and the completion of additional tasks designed to probe the subject's rational-number understandings. Some interview tasks were borrowed from clinical interview protocols developed in the testing component of the project. The interviews were individually tailored to probe each subject's errors and understandings.

From the clinical interview sample, eight students were chosen for instructional intervention. Each student met with the investigator in individual sessions of about 45 minutes. The number of sessions ranged from one session for two of the subjects to eight sessions for one of the subjects. The focus of the intervention sessions was the remediation of errors and misconceptions identified in the clinical interviews and written tests. The instructional methodology employed was borrowed or adapted from instructional routines developed in the project.

Perceptual Cues and the Quality of Children's Thinking: A Project Study

Although it is frequently recommended that children should learn mathematical ideas using concrete manipulative aids, very little is known about how manipulative aids affect a child's mathematics learning or conceptual development.

A number of research reviews (Fennema, 1972; Gerling & Wood, 1976; Kieren, 1969; Suydam & Higgins, 1977) provide evidence that the use of manipulatives does facilitate the learning of mathematical skills, concepts and principles. Existing research has dealt mainly with the superficial question of whether or not their use is effective, or which material is "best." The results have been equivocal. The literature contains little information about how manipulative aids affect children's cognitive functioning or why their use does or does not facilitate mathematical learning.

Data from three parallel 16–18-week teaching experiments recently conducted with fourth- and fifth-grade children by the Rational Number Project indicate considerable individual differences among children concerning (*a*) information they encode from a manipulative aid and (*b*) what features of an aid interfere with their logical–mathematical thinking. Concepts such as *partitioning, equivalence, order,* and *unit recognition* are basic thinking tools for understanding rational numbers. In our work it has been observed that certain components of a manipulative aid or pictorial display that are essential to illustrate one basic concept frequently impair the child's ability to use the aid for another concept. In particular, various types of perceptual cues can negatively influence children's thinking. In some cases, these perceptual cues act as distractors and overwhelm children's logical thought processes.

In the instructional component of the project, we found that children tend to assume that physical conditions within which problems are presented are relevant to and consistent with the task. This tendency is probably an artifact of their learning from a textbook- or worksheet-dominated instructional program that places little emphasis on manipulative materials. Within such a program, problem conditions are necessarily static in nature, providing little opportunity for children to manipulate problem conditions. Students expect that mathematical problem conditions (context) conform to the intended task and, therefore, are not in need of restructuring or rethinking. Children learn that one simply takes what is given, and proceeds directly to the solution.

Manipulative materials offer a mechanism for freeing children's thought processes because properly conceived and sequenced materials can provide for continual reconstruction of problem conditions and concurrently can permit a dynamic interaction between problem solver and problem conditions.

Meaningful understanding of mathematical ideas and of the mathematical symbolism for these ideas depends in part on an ability to demonstrate interactively the association between the symbolic and manipulative-aid modes of representation. Theoretically, as children deal with mathematical ideas, embodied in manipulative aids, the mathematical ideas are abstracted into logical–mathematical structures. As children's logical–mathematical structures expand, it is presumed that their dependence upon the concrete manipulative aids decreases. Ultimately, logical–mathematical thought becomes sufficiently strong so that it dominates the visual–perceptual information.

By using a series of tasks in which visual–perceptual distractors were deliberately introduced, we obtained information that indicates differences among children in their ability to put aside, overcome, or ignore the distractors and deal with the tasks on a logical–mathematical level. The extent to which a child is

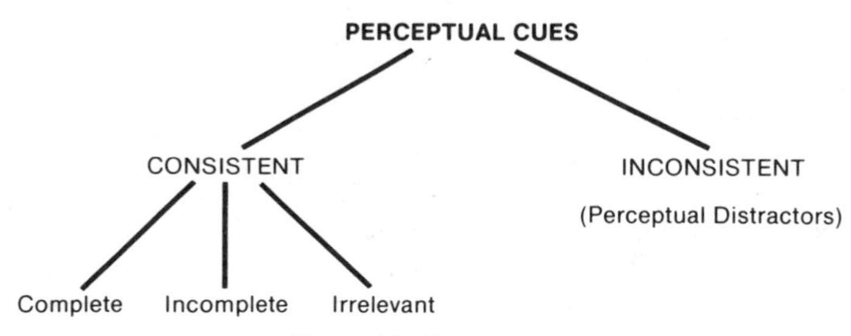

Figure 4.5. Perceptual cues.

able to do this—resolve conflicts between this perceptual processing of visual information and the child's cognitive processing of logical–mathematical relations—is viewed as one of several important indicators of the strength of the child's understanding of rational number concepts.

Definitions

By the term *visual–perceptual cues,* we mean the figures, models, or diagrams that accompany standard school tasks involving rational numbers. These cues are of two general types: those *consistent* with the task and those *inconsistent* with the task. Consistent cues may be further subdivided into three categories: *complete, incomplete,* and *irrelevant.* Inconsistent cues will be referred to as *perceptual distractors.* These relationships are depicted in Figure 4.5.

Consistent cues generally provide information to the student that can be used (perhaps with some modification) to aid in the solution of the task.

1. Complete cues contain all necessary information to aid in the solution of task or problem. Furthermore the information is presented in a form that is useful for the task at hand.
2. Incomplete cues require the student to add or modify existing features to complete the diagram or model.
3. Irrelevant cues contain extraneous but neutral information. Such cues require the solver to ignore certain information.

Inconsistent cues are those that conflict with the conceptualization of the task or problem and, therefore, must be reconciled prior to solution. Inconsistent cues provide visual information that may be misleading or may distract the subject from the intended task. Inconsistent cues are referred to here as *visual–perceptual distractors.*

An example illustrates these distinctions. The task in all cases is to shade three-fourths of the rectangle.

Consistent

Complete Cue: | Subject shades 3 of 4 parts.

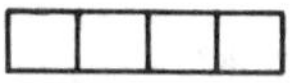

Incomplete Cue: 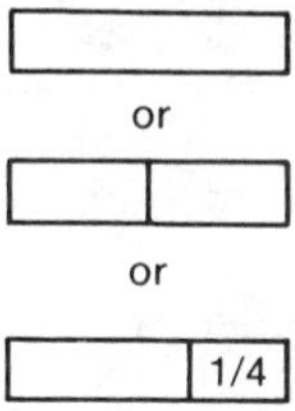 or or 1/4 | Subject adds required lines so that figure is divided into 4 equal parts, then shades 3 of them.

Irrelevant Cue: | Subject ignores every other line; clumps two $\frac{1}{8}$'s as $\frac{1}{4}$ and shades 3 such clumps.

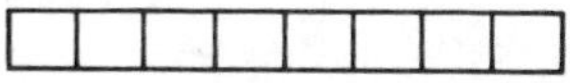

Inconsistent

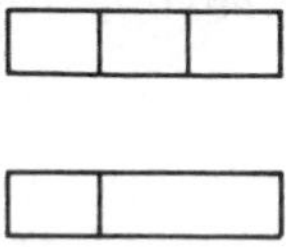

Subject ignores all lines, reconstructs diagrams and proceeds as in A.

SCHEMA 4.1

Further clarification about types of perceptual cues and solution strategies is provided in Tables 4.1 and 4.2.

Results from Perceptual Distractor Tasks

SURVEY DATA FROM A WRITTEN TEST

Table 4.3 provides information about the relative difficulty of various types of perceptually cued items. These data were collected in the early spring after children had undergone normal fourth-grade instruction dealing with fractions. Three fourth-grade classes ($N = 77$) from a suburban elementary school in St. Paul, Minnesota are represented.

Three major trends are immediately apparent. First, there is a disproportionate number of errors in number-line problems across all categories and specific fractions. This is true despite the fact that for 3 years the students' text series had employed the number-line model for whole-number interpretations of addition and subtraction. Children in this sample were generally incapable of concep-

TABLE 4.1

A Sample of Perceptual Cues for the Task: "Find $\frac{3}{4}$ of Each Region, Number Line, or Set"

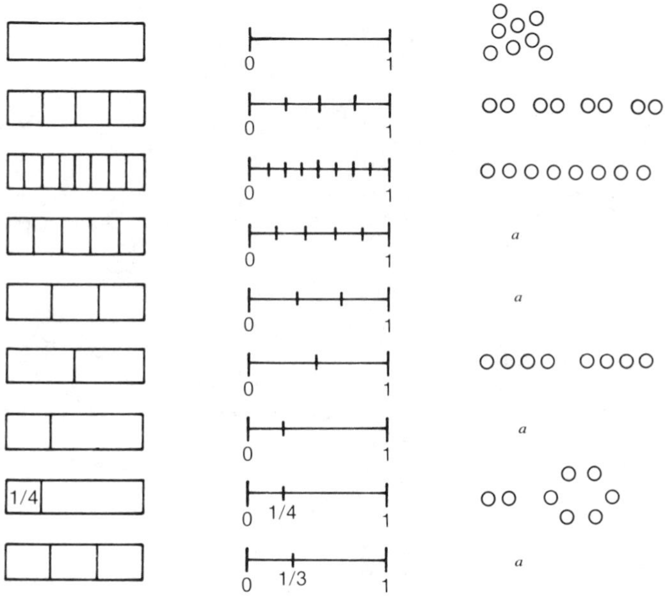

Category	Type of cue		
	Continuous	Number line	Discrete
Incomplete		0 1	
Complete		0 1	
Irrelevant		0 1	
Inconsistent		0 1	[a]
Inconsistent		0 1	[a]
Incomplete		0 1	
Inconsistent		0 1	[a]
Incomplete	1/4	0 1/4 1	
Inconsistent		0 1/3 1	[a]

[a]There is no discrete counterpart to the number line and area interpretation since a set of 8 cannot be divided in 5 (or 3) equal parts without subdividing individual elements.

TABLE 4.2

Solution Strategies for Various Types of Perceptual Cues

Type of visual cue	Recommended solution strategy
Consistent	
Complete	No modification necessary. Find required fractional part using part–whole strategies.
Incomplete	Identify appropriate unit fraction. Iterate unit fraction to complete diagram. Proceed as with consistent–complete cues.
Irrelevant	Highlight appropriate unit fraction. Ignore other partitions. Proceed as with consistent–complete cues.
Inconsistent (perceptual distractor)	Ignore all subdivisions. Reconstruct diagram so as to be complete. Proceed as with consistent–complete cues.

TABLE 4.3

Percentage of Student Errors by Type of Perceptual Cue, Interpretation of Fraction, and Specific Example[a]

Baseline	Specific fraction	Number line	Discrete	Circles	Rectangles
	$3/4$	61	18	0	1
Consistent–complete	$2/3$	68	22	5	0
	$5/3$	75	90		
	$3/4$	68	31	6	1
Consistent–incomplete	$2/3$	71	26	26	25
	$5/3$	79	91		
	$3/4$	74		19	21
Consistent–irrelevant	$2/3$	78		23	25
	$5/3$	82	88		
	$3/4$	87	31	48	57
Inconsistent	$2/3$	75	27	39	36
	$5/3$	81,82	87		

[a]These data were collected by Nadine Bezuk as part of an M.A. thesis at the University of Minnesota. The authors are indebted to Ms. Bezuk for her permission to include them here. Three fourth-grade classes; $N = 77$.

tualizing a fraction as a point on a line. This is probably due to the fact that the majority of their experiences had been with the part–whole interpretation of fraction in a continuous (area) context. These results are consistent with other findings (e.g., Novillis-Larson, 1980) that suggest that number line interpretations are especially difficult for children.

Second, in the discrete context, the fraction $\frac{5}{3}$ causes many more errors than do the fractions $\frac{3}{4}$ and $\frac{2}{3}$. This is due, perhaps, to the fact that school instruction has placed virtually all of its emphasis on fractions less than one, certainly a limited interpretation of rational number but hardly uncommon at this level.

Third, the percentage of errors increases as the type of perceptual cue changes from complete to incomplete to irrelevant to inconsistent. As would be predicted, the highest percentage of errors occurred with the inconsistent cues. This is true across all four physical interpretations of these three fractions. The discrepancy seems to be especially apparent in the continuous context, which was represented here by circles and rectangles. (Unfortunately, as of this writing, no data exist for the fraction $\frac{5}{3}$ in the continuous context.) The stability of this trend suggests the need for a closer examination of the cognitive processes involved in dealing with various types of perceptual cues across continuous, discrete, and number-line tasks.

CLINICAL DATA FROM ONE-ON-ONE INTERVIEWS

Children involved in the project teaching experiments were periodically given tasks involving perceptual distractors. These tasks were given along with others in one-on-one clinical interviews. The results and findings from the tasks are presented according to the type of embodiment on which they were based.

Continuous Embodiment Tasks

One task was designed to assess children's flexibility in regarding a part of a whole as an unpartitioned region and as a partitioned region. It requires the observation that two equivalent parts of a whole can each be named by the same fractions when one part is appropriately partitioned. In Figure 4.6, *b* and *cde,* equivalent parts, each can be named as *one-fourth* and *three-twelfths.*

Of interest was whether the child could ignore the partition lines in *cde* to consider it as one-fourth and imagine partition lines placed in *b* to consider it as three-twelfths. This was one of several contexts in which we found the presence of subpartitioning lines to be a distractor to children's logical–mathematical understanding of rational number concepts.

Children's responses to questions in this context reveal various degrees of flexibility in thinking about fractions. Children readily determined that it would take 4 pieces like *b* to cover the circle and concluded *b* to be one-fourth of the whole. Children also extrapolated beyond the boundaries of *cde* to determine that it would take 12 parts like *c, d,* or *e* to cover the entire circle. Children did this in various ways; the most common was iterating *cde* to count the total number of pieces in the whole, and naming *c, d,* and *e* each as one-twelfth.

Problems arose when students were asked to give more than one name to either *b* or *cde*. Portions of an interview sequence with a fourth-grade student indicate the general nature of these difficulties.

I: *b* Is what fraction of the whole?
S: One-fourth . . . [Why?] . . . Well this, [*b*] is that big and I measured with my eyes about that big again and then all the way around.
I: *cde* together is what fraction of the whole?
S: One-fourth . . . [Explain] . . . Well if you took all of them [*c, d,* & *e*] it would be one-fourth because that is as big as that [*b*].

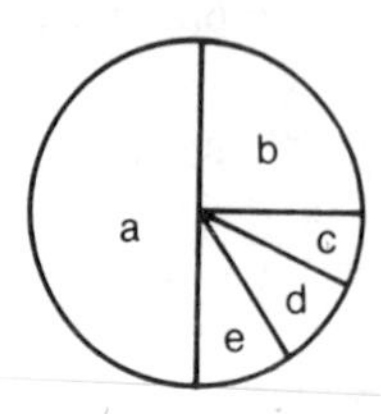

Figure 4.6. Unpartitioned and partitioned circular regions showing $\frac{1}{4} = \frac{3}{12}$.

I: Is there another way you can tell me what fraction this [*b*] is of the whole?
S: I don't think so.
S: [As *I* points to *c, d,* and *e* in turn] One-twelfth, one-twelfth, one-twelfth.
I: So what fraction is this [*cde*] altogether?
S: One-fourth.
I: Now count with me.
S: One–twelfth, two-twelfths, three-twelfths.
I. So what fraction is this of the whole circle?
S: One-twelfth . . . oh! Hold it . . . one-fourth.
I: Now count with me again [pointing in turn to *c, d,* and *e*].
S: One-twelfth, two-twelfths, three-twelfths.
I: Now how can I say another name besides one-fourth for all of this [*cde*]?
S: One-fourth, two-fourths, three-fourths [counting while pointing to *e, d,* and *c*]
I: Let's see, what was this [*e*] again?
S: One-twelfths, two-twelfths, three-twelfths [while *I* points to *e, d,* and *c*].
I: How can I describe the whole thing?
S: Three-twelfths . . . because there are three twelfths, so we can call it three-twelfths.
I: Is there another name for this [*b*]?
S: No.
I: What did you say about this [*cde*] compared to this [*b*] . . . and what did you call this [*cde*]?
S: Three-twelfths.
I: If they are the same size [*cde* and *b*] and this [*cde*] is [*S* says three-twelfths] then what's another name for this [*b*]?
S: Three-twelfths.
I: Explain to me.
S: Well these are the same size [each of *c, d,* and *e*] and this [*b*] and this [*c, d, e*] are the same size so you can call this [*b*] three-twelfths, *if you cut it in three pieces* [emphasis added].

Notice that even though the student shows increased flexibility in this sequence, he is still unable to label the piece with two names simultaneously. There appears to be some rigidity that seems to lead to confusion as evidenced when the subject mistakenly counts *cde* as "one-fourth, two-fourths, three-fourths." It has often been observed in the course of the teaching experiment that when children have difficulty with a new concept, skill with already learned tasks sometimes deteriorates temporarily. Identification of twelfths was considered routine for this subject, but dual labeling of fractions was somewhat novel. Although the subject did eventually resolve the discrepancy, it was evident that some further instruction was needed.

The results of eight such interviews suggest a trend in the development of the ability to identify fractional parts from multiple perspectives. At one level children were able to label *b* and *cde* with a single label only, centering either on the size of the parts ($b = \frac{1}{4}$, $cde = \frac{1}{4}$) or the partitioning lines ($b = \frac{1}{4}$, $cde = \frac{3}{12}$). At a transitional level, subjects showed increasing flexibility with respect to only one region. For example, some subjects first acknowledged that partition lines could potentially be drawn in *b*. Thus, *b* could be named either *one-fourth* or *three-*

twelfths (but not both at the same time). They persisted in naming region *cde* *three-twelfths* because the partition lines had already been drawn. Other subjects said that *cde* could be named *one-fourth* if the partition lines were removed, but maintained that *b* must be named *one-fourth*. At the last level, the subjects' flexibility increased to the point where they could label both regions with either name.

Discrete Embodiment Tasks

When a continuous object, such as a sheet of paper, is partitioned into *n* equal-sized parts, each part is also a *single* continuous piece. When a set of discrete objects is used as a unit, partitioning of the unit into *n* equal-sized parts frequently results in subsets, each with *several* objects. This characteristic of the discrete set embodiment forces an extension of the child's part–whole rational-number schema.

To investigate the strength of children's logical–mathematical thinking about rational number in the context of discrete embodiments, several tasks involving perceptual distractors were developed. The distractor was created by transforming the arrangement of the objects in the initial unit set from "consistent" to "inconsistent."

Task 1 involved an initial presentation of 6 paper clips arranged as !!! !!! and transformed to !! !! !!; Task 2 involved an initial presentation of 10 paper clips arranged as !!!!! !!!!! and transformed to !!! !!!! !!!. For each part of each task, the subject was asked to produce a set of paper clips equal in number to three-halves the number of clips in the stimulus set. Task 3 involved a set of 12 paper clips; for the initial presentation they were arranged in 3 groups of 4 as !!!! !!!! !!!! and transformed to 2 groups of 6 as !!!!!! !!!!!!. The problem for the subject in each case in Task 3 was to present a set of clips equal in number to five-thirds the number of clips in the stimulus set.

In the final interview of the teaching experiment, the children were given two tasks, each of which involved a unit of six counting chips. The unit was first presented in a row: ••••••; then while the children observed, the set was transformed to •• •• ••; finally, again while the children observed, the set was transformed to 2 groups of 3 as ••• •••. For the first of these tasks (Task 4) the child was asked, after the initial presentation, and after each transformation, to show a set of chips equal in number to two-thirds of the stimulus set; in Task 5, the child was asked to show three-halves of the stimulus set.

For each of the tasks, some leading questions were asked if the child initially failed at showing the requested fraction of the consistently arranged set; the purpose was to determine whether the child conserved the quantitative aspect of the fraction in spite of the distraction induced by transforming the set to an inconsistent arrangement. For example if a child had difficulty in showing three-halves of !!! !!!, it was suggested that the child show one-half, two-havles, and finally three-halves. Children were invariably successful after this intervention. Therefore, the interest in this set of subject responses is in the disparity, where it exists, between subjects' performance on the task in the absence of a distracting arrangement (consistent) as compared with performance in the presence of a distracting arrangement (inconsistent).

Successful performance for the second part of the first three tasks was characterized by returning to the consistent arrangement and repeating the previous correct solution. Two subjects succeeded immediately; two others improved during the course of the interview and were eventually successful. All four of these subjects commented on the fact that moving the objects did not change the problem. On Tasks 4 and 5 these children had no trouble transforming each display into a consistent arrangement and solving correctly.

The most typical error on Tasks 1 and 2 for the four subjects mentioned above, and on Task 3 as well, for the remaining two subjects was of the following type: For three-halves of !! !! !!, the solution was !! !! !!, with the subject naming each pair of clips as one-half, two-halves, and three-halves. A second common error resulted from confusing the numerator of the requested fraction with the number of objects in each subset. For example, for five-thirds of !!!!!! !!!!!!, two subjects showed five sets of five clips. One child called them "one-fifth, two-fifths, three-fifths, four-fifths, five-fifths;" the other child said "one-third, two-thirds, three-thirds, four-thirds, five-thirds." The inconsistent arrangements for Tasks 4 and 5 elicited the same types of errors from unsuccessful subjects.

It is possible that the similarity between the numerals used in the fraction and the numbers that describe the arrangement of the visual stimulus presented may have caused difficulties for many students. For example, some children had more trouble with the problem "Find three-halves of •• •• ••" than they did with the problem "Find five-thirds of •••••• ••••••." It is possible that the difficulty in the first problem involved the numerical similarity between the fraction (3/2) and the arrangement of the chips presented (3 groups of 2). This similarity apparently overwhelmed some students, causing them to abandon their customary solution process in favor of an illogical process. For example, some students said that the answer to the problem "Find three-halves of •• •• ••" was •• •• ••. Some of these same students, however, correctly stated that five-thirds of •••••• •••••• was 20 chips, not 5 groups of 3 as their previous strategy might have suggested.

Number-line Tasks

One of the subconstructs of rational numbers which Kieren (1976) identified is the measure subconstruct, for which some unit of measure is involved, as well as the subdivision of units into smaller components. The measure (number) associated with an object is then the number of units or subunits that "equal" the object measured. A common concrete embodiment of the measure subconstruct of rational number is the number line. In this context, a unit is represented by a length, in contrast with the part–whole subconstruct in which the unit is most often an area or a set of discrete objects.

Some research (e.g., Novillis-Larson, 1980) indicates that children as late as seventh grade have difficulty interpreting the unit on the number line. Another variable that was found to cause children difficulty was whether or not the unit subdivisions were equal in number to the denominator of the fraction in question.

We investigated fourth- and fifth-grade children's ability to deal with rational-number situations on a number line for which the number of subunits of each unit was, with respect to the fraction in question, equal to the denominator, one-half the denominator, twice the denominator, or neither a divisor nor multiple of the denominator.

According to the definition of perceptual cues used in this chapter, the number-line problems represented either complete, incomplete, irrelevant, or inconsistent perceptual cues.

The types of number-line problems that are discussed here are (*a*) locating the point on the number line corresponding to a given rational number; and (*b*) using the number line to generate a fraction equivalent to a given fraction, or using the number line to justify answers given abstractly.

Our results were similar to those reported by Novillis-Larson:

1. Children differed in how they identified the unit on the number line.
2. Problems in which the subdivisions of the unit did not equal the denominator of the fraction were harder to solve than were problems in which subdivisions equalled the denominator.
3. Problems with perceptual distractors (inconsistent cues) were harder to solve than were problems in which subdivisions of the unit were factors or multiples of the denominator or the fraction (incomplete cues or irrelevant cues).

In the series of number-line tasks, with cues ranging from complete to inconsistent, given to two groups of fourth graders ($N = 11$) after instruction, we found differences in the levels of success, the strategies used to solve the problems, and the amounts of assistance needed to reach a solution. Several of the tasks and the results obtained from the interviews are discussed subsequently.

All 11 children were able to locate $\frac{2}{3}$ on a 4-unit-long number line on which each unit was partitioned into thirds. They were also successful with a similar problem using fractions greater than one.

The related problem with inconsistent cues required the subjects to locate $\frac{2}{4}$ on a 4-unit-long number line on which each unit was partitioned into thirds. Only four subjects solved this problem easily. Successful strategies were to ignore the markings for thirds and draw in fourths or to simplify $\frac{2}{4}$ to $\frac{1}{2}$ and locate $\frac{1}{2}$ midway between zero and one. Another subject was able to respond correctly after being told to ignore the one-third markings.

One subject located $\frac{2}{4}$ at the $\frac{2}{3}$ point, realized his mistake, but could not correct it. Three subjects changed the length of the unit: one located $\frac{2}{4}$ at 2, changing the unit to 4, whereas the others located $\frac{2}{4}$ at $\frac{2}{3}$, indicating that they had changed the unit to $1\frac{1}{3}$.

An external ruler on which $\frac{1}{4}$ was approximated was developed by one subject who then measured off his estimated $\frac{1}{4}$ length twice to locate $\frac{2}{4}$. The remaining subject squeezed an additional partition point between 0 and $\frac{1}{3}$, then located $\frac{2}{4}$ at the second of the four partition points (actually at $\frac{1}{3}$).

The second number-line task involved equivalent fractions. Subjects were asked to use the number line to find $\frac{5}{3}$ = [ ]/12 on a number line 4 units long, divided into thirds. The students had no trouble locating $\frac{5}{3}$ on the number line; identifying twelfths did present a problem.

Four of the nine fourth graders solved the problem without using the number line. They observed that $3 \times 4 = 12$, $5 \times 4 = 20$, so $\frac{5}{3} = \frac{20}{12}$. Three of these subjects then used the number line to verify the solution. The fourth subject could not successfully reconcile her correct symbolic result with attempts to subdivide the number line into twelfths.

Four subjects actually attempted to solve the problem using the number line, without first finding the result symbolically. One successfully divided each third from 1 to 2 into four equal parts to obtain twelths, then counted $\frac{20}{12}$ without partitioning the segment from zero to one. One child was able to complete the solution after the interviewer divided the number line into twelfths and highlighted the thirds. The other two subjects divided each third in half, then labeled $\frac{5}{6}$ and $\frac{10}{6}$ as $\frac{10}{12}$ and $\frac{20}{12}$, respectively. The remaining subject was unable to obtain an answer to the problem.

These results illustrate the fact that, faced with a representation that is not immediately useful in solving the required task, a number of children prefer to translate the problem into a different mode of representation. In this case, the pictorial number-line representation did not include markings for the required twelfths; four of the nine subjects solved the problem using a symbolic representation. A similar preference for symbolic representations is reported by Lesh, Landau, and Hamilton (Chapter 9, this volume) for the solution of problems presented in the context of real-world and manipulative-aid modes.

Discussion

Perceptual distractors represent one class of instructional conditions that make some types of problems more difficult for children to solve. Knowledge of their impact will be helpful in the design of more effective instructional sequences for children. It seems reasonable to suggest that initial examples might be given wherein the potential impact of perceptual distractors is minimized, but that later examples should deliberately provoke children to resolve conflicts that arise in association with perceptual distractors.

Although performance with rational numbers is affected by the presence of distractors, children can be taught to overcome their influence. Furthermore, the strategies generated by children to overcome these distractors lead to more stable rational-number concepts.

Our research raises questions about the nature of an appropriate role for perceptual and other distractors in the learning process. Distractors that initially caused children problems did not affect them later when the concept had become internalized. When a distractor is coupled with the introduction of a new subconcept, the resulting situation is contaminated with unnecessary and irrelevant cues and causes difficulty for the child. We believe that the process of developing new or extended concepts requires the student to identify relevant and irrelevant variables and to view problem conditions more critically. This is, in fact, the process of discriminating what is and what is not relevant to the concept in question. Such discrimination is continually referred to in the concept formation literature as an integral component of what is meant by "knowing a concept."

Directions for Future Research

Refinement of the Theoretical Models

This section provides a brief discussion of three broad areas of research interest that have been identified in the work of the Rational Number Project. One interest is the question of whether emphasis on oral language can serve a facilitating intermediary role to bridge the apparently large gap between a learner's ability to represent mathematical ideas with manipulative aids and the ability to represent the same ideas with mathematical symbolism. A second issue deals with how manipulative aids might be used to facilitate a learner's ability to develop appropriate mathematical models for real-world problem situations. These issues involve refining two traids of the theoretical model depicted in Figure 4.3. The first issue involves the manipulative aids–oral language–written symbols triad; the second involves the real-world problems–manipulative aids–written symbols triad.

The third broad area of interest concerns which subconstruct of rational number might best serve as the *fundamental subconstruct* for developing initial fraction-concepts in children. The importance of unit fractions (of the form $1/n$) is discussed in this context.

The Role of Oral Language in Facilitating Mathematics Learning

How is it that learners make a meaningful connection between a mathematical idea represented with concrete manipulative aids and the appropriate mathematical symbolism for that idea? Behr (1977) indicated that the gap between children's ability to represent mathematical ideas in these two modes is much greater than usually perceived and that the mental schemata necessary to bridge this gap are apparently much more complex than expected. There is an obvious need for research, via teaching experiments, to investigate instructional situations with the objectives of (*a*) gaining insights into the source of children's difficulty, and (*b*) providing experience to help children overcome the difficulty.

The problem of children's learning of mathematical symbolism has not gone completely unnoticed in mathematics education research. Three studies (Coxford, 1965; Hamrick, 1978; Pinchback, 1970) have dealt with the question. Of particular interest is Hamrick's readiness criterion for symbolism: a subject's ability to *state orally* the mathematical sentence of concern. She found that children who met this criterion outperformed other children on her test of symbolic addition and subtraction. Ellerbruch and Payne (1978) reported that children who, first, say aloud the fraction represented by a display, then transcribe the oral sound as *three-fifths,* for example, before going to the symbolic form of $\frac{3}{5}$, seldom make the common reversal error of writing $\frac{5}{3}$.

Two observations from our current work bear upon the question:

1. One child having difficulty writing "mixed numerals" to correspond with a fraction display was asked each time to say orally the fraction shown, and then write what he said. Repetitions of this, along with an instruction always to say the fraction to himself before writing it, alleviated the problem.

2. We often notice when children are doing worksheets where either pictures or manipulatives are used, they spontaneously vocalize or subvocalize the fraction represented before writing it down. When asked to show and write a fraction using our colored-parts model, one subject was observed first to place the pieces down, then orally to count them before writing the fraction.

The Role of Manipulative Aids in Developing Problem Modeling

Past research indicates that instruction using manipulative aids is at least as effective, but perhaps less efficient, then other forms of instruction. Unfortunately, the learning outcomes that have been assessed have usually been restricted to initial learning or short-term retention. Less attention has been given to the transferability and usefulness of the learning in real problem-solving situations—precisely what learning from concrete materials might be expected to facilitate.

Uninvestigated is the role that manipulative materials play in modeling real-world problem situations, which require both (*a*) a translation from the real-world situation to the realm of mathematics and (*b*) a representation of the real-world situation with mathematical symbolism and assumptions. Manipulative aids are an intermediary between the real world of problem situations and the world of abstract ideas and written symbols. They are symbols, in that they can be used to represent several different real-world situations, and they are concrete, in that they involve real materials. A manipulative aid, such as poker chips, can easily model certain real-world problems. For example, consider the problem:

> $\frac{5}{7}$ of a group of children will receive a prize. There are 35 children. How many will receive a prize?

If counting chips are used to represent people, then the problem situation is easily modeled and the answer determined. If the children have had prior experience associating the sentence $\frac{5}{7} \times 35 = [\ ]$ with the chip demonstration, then the fact that the same mathematical sentence is a model for the problem situation is easier to see. Experiences of this kind may help a child to move gradually from a manipulative model of the problem situation to a symbolic model.

Research questions of interest concern whether or not children are able to display a model via a manipulative aid for a real-world problem situation, and whether or not they are able to solve the problem. Also of interest are questions concerning children's ability, following modeling experiences such as those suggested above, to relate symbolic mathematical statements to models and to real-world situations.

The Importance of Unit Fraction Approach in Rational-Number Learning

Curriculum materials currently used in schools develop the rational-number concept predominantly from the part–whole subconstruct. The question of which subconstruct should play a central role in the development of the *basic* rational-number concept and in the development of the *basic* concepts underlying rational-number relations and operations is open to empirical investigation. A strand of data arising from our teaching experiments bears on this question.

The data strand concerns children's development of a quantitative notion of rational number. (By *quantitative notion* of rational number we mean children's ability to demonstrate the size of rational numbers.) Our observations suggest that this notion is fundamental in children's development of rational-number concepts, relations, and operations. It apparently underlies children's ability to order rational numbers, to internalize the concept of equivalent fractions, and to have a meaningful grasp of addition and multiplication of fractions. What meaning does the addition of $\frac{3}{8}$ and $\frac{4}{8}$ have for a child without an internalized notion of the "bigness" of each addend and the sum?

Although the historical and cognitive importance of *unit fractions* has been recognized (Gunderson & Gunderson, 1957; Kieren, 1976), curriculum materials do not exploit this notion in the development of rational-number concepts. An important hypothesis that arose from our current work is that children develop a stronger quantitative notion of rational numbers when the development of basic rational-number concepts arises from iteration of unit fractions. In this context, nonunit fractions would develop through counting or adding related unit fractions (i.e., $\frac{3}{4}$ is $\frac{1}{4}$ and $\frac{1}{4}$ and $\frac{1}{4}$, rather than 3 of 4 parts). The addition, $\frac{3}{8} + \frac{4}{8}$, could be processed as $\frac{3}{8}$, $\frac{4}{8}$, $\frac{5}{8}$, $\frac{6}{8}$, $\frac{7}{8}$ by counting unit fractions.

Some observations from our current work give credibility to this hypothesis: Children have had no difficulty doing symbolic addition problems, such as $\frac{3}{7} + \frac{4}{7}$ when instruction emphasizes counting sevenths on a manipulative display. The progression from fractions less than one to fractions greater than one is also facilitated. Similarly, children who mistakenly interpret

SCHEMA 4.2

as $\frac{7}{8}$ rather than $\frac{7}{4}$ are helped when emphasis is placed on identifying the fraction represented by one part ($\frac{1}{4}$) and then counting the numbers of fourths shaded.

Other Critical Issues in Rational-Number Research

The questions of how and why manipulative aids facilitate the learning of mathematical ideas for children have not been adequately addressed by research. We present here some observations about the use of manipulative aids as hypotheses for further investigation.

1. Which manipulative aids should be used to teach which concepts, and in what order should they be introduced? Manipulative aids that are used must differ from one another in perceptual features and in the mathematical way they embody the concept. We have observed that when children face the representation of a tenuously familiar concept in the context of a new manipulative (which differs in a nontrivial way from its predecessors) they are forced to rethink the concept. Thus, whereas the representation of a concept with a first manipulative aid may be characterized as a bottom-up interpretation (i.e., the manipulative provides an interpretation of the concept for the child), subsequent manipulative representations provide the child with an opportunity for a top-down interpretation (i.e., the child uses the concept to interpret how the manipulative represents the concept). It seems likely that it is a series of such top-down interpretations that provides for the mathematical generalization and abstraction of the concept (Dienes, 1967).

2. Children first learn to represent mathematical ideas with a first manipulative by imitation of teacher demonstration. Subsequent manipulatives are introduced in one of two ways: (*a*) the teacher demonstrates with the familiar aid and students use the new aid to interpret the teacher's demonstration, or (*b*) the teacher demonstrates with the new aid and students interpret with the familiar one.

3. Contrary to the prevailing opinion among mathematics educators, we have learned that a "good" manipulative aid is one that *causes* a certain amount of confusion. The resultant cognitive disequilibrium leads to greater learning.

4. We have gained insights into what kinds of rational number tasks children can learn. We would argue that more attention needs to be given certain fundamental rational-number concepts in earlier grades. In particular, children should be given experiences in partitioning activities before Grade 3. Through the use of appropriate manipulative materials, children should begin to observe the compensatory relationship between the size and number of parts into which a whole is partitioned.

Acknowledgments

Our sincere thanks go to the following people who assisted us during this research: Nik Pa Nik Azis, Nadine Bezuk, Diane Briars, Kathleen Cramer, Issa Feghali, Eric Hamilton, Cherl Hoy, Leigh McKinlay, Marsha Landau, Roberta Oblak, Mary Patricia Roberts, Robert Rycek, Constance Sherman, and Juanita Squire. We are also indebted to Nadine Bezuk, Kathleen Cramer, Marsha Landau, Mary Patricia Roberts, Robert Rycek, and Juanita Squire for their help in preparing this manuscript.

Reference Note

1. Behr, M., Post, T., & Lesh, R. *Construct analysis, manipulative aids, representational systems, and learning of rational numbers.* NSF RISE Proposal, 1981.

References

Behr, M. *The effects of manipulatives in second graders' learning of mathematics* (Vol. 1). PMDC Technical Report No. 11, Tallahassee, FL: PMDC, 1977.

Bruner, J. S. On cognitive growth. In J. S. Bruner, R. R. Oliver, & P. M. Greenfield (Eds.), *Studies in cognitive growth.* New York: Wiley, 1966.

Carpenter, T. P., Coburn, T. G., Reys, R. E., & Wilson, J. W. Notes from national assessment: Addition and multiplication with fractions. *Arithmetic Teacher,* 1976, *23*(2), 137–141.

Carpenter, T., Coburn, T. G., Reys, R. E., & Wilson, J. W. *Results from the first mathematics assessment of the National Assessment of Educational Progress.* Reston, Virginia: National Council of Teachers of Mathematics, 1978.

Carpenter, T. P., Corbitt, M. K., Kepner, H. S., Jr., Lindquist, M., & Reys, R. E. National assessment: Prospective of students' mastery of basic skills. In M. Lindquist (Ed.), *Selected issues in mathematics education.* Berkeley, California: McCutchan, 1980.

Cobrun, T. G., Beardsley, L. M., & Payne, J. N. Michigan educational assessment program. Mathematics Interpretive Report, 1973. grades 4 and 7 tests. *Guidelines for quality mathematics teaching monograph series, No. 7.* Birmingham, Michigan: Michigan Council of Teachers of Mathematics, 1975.

Coxford, A. F. *The effects of two instructional approaches on the learning of addition and subtraction concepts in grade one.* Unpublished doctoral dissertation, University of Michigan, 1965.

Dienes, Z. P. *Building up mathematics* (Rev. ed.). London: Hutchinson Educational, 1967.

Ellerbruch, L. W., & Payne, J. N. A teaching sequence for initial fraction concepts through the addition of unlike fractions. In M. Suydam (Ed.), *Developing computational skills.* Reston, Virginia: National Council of Teachers of Mathematics, 1978.

Fennema, E. H. Models and mathematics. *Arithmetic Teacher,* 1972, *19,* 635–640.

Gagne, R. M., & White, R. T. Memory structures and learning outcomes. *Review of Educational Research,* 1978, *48,* 187–222.

Ganson, R. E., & Kieren, T. Operator and ratio thinking structures with rational numbers – A theoretical and empirical exploration. *The Alberta Journal of Educational Research,* 1980, in press.

Gerling, M., & Wood, S. *Literature review: Research on the use of manipulatives in mathematics learning* (PMDC Technical Report No. 13). Tallahassee, Florida: Florida State University, 1976.

Gunderson, A. G., & Gunderson, E. Fraction concepts held by young children. *Arithmetic Teacher,* 1957,*4*(4), 168–174.

Hamrick, A. K. *An investagation of oral language factors in reading for the written symbolization of addition and subtraction.* Unpublished doctoral dissertation, University of Georgia, 1976.

Hawkins, P. *An approach to science education policy.* Background paper for a program of research in mathematics and science education. Washington, D.C.: National Science Foundation, 1979.

Hiebert, J., & Tonnessen, L. H. Development of the fraction concept in two physical contexts: An exploratory investigation. *Journal for Research in Mathematics Education,* 1978, *9*(5), 374–378.

Karplus, R., Karplus, E., Formisano, M., & Paulsen, A. C. Proportional reasoning and control of variables in seven countries. In J. Lochhead & J. Clements (Eds.), *Cognitive process instruction.* Philadelphia, Pennsylvania: Franklin Institute Press, 1979.

Karplus, R., Karplus, E. F., & Wollman, W. Intellectual development beyond elementary school: Ratio, the influence of cognitive style (Vol. 4). *School Science and Mathematics,* 1974, *76*(6), 476–482.

Karplus, R., Pulos, S., & Stage, E. K. *Proportional reasoning of early adolescents.* Paper presented at the meeting of the Fourth Annual MERGA Conference, Hobart, Australia, May 1980.

Kieren, T. E. Activity learning. *Review of Educational Research,* 1969, *39,* 509–522.

Kieren, T. E. On the mathematical, cognitive, and instructional foundations of rational numbers. In R. Lesh (Ed.), *Number and measurement: Papers from a research workshop.* Columbus, Ohio: ERIC/SMEAC, 1976.

Kieren, T. E. *Five faces of mathematical knowledge building.* Edmonton: Department of Secondary Education, University of Alberta, 1981.

Kieren, T. E., & Nelson, D. The operator construct of rational numbers in childhood and adolescence – An exploratory study. *The Alberta Journal of Educational Research,* 1978, *24*(1).

Kieren, T. E., & Southwell, B. Rational numbers as operators: The development of this construct in children and adolescents. *Alberta Journal of Educational Research,* 1979, *25*(4), 234–247.

Klahr, D., & Siegler, R. S. The representation of children's knowledge. *Advances in Child Development and Behavior,* 1978, *12,* 62–116.

Kurtz, B., & Karplus, R. Intellectual development beyond elementary school: Teaching for proportional reasoning (Vol. 7). *School Science and Mathematics,* 1979, *79*(5), 387–398.

Lankford, F. G., Jr. *Some computational strategies of seventh grade pupils.* U.S. Office of Education, Project No. 2–C–013. Washington, D.C.: Government Printing Office, 1972.

Lesh, R. Mathematical learning disabilities: Considerations for identification, diagnosis, and remediation. In R. Lesh, D. Mierkiewicz, & M. G. Kantowski (Eds.), *Applied mathematical problem solving.* Columbus, Ohio: ERIC/SMEAC, 1979.

Lesh, R., & Hamilton, E. *The Rational Number Project Testing Program.* Paper presented at the American Educational Research Association Annual Meeting, Los Angeles, California, April 1981.

Lesh, R., Landau, M., & Hamilton, E. Rational number ideas and the role of representational systems. In R. Karplus (Ed.), *Proceedings of the Fourth International Conference for the Psychology of Mathematics Education.* Berkeley, California: Lawrence Hall of Science, 1980.

Noelting, G. The development of proportional reasoning in the child and adolescent through combination of logic and arithmetic. In E. Cohors-Fresenborg & I. Wacksmuth (Eds.), *Proceedings of the Second International Conference for the Psychology of Mathematics Education.* Osnabruck, West Germany: University of Osnabruck, 1978.

Noelting, G. *The development of proportional reasoning and the ratio concept (the orange juice experiment).* Ecole de Psychologie Universite Laval, Quebec, November, 1979.

Novillis, C. An analysis of the fraction concept into a hierarchy of selected subconcepts and the testing of the hierarchy dependences. *Journal for Research in Mathematics Education,* 1976, *7,* 131–144.

Novillis-Larson, C. Locating proper fractions. *School Science and Mathematics,* 1980, *53*(5), 423–428.

Owens, D. T. Study of the relationship of area concept and learning concepts by children in grades three and four. In T. E. Kieren (Ed.), *Recent research on number concepts.* Columbus, Ohio: ERIC/SMEAC, 1980.

Piaget, J., Inhelder, B., & Szeminska, A. *The child's conception of geometry.* New York: Basic Books, 1960.

Pinchback, Carolyn L. Relating Symbolism to Mathematical Concepts by 10–11 year olds. Unpublished Docotral Dissertation. University of Texas at Austin, 1978.

Polkinghorne, A. R. Young children and fractions. *Childhood Education,* 1935, *11,* 354–358.

Post, T. R., & Reys, R. E. Abstraction, generalization, and design of mathematical experiences for children. In K. Fuson & W. Geeslin (Eds.), *Models of mathematics learning.* Columbus, Ohio: ERIC/SMEAC, 1979.

Rappaport, D. The meaning of fractions. *School Science and Mathematics,* 1962, *62,* 241–244.

Riess, A. P. A new approach to the teaching of fractions in the intermediate grades. *School Science and Mathematics,* 1964, *54,* 111–119.

Sambo, Abdussalami, A. *Transfer effects of measure concepts on the learning of fractional numbers.* Doctoral dissertation, The University of Alberta, 1980.

School Mathematics Study Group. In James W. Wilson, Leonard S. Cohen, and Edward G. Begle (Eds.), description and statistical properties of X-population scales. National Longitudinal Study of Mathematics Abilities Reports: No. 4 Stanford, California: The Board of Trustees of the Leland Stanford Junior University, 1968.

Suydam, M. N., & Higgins, J. L. *Activity-based learning in elementary school mathematics: Recommendations from research.* Columbus, Ohio: ERIC/SMEAC, 1977.

Tulving, E. Episodic and semantic memory. In E. Tulving (Ed.), *Organization of memory.* New York: Academic Press, 1972.

Usiskin, Z. P. The future of fractions. *The Arithmetic Teacher,* 1979, *26,* 18–20.

Vygotsky, L. S. *Mind in society.* Cambridge, Massachusetts: Harvard University Press, 1976.

CHAPTER 5

Multiplicative Structures

*Gerard Vergnaud**

History teaches us that science and technology have developed with the aim of solving problems. One of the most challenging points in education is probably to use meaningful problems so that knowledge, both in its theoretical and its practical aspects, may be viewed by students as a genuine help in solving real problems. However, this condition, that knowledge be both operational and interesting, cannot easily be satisfied.

Piaget has demonstrated that knowledge and intelligence develop over a long period of time, but he has done this by analyzing children's development in terms of general capacities of intelligence, mainly logical, without paying enough attention to specific contents of knowledge. It is the need to understand better the acquisition and development of specific knowledge and skills, in relation to situations and problems, that has led me to introduce the framework of conceptual fields. A *conceptual field* is a set of problems and situations for the treatment of which concepts, procedures, and representations of different but narrowly interconnected types are necessary.

Why is such a framework necessary?

1. It is difficult and sometimes absurd to study separately the acquisition of interconnected concepts. In the case of multiplicative structures, for instance, as we see in this chapter, it would be misleading to separate studies on multiplication, division, fraction, ratio, rational number, linear and n-linear function, dimensional analysis, and vector space; they are not mathematically independent of one another, and they are all present simultaneously in the very first problems that students meet.

*Other members of the research group whose experiments were reported are A. Rouchier, G. Ricco, P. Marthe, C. Landré, A. Viala, R. Metregiste and, recently, J. Rogalski and R. Samurcay.

ISBN No. 0-12-444220-X

2. It is also wise, in a psychogenetic approach to the acquisition of specific ideas, to delineate rather large domains of knowledge, covering a large diversity of situations and different kinds and levels of analysis. This enables one to study their development in the student's mind over a long period of time.

3. Finally, there are usually different procedures and conceptions and also different symbolic representations involved in the mastery by students of the same class of problems. Even though some of these conceptions and representations are weak or partially wrong, they may be useful for the solution of elementary subclasses of problems and for the emergence of stronger and more nearly universal solutions.

The conceptual-field framework makes it possible to study the organization of these interconnected ideas, conceptualizations, and representations over a period of time long enough to make the psychogenetic approach meaningful.

I have been interested in two main conceptual fields, additive structures and multiplicative structures, viewing them as sets of problems involving arithmetical operations and notions of the additive type (such as addition, subtraction, difference, interval, translation) or the multiplicative type (such as multiplication, division, fraction, ratio, similarity). Of course, multiplicative structures rely partly on additive structures; but they also have their own intrinsic organization which is not reducible to additive aspects. See Vergnaud, 1981; Vergnaud, 1982; Vergnaud, 1983 for additive structures.

Other important conceptual fields, interfering with these two, include (*a*) displacements and spatial transformations; (*b*) classifications of discrete objects and features, and Boolean operations; (*c*) movements and relationships among time, speed, distance, acceleration, and force; (*d*) parenthood relationships; and (*e*) measurement of continuous spatial and physical quantities.

Some of these last conceptual fields play an important part in the meaning and understanding of additive and multiplicative structures; reciprocally, the development of additive or multiplicative structures is necessary for the mastery of certain relationships involved in other conceptual fields. It is a complex landscape. Still, I find it fruitful to delineate distinct domains, if they can be consistently described, even though these domains are not independent.

Preliminary Analysis

Looking at multiplicative structures as a set of problems, I have identified three different subtypes: (*a*) isomorphism of measures, (*b*) product of measures, and (*c*) multiple proportion other than product.

Isomorphism of Measures

The isomorphism of measures is a structure that consists of a simple direct proportion between two measure-spaces M_1 and M_2. It describes a large number of situations in ordinary and technical life. These include: equal sharing (persons and objects), constant price (goods and cost), uniform speed or constant average speed (durations and distances), constant density on a line (trees and distances); on a surface, or in a volume. Four main subclasses of problems can be identified:

MULTIPLICATION

Schema 5.1 illustrates the isomorphism of measures for multiplication.

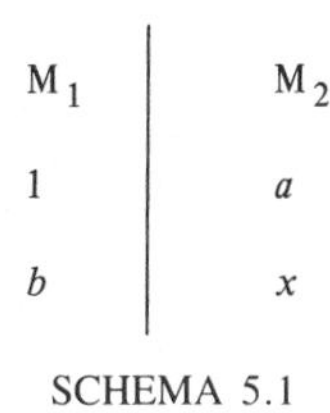

SCHEMA 5.1

Example 1. Richard buys 4 cakes priced at 15 cents each. How much does he have to pay?

$a = 15, \quad b = 4, \quad M_1 = [\text{numbers of cakes}], \quad M_2 = [\text{costs}].$

Example 2. A farm of 45.8 ha produces 6850 kg of corn per ha. What will be the yield?

$a = 6850, \quad b = 45.8, \quad M_1 = [\text{areas}], \quad M_2 = [\text{weights of corn}].$

Multiplication problems do not consist of a three-term relationship but of a four-term relationship from which children have to extract a three-term relationship. They can do it by extracting either a binary law of composition or a unary operation. Each method implies different operations of thinking, as shown below.

Binary Law of Composition

From Schema 5.1 children can extract $a \times b = x$. In Example 1, for instance, the child recognizes the situation to be multiplicative, and therefore multiplies 4 $\times$ 15 or 15 $\times$ 4 to find the answer. This binary composition is correct if a and b are viewed as numbers. But, if they are viewed as magnitudes, it is not clear why 4 cakes $\times$ 15 cents yields cents and not cakes.

Unary Operation

It is most likely that children, especially young ones, do not extract a binary law of composition but rather a unary operation. This can be done in two different ways. Children can (*a*) use a scalar operator ($a \overset{\times b}{\rightarrow} x$) that consists of transposing in M_2, from a to x, the operator that links 1 to b in M_1.

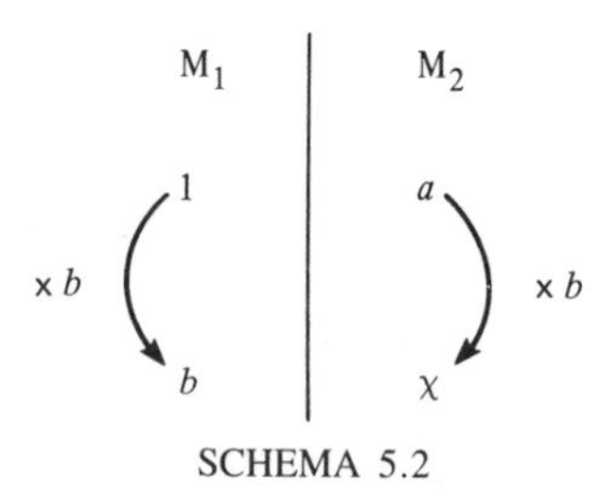

SCHEMA 5.2

In Schema 5.2, $\times b$ is a scalar operator because it has no dimension, being a ratio of two magnitudes of the same kind; b cakes is b times as much as 1 cake, and the cost of b cakes is also b times as much as the cost of 1 cake. Or, (*b*) children can use a function operator ($b \overset{\times a}{\rightarrow} x$) that consists of transposing on the lower line, from b to x, the operator that links 1 to a on the upper line.

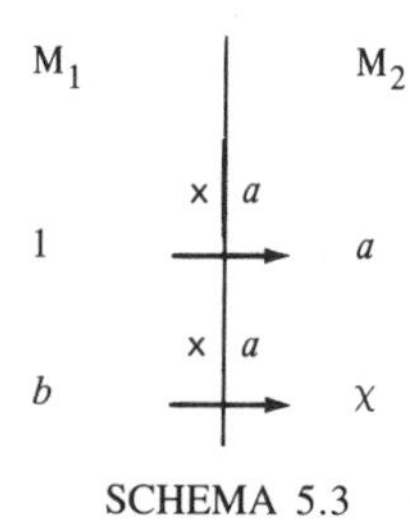

SCHEMA 5.3

In Schema 5.3, $\times a$ is a function operator because it represents the coefficient of the linear function from M_1 to M_2. Its dimension is the quotient of two other dimensions (e.g., cents per cake, kg per ha).[1]

FIRST-TYPE DIVISION

Schema 5.4 illustrates the first-type division, which is to find the unit value f(1).

[1]Another procedure for solving multiplication problems consists of adding $a + a + a$ . . . (b times), but it is not a multiplicative procedure. It only shows that the scalar procedure relies upon iteration of addition. One does not find, in young children, the symmetric procedure $b + b + b$ . . . (a times) because it is not meaningful.)

M_1	M_2
1	$\chi = f(1)$
a	$b = f(a)$

SCHEMA 5.4

Example 3. Connie wants to share her sweets with Jane and Susan. Her mother gave her 12 sweets. How many sweets will each receive?

$$a = 3, \quad b = 12, \quad M_1 = \text{[numbers of children]}, \quad M_2 = \text{[numbers of sweets]}$$

Example 4. Mrs Johnson bought some large peaches. Nine peaches weigh 2 kg. How much does one peach weigh, on the average?

$$a = 9, \quad b = 2, \quad M_1 = \text{[numbers of peaches]}, \quad M_2 = \text{[weights]}$$

This class of problems can be solved by applying a scalar operator $/b$ to the magnitude c (see Schema 5.5.).

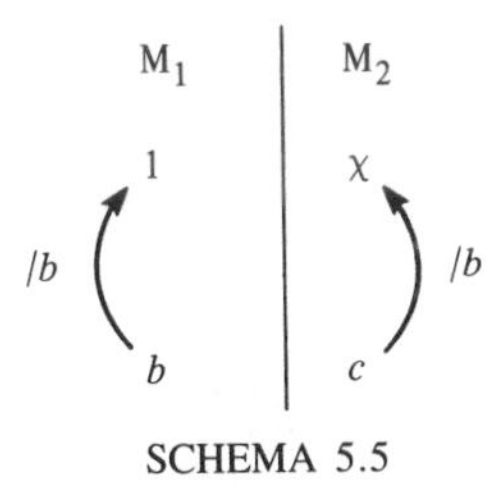

SCHEMA 5.5

Some children, because mental inversion of the relationship $\times$b into $/b$ is difficult, prefer to try to find x such that $x \times b = c$ (eventually by trial and error). This *missing factor procedure*, which is similar to *missing addend procedures* in subtraction problems, avoids the conceptual difficulty raised by inversion. But it is of value only for small whole numbers. Adults also use this missing factor procedure, when b and c are entries in the familiar multiplication table, for instance. But whereas they are able to shift to the canonical procedure c/b when necessary, young children usually fail.

Another procedure is available in the case of sharing objects: delivering them one by one to the participants or to different places in space. This can also be done mentally, even in other cases (by analogy), but it is inefficient and has no multiplicative character.

SECOND-TYPE DIVISION

Schema 5.6 illustrates second-type division, which is to find x knowing $f(x)$ and $f(1)$.

M_1	M_2
1	$a = f(1)$
x	$b = f(x)$

SCHEMA 5.6

Example 5. Peter has $15 to spend and he would like to buy miniature cars. They cost $3 each. How many cars can he buy?

$$a = 3, \quad b = 15, \quad M_1 = \text{[numbers of cars]}, \quad M_2 = \text{[costs]}.$$

Example 6. Dad drives 55 miles per hour on the freeway. How long will it take him to get to his mother's house, which is 410 miles away?

$$a = 55, \quad b = 410, \quad M_1 = \text{[durations]}, \quad M_2 = \text{[distances]}.$$

This class of problems can usually be solved by inverting the direct function operator and applying it to b, as shown in the Schema 5.7.

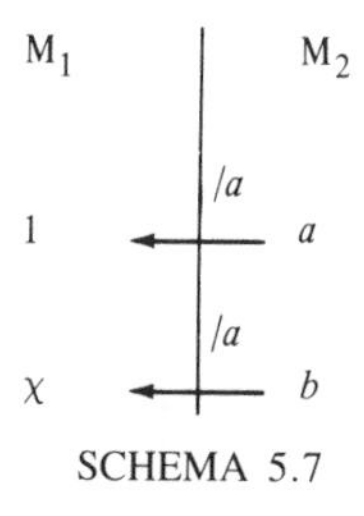

SCHEMA 5.7

This procedure is difficult for children, not only because of the inversion problem, but also because the inverse operator has a dimension (inverse of the direct one) that is unusual and harder to conceive (e.g., cars per dollar, hours per mile). Frequently, especially when the numbers are not small whole numbers, children prefer to find out how many times a goes into b, get the scalar operator, and transpose it in M_1. This avoids reasoning on inverse quotients of dimensions.

Children can also attempt additive procedures $a + a + a, \ldots$ until they get to b, then count the number of times they have added a.

RULE-OF-THREE PROBLEMS: GENERAL CASE

Schema 5.8 illustrates rule-of-three problems in the general case.

M_1	M_2
a	b
c	x

SCHEMA 5.8

Example 7. The consumption of my car is 7.5 liters of gas for 100 km. How much gas will I use for a vacation trip of 6580 km?

$$a = 100, \quad b = 7.5, \quad c = 6580, \quad M_1 = [\text{distances}], \quad M_2 = [\text{gas consumptions}].$$

Example 8. When she makes strawberry jam, my grandmother uses 3.5 kg of sugar for 5 kg of strawberries. How much sugar does she need for 8 kg of strawberries?

$$a = 5, \quad b = 3.5, \quad c = 8, \quad M_1 = [\text{strawberry weights}], \quad M_2 = [\text{sugar weights}].$$

This class of problems can be solved by different procedures, using different properties of the four-term relationship. They will be examined later in this chapter, under procedures for rule-of-three problems in the experiments.

It should already be clear that multiplication and division problems are simple cases of the more general rule-of-three class of problems in which four terms are involved, one of which is equal to one. In solving problems in this structure, students naturally use the isomorphic properties of the linear function:

$$f(x + x') = f(x) + f(x')$$
$$f(x - x') = f(x) - f(x')$$
$$f(\lambda x) = \lambda f(x)$$
$$f(\lambda x + \lambda' x') = \lambda f(x) + \lambda' f(x')$$

It is less natural for them to use the standard properties of the proportion coefficient:

$$f(x) = ax$$
$$x = 1/a\, f(x)$$

Because the isomorphism properties appear to be more natural than the proportional coefficient properties, the expression *isomorphism of measures* is used to name and describe the simple direct proportion structure. This term enables us to distinguish very clearly this structure from the next ones, the *product of measures* and the *multiple proportion*. Whereas the product of measures structure and the multiple proportion structure involve three (or more) variables and a bilinear (or n-linear) function's model, the isomorphism of measures involves only two variables and is properly modelled by the linear function.

Product of Measures

The product of measures is a structure that consists of the Cartesian composition of two measure-spaces, M_1 and M_2, into a third, M_3. It describes a fair number of problems concerning area, volume, Cartesian product, work, and many other physical concepts.

Because there are (at least) three variables involved, this structure cannot be represented by a simple correspondence table like the one used for the isomorphism of measure structure. Rather it is represented by a double-correspondence table. For example, in the case of the area of a rectangle:

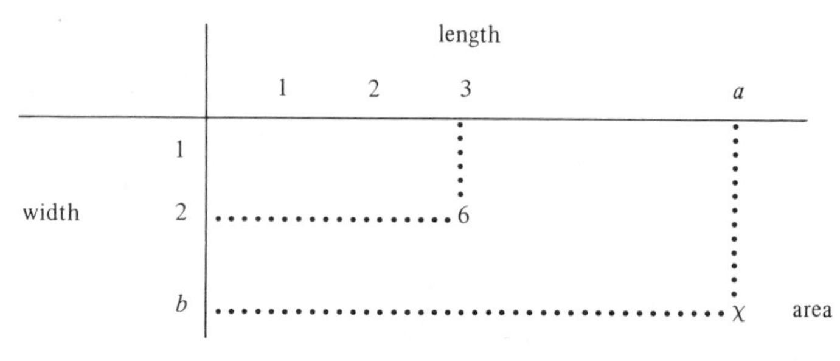

SCHEMA 5.9

Schema 5.9 reflects the double proportion of area to length and width independently. A similar relationship exists in the next structure (multiple proportion) but the choice and the expression of units do not obey the same rules. In the product of measures, there is a canonical way of choosing units. That is, $f(1,1) = 1$; or in the case of finding the area of a square,

$$(1 \text{ unit of length}) \times (1 \text{ unit of length}) = (1 \text{ unit of area}).$$

The units of the product are expressed as products of elementary units; for example, square centimeters, cubic centimeters or, in Example 9:

$$f(1 \text{ boy} \times 1 \text{ girl}) = 1 \text{ couple}.$$

In the multiple proportion (to be described later), units do not generally have these properties.

Example 9. Four girls and 3 boys are at a dance. Each boy wants to dance with each girl, and each girl with each boy. How many different boy–girl couples are possible?

The different possible couples can be easily generated and classified by a double-entry table, and the proportion of the number of couples to the number of boys and the numbers of girls independently can also be made visible by a double correspondence table: the number of couples is proportional to the number of boys when the number of girls is held constant (parallel columns), and to the number of girls when the number of boys is held constant (parallel lines) (see Schema 5.10).

		Girls			
		L	M	N	O
	A	AL	AM	AN	AO
Boys	B	BL	BM	BN	BO
	C	CL	CN	CN	CO

Couples (Ex.9)

	Number of girls					
	1	2	3	4	5	. . . n
1	1		3			
Number of boys 2	2	4	6	8	10	$2n$
3			9			
m			$3m$			mn

Number of couples

SCHEMA 5.10

The Cartesian product is so nice that is has very often been used (in France anyway) to introduce multiplication in the second and third grades of elementary school. But many children fail to understand multiplication when it is introduced this way. The arithmetical structure of the Cartesian product, as a product of measures, is indeed very difficult and cannot really be mastered until it is analyzed as a double proportion. Simple proportion should come first.

Two classes of problems can be identified, multiplication and division, the first of which is illustrated in Schema 5.11. Given the value of the elementary measures, find the value of the product-measure.

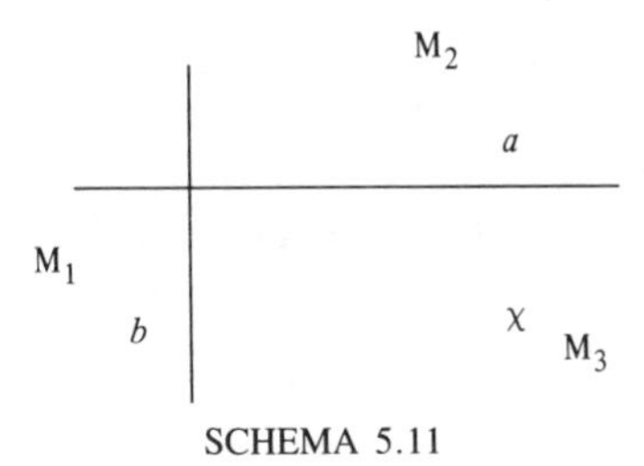

SCHEMA 5.11

Example 10. What is the area of a rectangular room that is 7 m long and 4.4 m wide?

$a = 7$, $b = 4.4$, M_1 = [widths], M_2 = [lengths], M_3 = [areas].

Example 11. What is the volume of a pipe that is 120 cm long and has a cross-sectional area of 15 cm^2?

$a = 120$, $b = 15$, M_1 = [section areas], M_2 = [heights], M_3 = [volumes].

The solution $a \times b = x$ is not so easy to analyze in terms of scalar and function operators. It is a product of two measures both in the dimensional and the numerical aspects:

$$\text{area (m}^2) = \text{length (m)} \times \text{width (m)}$$
$$\text{volume (cm}^3) = \text{height (cm)} \times \text{section area (cm}^2)$$

In the first structure (i.e., isomorphism of measures), we could not explain why multiplying cents by cakes would output cents and not cakes. This could only be explained either by the scalar operator (cents $\rightarrow$ cents) or by the function operator (cakes $\xrightarrow{\text{cents/cake}}$ cents).

In this second structure (i.e., product of measures), the landscape is different, and multiplying meters by meters outputs square meters; multiplying girl-dancers by boy-dancers outputs mixed couples of dancers.

The second class of problems, division, is illustrated by Schema 5.12. Given the value of the product measure and the value of one elementary measure, find the value of the other one.

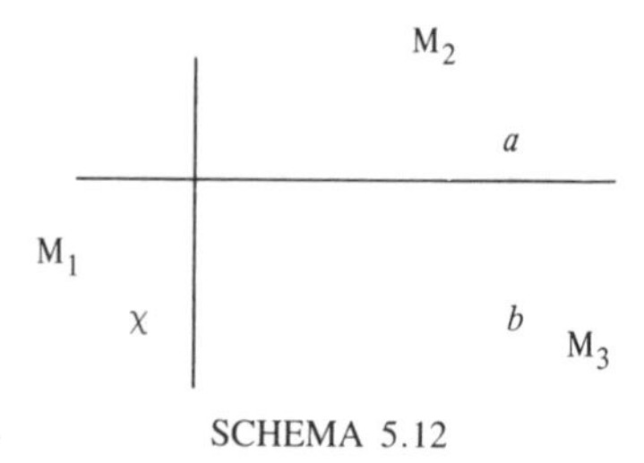

SCHEMA 5.12

Example 12. The area of a pool is 150 m^2. Filling it up requires 320 m^3 water. What is the average height of water?

Here again, the division procedure cannot easily be described by a scalar or function operator. The dimension of the quantity to be found is the quotient of the dimension of the product by the dimension of the other "elementary measure."

$$\text{volume (m}^3) \;/\; \text{area (m}^2) = \text{height (m)}$$

One way to explain the structure of the product is to see it as a double isomorphism or double proportion. Let us take the example of the volume of straight prisms.

If height is multiplied by 2, 3 or λ, volume is multiplied by 2, 3 or λ (provided basic area ia held constant) (see Schema 5.13.)

Similarly, if basic area is multiplied by 2, 3 or λ', volume is multiplied by 2, 3 or λ' (provided height is held constant). If one adds basic areas of different prisms, volumes are also added (provided height is the same).

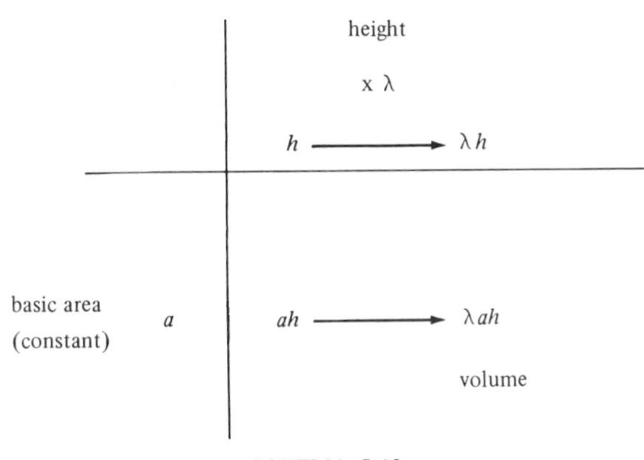

SCHEMA 5.13

These properties issue directly from the isomorphism properties of the linear function (scalar aspect, additive aspect). When height is held constant, ($\times h$) can be viewed as a function operator linking basic area to volume (see Schema 5.14.)

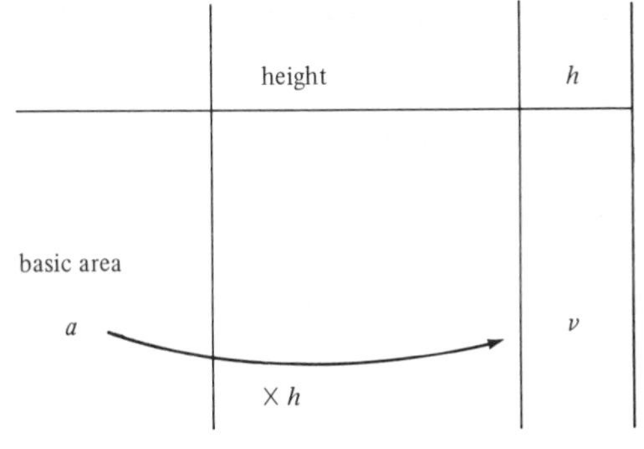

SCHEMA 5.14

The same can be said for basic area when it is held constant. Although this analysis is a bit sophisticated, it shows that although product is different from isomorphism it can also be considered as a double isomorphism.

It follows that if height is multiplied by λ and basic area by λ', volume is multiplied by $\lambda\lambda'$.

Reciprocally, isomorphism can also be viewed as a product. For instance:

time × speed = distance
volume × volumic mass = mass.

These relationships are well illustrated by the function operator (see Schema 5.15.)

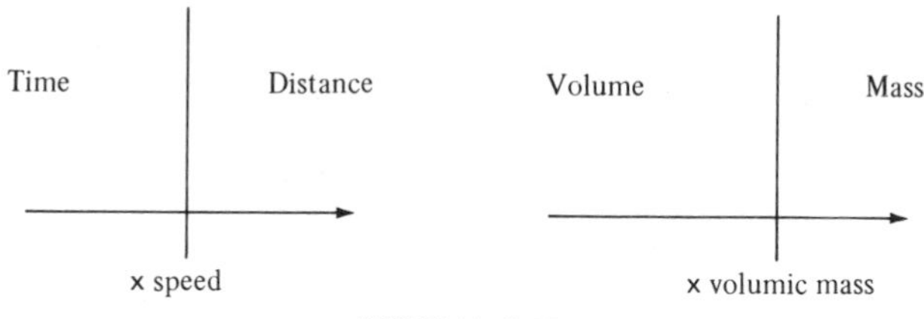

SCHEMA 5.15

Still, one should note that, in this case, speed and volumic mass are considered as constants and not as variables, whereas in the product (volume, for instance) both elementary measures (basic area and height in the case of volume) are variables. One must also keep in mind that, in the isomorphism structure, the quotient of dimensions is a derived magnitude and not an elementary one. If time × speed = distance, it is because speed = distance/time. If volume × volumic mass = mass, it is because volumic mass = mass/volume. However, in the product structure, at least in the product met by children at the primary and secondary levels (e.g., area, volume, Cartesian product), the elementary measures are really elementary and not quotients.

Multiple Proportion

The multiple proportion is a structure very similar to the product from the point of view of the arithmetic relationships: a measure-space M_3 is proportional to two different independent measure-spaces M_1 and M_2. For example:

1. The production of milk of a farm is (under certain conditions) proportional to the number of cows and to the number of days of the period considered.
2. The consumption of cereal in a scout camp is proportional to the number of persons and to the number of days.

Time is very often involved in such structures because it intervenes in many phenomena as a direct factor of proportionality (e.g., consumption, production, expense, outcome. But there are other factors; in physics, for instance:

$$P = \mathrm{k}RI^2 \qquad \text{(Power, Resistance, Intensity)}$$

Whereas in physics multiple proportion phenomena can often be interpreted as products, this is not always possible in multiple proportion problems. Most of the time, no natural choice of units can provide $f(1,1) = 1$. For instance, there is no reason why a cow should produce 1 liter of milk per day, or a person eat 1 kg of cereal per day or per week. Usually there exists a coefficient k not equal to 1; $f(1,1) = \mathrm{k}$.

In multiple proportion, the magnitudes involved have their own intrinsic meaning, and none of them can be reduced to a product of the others. There is no reason to interpret the double proportionality of the consumption of cereal to the number of persons and to the number of weeks as a dimensioned operation

(i.e., cereal = persons × weeks).

Here again several classes of problems can be identified. I will only give examples, the analysis being similar to what has been explained earlier.

MULTIPLICATION

Example 13. A family of 4 persons wants to spend 13 days at a resort. The cost per person is $35 per day. What will be the expense?

FIRST-TYPE DIVISION

First-type division involves finding the unit value $f(1,1)$.

Example 14. A farmer tries to calculate the average production of milk of his cows during the 180 best days of the year. With 17 cows, he has produced 70,340 liters of milk during that period. What is the average production of milk per cow per day? (See Schema 5.16.)

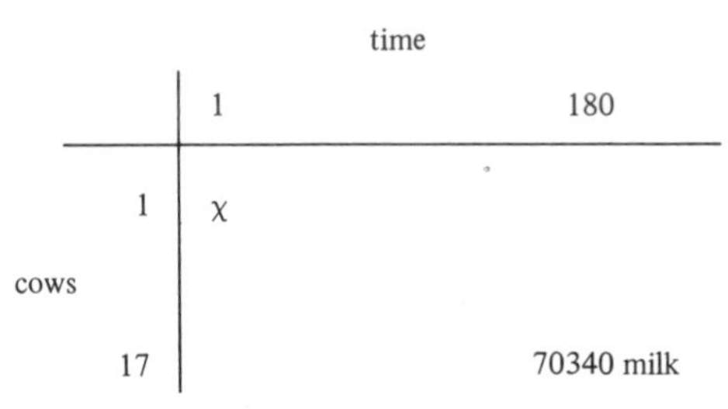

SCHEMA 5.16

This division does not usually exist in the product of measures, because $f(1,1) = 1$, at least in the metric system.

SECOND-TYPE DIVISION

Second-type division involves finding x knowing $f(x,a) = b$ and $f(1,1)$.

Example 15. A scout camp has just received 500 kg of cereal. The allowed distribution of cereal is 0.6 kg per person per week. There are 236 persons in the camp. How long will the cereal last? (See Schema 5.17.)

	time 1	x
persons 1	0.6	
$a = 236$		$b = 500$ cereal

SCHEMA 5.17

The bilinear function is an adequate model for both the product of measures and the multiple proportion. One hypothesis is that it implies more complex operations of thinking than does the linear function. Another hypothesis is that the product of measures raises its own difficulties that are not reducible to those of the multiple proportion.

In the next part of this chapter, I report experiments on problems that can be analyzed by these structures. Sometimes problems are combinations of different structures. The above analysis is a first approach to these problems. A complementary analysis will be made later. The value of the magnitudes involved, the concept of average, and the reference context are also important characteristics of problems.

Experiments

This section describes several experiments, conducted during the past 4 or 5 years, showing either results on the comparative complexity of problems and procedures, or the evolution in the classroom of conceptualizations and procedures in a dialectical relationship with situations.

Isomorphism, Product, and Multiple Proportion

The very first experiment performed by our research group (Vergnaud, Ricco, Rouchier, Marthe, & Metregiste, 1978) aimed at comparing the difficulty of different problems and evaluating the stability of procedures used by students. It consisted of different versions of the same three problem-structures:

1. Volume: calculation of the volume of a right parallelipiped: $x = a \times b \times c$.
2. Direct proportion: among three measure-spaces; calculation of x knowing $f \circ g(x) = c$ $\quad x = c/(a \times b)$ (See Schema 5.18.)

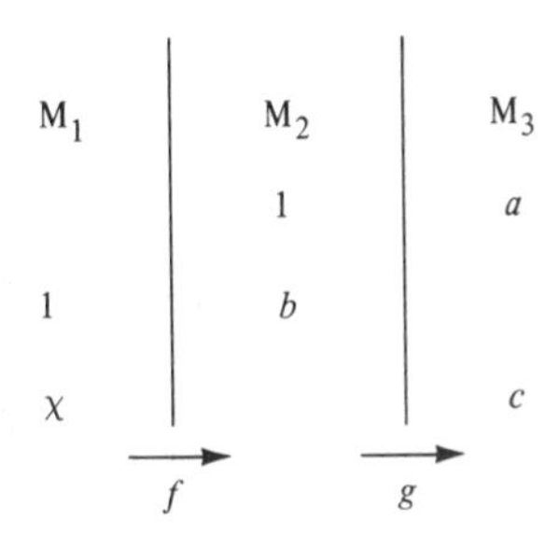

SCHEMA 5.18

3. Double proportion: calculation of a magnitude proportional to time and to another magnitude. (See Schema 5.19.)

		time	
		1	c
	1	a	
other magnitude	b		x

SCHEMA 5.19

The first version of these structures was a complex problem (with several questions), chosen from a handbook for 11–12-year-olds, involving all three structures.

> ‘‘Central heating is being installed in a house; the dimensions are length = 18 m, width = 6 m, height = 4 m.
>
> 1. A radiator is made of 8 elements. Each element can heat 6 m^3. How many radiators are needed?
>
> 2. The average consumption is 4 kg coal per day for each radiator. The heating period runs from October 1 to April 15. How much coal will be used?’’

The other versions were single questions, built ad hoc for the sake of comparison:

Volume: ‘‘What is the volume of water used to fill up a rectangular pool: Length = 17 m, width = 8 m, depth = 3 m.’’

Direct proportion: ‘‘A luxury train should contain 432 first-class seats. Each car has 8 compartments and each compartment has 6 seats. How many cars are needed?’’ Another version was also used.

Double proportion: "A farmer owns 5 cows. They each yield an average of 23 liters of milk per day during the 180 best days of the year. How much milk does the farmer produce during this period?"

Among 84 students (11–12 years old) we could find the following hierarchy:

The easiest problems were the direct proportion and the double proportion ones, although the direct proportion problem needed either two divisions, or a multiplication and a division. Both problem types were solved by $\frac{2}{3}$ of the students in the simple version and $\frac{1}{3}$ in the complex version.

The most difficult problem had to do with volume, both in the simple and the complex versions (success rate of $\frac{38}{84}$ and $\frac{28}{94}$), although the calculations were the simplest.

We analyzed the different procedures used and categorized them, so as to compare the stability of these categories on two (or three) versions of the same problem-structure. Many students had a *perimetric* representation of volume, adding lengths together and trying to take into account as many sides as possible (either by multiplying 2, 3 or 4 times the sum $L + W + H$, or by adding 1, 2 or 4 times the height to the perimeter of the base, or by any other combination). Some students also had a *surface* representation (adding areas) or a *mixed* representation (multiplying perimeter by height, for instance). We found only 50% stability on classes of procedures, which is far beyond random coincidence; but still rather weak.

For the direct proportion we also found different classes of procedures. Most errors involved partially correct procedures that were not carried out to the solution. The stability was quite good for one of the correct procedures, which had the easiest physical meaning, and comparatively weak for the others.

As for the double proportion, we found better performance in relation to the time factor. Both consumption of coal and production of milk are conceived as proportional to time, even though the other factor involved (number of radiators, number of cows) did not appear to be difficult.

Another finding of this experiment was that most procedures used by students, even when they were wrong, had a physical meaning. We very rarely found meaningless calculations.

A Variety of Procedures for Rule-of-Three Problems

The experiment (Vergnaud, Rouchier, Ricco, Marthe, Metregiste, & Giacobbe, 1979; Vergnaud, 1980) was designed to test the hypothesis of a better availability of scalar procedures compared with function procedures. It also permitted us to make an extensive description of procedures (correct or incorrect) used by students. The problem was the following:

"In a hours the central heating consumption is b liters of oil. What is the consumption in c hours?" (See Schema 5.20.)

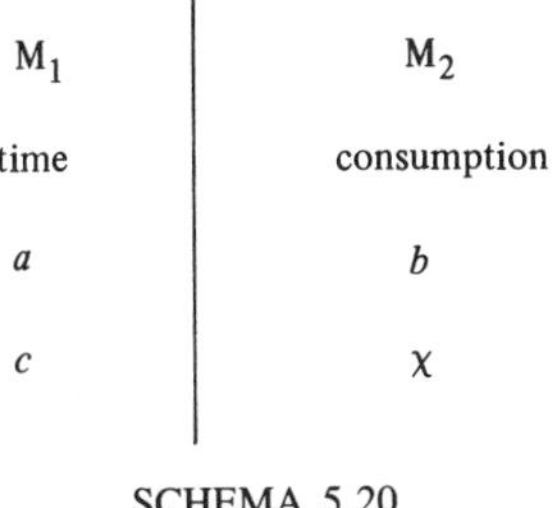

SCHEMA 5.20

By choosing adequate values for a, b and c, it is possible to simplify either the scalar ratio $(\frac{c}{a})$ or the function ratio $(\frac{b}{a})$, or both, or none.

In order to test the hypothesis that scalar procedures would be more easily and frequently used than function procedures, we used problems with a simple scalar ratio (3 or 4) and a complex function ratio (12, 13) and problems with a simple function ratio and a complex scalar ratio. This was done for multiplication and division, resulting in four cases in all. (See Schema 5.21.)

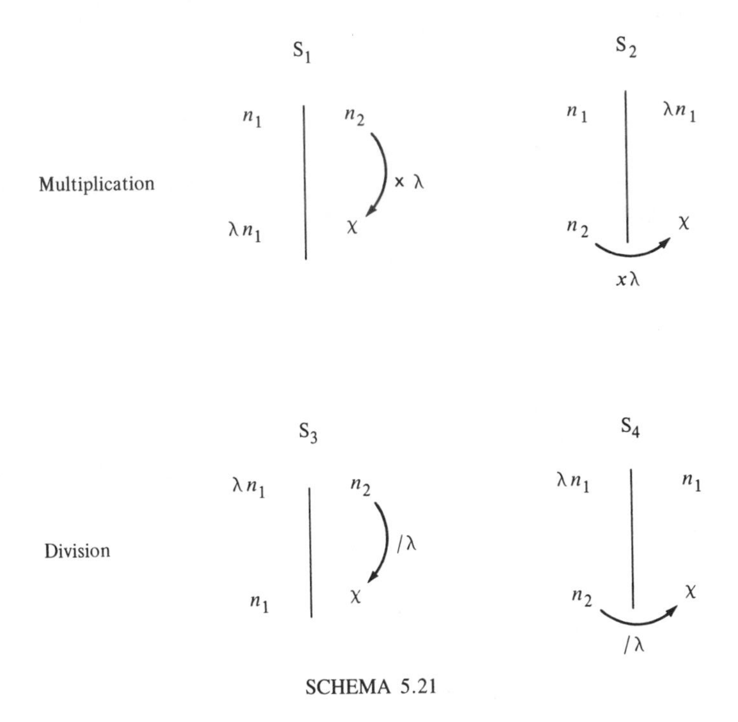

SCHEMA 5.21

We used a total of 16 problems. Examples are as follows. (See Schema 5.22.)

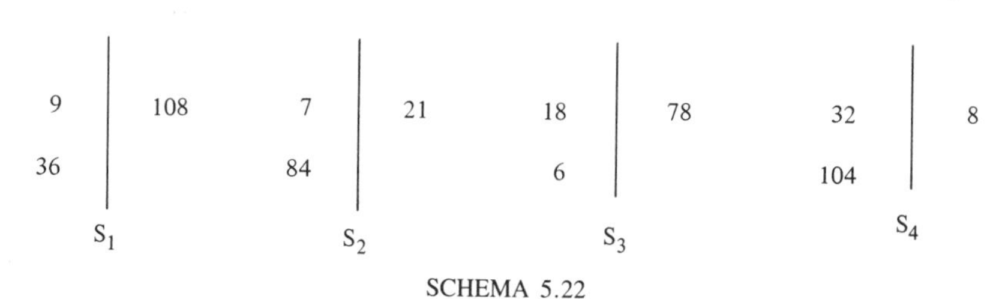

SCHEMA 5.22

We counterbalanced the order of the four problems that each student had to solve. Four groups of 25 students each (one group for each secondary comprehensive school grade, from 11–12-year-olds to 14–15-year-olds) participated in the experiment.

According to our hypothesis, we expected S_1 to be easier than S_2, and S_3 to be easier than S_4. We also expected to find differences in the use of the different possible procedures. We also meant to describe the evolution of success rates and procedures along the four grades of early secondary school.

Table 5.1 shows very clearly that there is no difference between the S_1 and S_2 problems. This result contradicts our hypothesis that an easy scalar ratio would enable children to solve S_1 problems more easily than S_2 problems (easy function ratio).

The situation is different for S_3 and S_4. Whereas S_3 problems are mastered nearly as well as S_1 and S_2, there is a big drop in the success rate for S_4. Unfortunately, this drop may be due to two different factors: the difficult scalar division on one hand, and the fact that S_4 is the only situation where $f(x) < x$ (as can be seen in the numerical examples above).

One interesting thing is the regular, slow evolution from younger students to older ones. This shows that a psychogenetic approach is useful in studying the acquisition of mathematical skills at the secondary school level. Even when these

TABLE 5.1

Success Rates for Rule-of-Three Problems (%)

	Problems			
	S_1	S_2	S_3	S_4
Grades	n_1 n_2 λn_1	n_1 λn_1 n_2	λn_1 n_2 n_1	λn_1 n_1 n_2
6th graders (11–12 years old)	39	39	29	16
7th graders (12–13 years old)	64	55	59	36
8th graders (13–14 years old)	65	69	69	35
9th graders (14–15 years old)	82	85	74	56
	63	63	58	36

skills are taught, the development is slow, and it takes students a few years to deal successfully with the different numerical cases. Some students still fail to handle the simplest situations even at the end of early secondary school, but most of them progress regularly. The most difficult case used here, which is not the most difficult that one can meet, even with whole numbers, is mastered by a majority of students only at the last early secondary school level (14–15-year-olds).

Procedures

More interesting is the variety of procedures we observed (over 25 kinds used by students). We classified correct procedures into five subcategories; we tried to classify incorrect ones into meaningful subcategories, but this was not always possible.

In the following description, we use letters *a, b, c* and *x: a* and *c* are time-measures, *b* and *x* are consumption measures as shown in Schema 5.23.

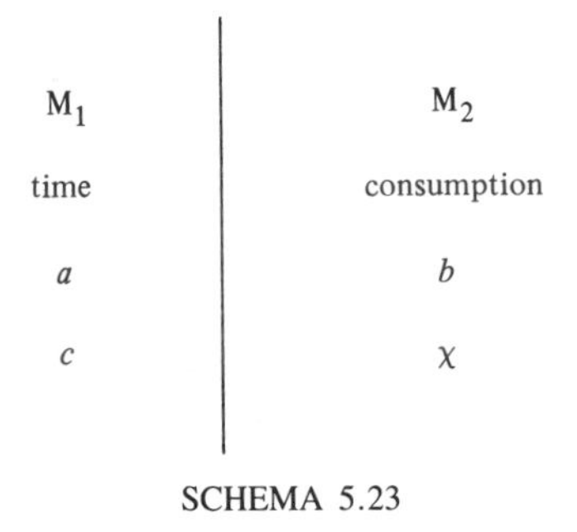

SCHEMA 5.23

Correct Procedures

S *Scalar:* The student calculates ($c/a = \lambda$) (in S_1 and S_2) or ($a/c = \lambda$) (in S_3). This calculation can be made explicitly either by dividing or by using the missing factor procedure (see Schema 5.23). It can also be performed mentally. Afterwards, the student calculates $x = \lambda \times b$ or $x = b \times \lambda$ (in S_1 and S_2) or b/λ (in S_3).

F *Function:* The student calculates ($b/a = \lambda$) (in S_1 and S_2) or ($a/b = \lambda$) (in S_4), either explicitly or mentally and then $x = \lambda \times c$ (S_1 and S_2) or c/λ (S_4).

U *Unit value:* The student performs the same calculations as in F, but he explains that (b/a) is the unit value $f(1)$. This procedure is scalar in character, although the calculations are the same as in F. (See Schema 5.24.)

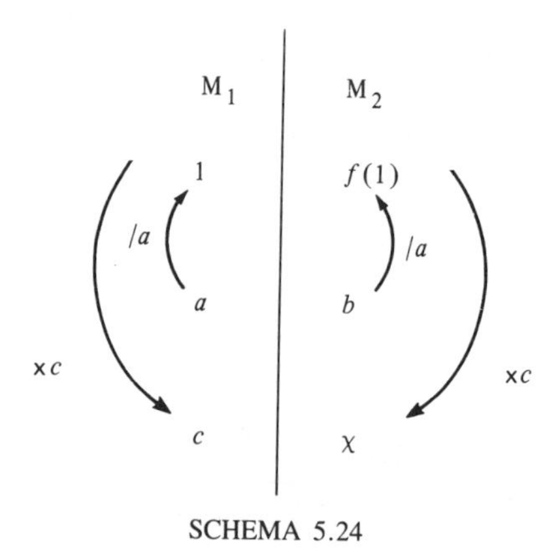

SCHEMA 5.24

R *Rule of three:* The student calculates $(b \times c)/a$ or $(c \times b)/a$ (multiplication first). This well-known algorithm is rarely used. We explain why later.

SD *Scalar decomposition:* The student tries to decompose magnitude c as a linear combination of different other magnitudes: multiples of a or fractions of a.

Example: $a = 32$, $b = 8$, $c = 104$ (boy; 14 years old).

Protocol	Comment
$32 \times 3 = 96 + 8 = 104$	$104 = (3 \times 32) + (\frac{1}{4} \times 32)$
	$\Downarrow$
$86 = 24$ liters $+ 8/4$	$x = (3 \times 8) + (\frac{1}{4} \times 8)$
$= 26$ liters	$= 24 + 2$

Although the equations in the protocol are all wrong, the procedure is efficient and shows the use of a powerful theorem (see comment):

$$(f(\lambda a + \lambda' a) = \lambda f(a) + \lambda' f(a)$$

This procedure is very often used by students (even at the primary school level) when they cannot think of the function-operator. The decomposition can also be multiplicative:

$$f(\lambda\lambda' a) = \lambda\lambda' f(a)$$

The properties of numbers are of course very important in the emergence of such procedures but it is also important to notice that these procedures cannot be explained by pure numerical properties. Numbers are magnitudes. As a matter of fact, if the numbers did not represent magnitudes of qualitatively different types of quantities, then one should also find function decomposition procedures ($b = \lambda a + \lambda' a$). This is never the case; b cannot be conceived as a linear combination of magnitudes of a different kind.

Incorrect Procedures

Many incorrect procedures are based on some aspects of the indicated real situation. We thought it might be interesting to classify these incorrect procedures in order to see what sort of features were the most salient.

S *Erroneous scalar:* The student uses a scalar ratio or difference, either (c/a) or $c - a$ or (a/c) or $a - c$, and either gives it as an answer, or multiplies it by b, or adds it to b, or divides b by it, or subtracts it from b.

Example: $a = 7$, $b = 21$, $c = 84$ (girl; 16 years old).

$$84h - 7h = 77\ h$$

The consumption is 77×21 liters $= 1617$ liters.
We also considered that multiplying b by an arbitrary number, approximately equal to c/a, could be classified in this category.

F *Erroneous function:* The student uses a function ratio or difference, either (b/a), or $b - a$, or (a/b), or $a - b$, and either gives it as an answer or applies it to c.

S′F′ *Erroneous scalar and function:* The student makes a calculation $b \times c$, forgetting or cancelling division by a or makes a combination of erroneous scalar and function operations.

Example: $a = 8$, $b = 32$, $c = 104$ (girl; 13 years old). (See Schema 5.25.)

$$8\,\overline{)\,32}\ \to 4 \qquad 8\,\overline{)\,104}\ \to 13 \ \ (8,\ 24,\ 24) \qquad 13 \times 4 = 52$$

SCHEMA 5.25

In 104h., the consumption is 52 litres.

I *Inverse:* The student uses the inverse ratio a/c instead of c/a, or c/a instead of a/c, or b/a instead of a/b, etc.

Example: $a = 21$, $b = 7$, $c = 90$ (girl; 14 years old).

$$21:7 = 3$$

$90 \times 3 = 270$ liters.

P *Erroneous product:* The student multiplies c and a, or b and a, which has no physical meaning at all.

Q *Erroneous quotient:* The student divides c by b, or b by c, which again has no meaning.

O *Others:* Procedures that could not be classified elsewhere.

Looking at these incorrect procedures, one can see some differences. Erroneous scalar and function, and inverse procedures are less "silly" than erroneous products and quotients. One can make better sense of them. Comparing two magnitudes of the same kind by looking at the difference $c - a$ instead of the ratio c/a also makes better sense than looking at $b - a$ instead of b/a.

Table 5.2 shows the distribution of procedures on each problem-structure for all grades together. The most striking fact is that scalar procedures are more frequently used than are function procedures, even for problems in which the function-operator is very simple (S_2 and S_4). This is true for procedure S alone, but it is still more striking if one considers S, V, SD and S′ together, compared with F, FD and F′.

This fact has been found or observed by other authors (Freudenthal, 1978; Lybeck, 1978) and discussed by others (see Karplus, chapter 3, this volume. Noelting, 1980 a,b) under the distinction between *Within* ratios and *Between* ratios. The discussion may be confused if one mixes up problems of comparison and problems of calculation, in which the answer is a certain magnitude. I return to this point when considering problems of fractions and ratios. For the time being, I just stress the convergence of Lybeck's (1978) results with ours, and the fact that the analysis in terms of *isomorphism of measures* is the most powerful and the most general one.

TABLE 5.2

Distribution of Procedures for Rule-of-Three Problems (%)

Procedures	Problems			
	S_1	S_2	S_3	S_4
Correct				
S Scalar	41	32	38	16
F Function	11	14	6	8
U Unit value	9	14	10	5
R Rule of three	1	1	2	1
SD Scalar decomposition	1	0	0	4
FD Function decomposition	0	0	0	0
Incorrect				
S′ Erroneous scalar	8	6	20	5
F′ Erroneous function	0	1	4	10
S′F′ Erroneous scalar and function	10	9	0	5
I Inverse	2	2	1	11
P Erroneous product	3	4	1	3
Q Erroneous quotient	3	4	2	1
O Others	11	13	16	31
	100	100	100	100

Another striking fact is that only 1% of the students use the rule-of-three algorithm. This also has been found by others (Hart, 1981), but it must be explained. My view is that it is not natural for students to multiply b by c and then divide by a if one considers b, c and a as magnitudes. It is fair to do this in the set of numbers because of the equivalence of different calculations $(b \times c)/a$, $c/a \times b$, $b/a \times c$. But children do not think of b, c and a as pure numbers. They see them as magnitudes and there is no meaning for them in multiplying b liters of oil by c hours, whereas they can more easily figure out scalar ratios c/a or even function ratios b/a.

We also found that the older students, although they had studied the linear function $f(x) = ax$ and the proportion coefficient, used the scalar procedure (in its different versions S, V, and SD) more often than did the younger students. For details on the first ideas of children on linear functions, see Ricco, 1978.

Volume: A Difficult Concept

This experiment, which was completed recently and will be published with more details at a later date, consisted of 80 individual interviews with secondary school students: 10 boys and 10 girls in each of the four grades, from 11–12 to 14–15-year-olds. Its aim was to obtain a varied and meaningful picture of students' skills and representations so as to be in a better position to make a series of didactic situations on volume for seventh graders (12–13-year-olds). Most of the questions concern trilinear properties of volume but we also tested the availability of the formula for parallelepipeds, and tried to obtain definitions of volume.

Because we intended to explore aspects varying in complexity, we used a branching program of items, posing more difficult questions to successful students and easier questions to those who had failed.

THE INTERVIEW

The interview started with the estimation of the volume of an aquarium, placed on the table. The student could use a meter stick to measure the dimensions of the aquarium, which were not given. We recorded what dimensions he (or she) did actually measure, and the sort of calculations he (or she) performed. The student was then asked to estimate the volume of the classroom (rectangular); no dimensions were given. We were not so much interested in the estimation of length, width, and height as in the calculations. The students were asked to explain how they arrived at their answers and were also asked, "What is volume for you?" This could also be repeated as, "If you had to explain to a younger fellow, what would you tell him?"

Students who failed to find a correct multiplicative calculation for the aquarium and the classroom were then asked a simpler question with blocks. A large quantity of blocks was displayed on the table, and then was hidden away in the experimenter's bag. The student was asked: "How many blocks should I give you, for you to make a straight box, as full as a box of sugar pieces, 3 blocks wide, 4 blocks long and 2 blocks high?" This item was obviously aimed at helping students find the correct calculation by using the blocks as a model (paving space with unit blocks).

Students who did not succeed on these items did not continue. The others were then asked the following problem: "Mr. Dupont has a small aquarium in his kitchen, and a large one in his den. The den one is twice as long, three times as wide and twice as deep as the kitchen one. How many times is the den one larger than the kitchen one?" We recorded the answers and the explanations.

There were two more items on the trilinear aspects of volume:

1. Two spheres ($D = 4$ cm and $d = 2$ cm) were shown, and the first question was: "How many times is the volume of the big one larger than the volume of the little one?" The second question concerned the weight. The student was then given several little spheres and a plastic toy scale. The big sphere was placed on one plate: "How many little ones should you put on this side to get the equilibrium?" (All spheres were solid and made of the same wood.)

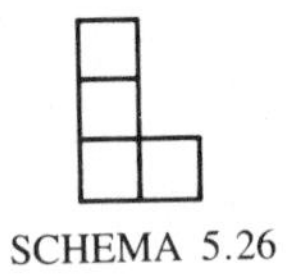

SCHEMA 5.26

2. An L-shape, made of 4 blocks (see Schema 5.26), and an enlarged version of the same shape (twice as long, twice as wide, twice as thick) were shown. The little L was permanently on the table; the big one was shown and immediately hidden. The question was: "How many blocks are there in the big L?" The difficulty of this item was expected to be intermediary between the two-aquarium item and the sphere item.

Finally a question was asked on division: "The volume of a box is 60 blocks. It is 3 blocks wide and 4 blocks high. How long is it?"

RESULTS

The response patterns are compatible with the hierarchy of items summarized in Schema 5.27.

Notice that 19 students failed completely and that only a small minority were able to handle trilinear aspects of volume. Most students just used the formula for the calculation of the volume of parallelepipeds, or were reminded of it through the block model. Only 4 students were successful in the sphere item, and the L-

Hierarchy of Patterns (Volume)

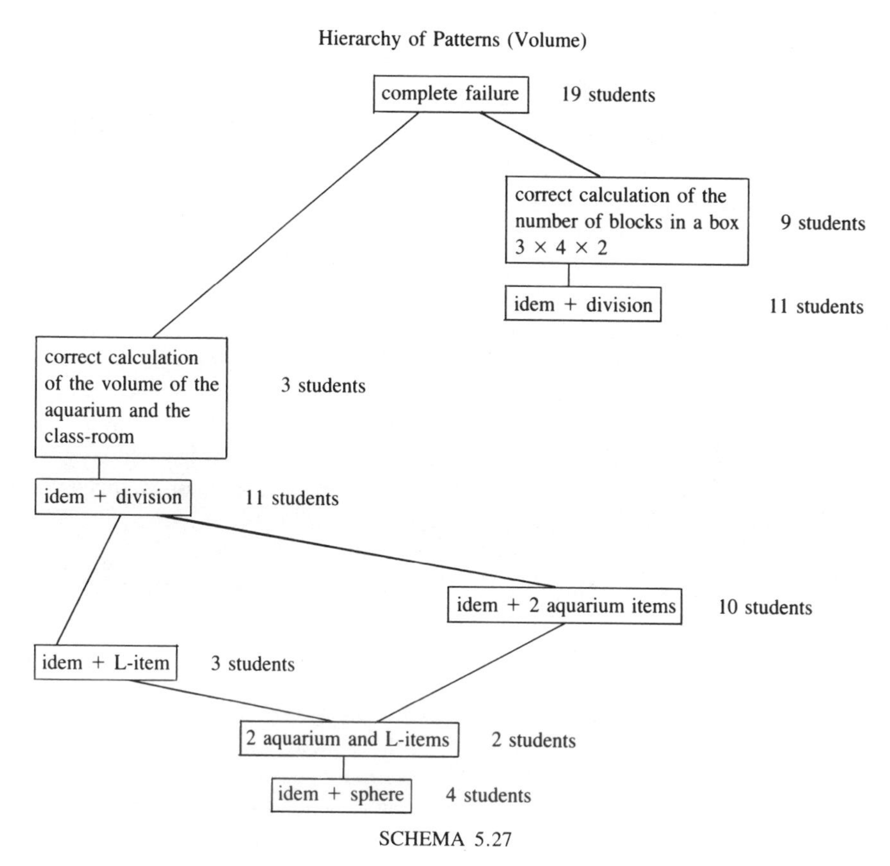

SCHEMA 5.27

item was also very difficult. There was no difference between the two first items; it was not easier to calculate the volume of the room in cubic meters with small numbers ($6 \times 4 \times 3$) than the volume of the aquarium in cubic centimeters with larger numbers ($40 \times 17 \times 20$).

Most 11–12-year-olds failed completely, but there were also total failures among older students. The jump is important from 11–12 to 12–13-year-olds (volume is taught again to 12–13), but the two-aquarium item was very difficult until ninth grade. Although volume is not taught any longer after the seventh grade (12–13) and is supposed to be well known by students (which is not the case), some skills go on improving, slowly.

We classified answers and procedures used by students and found again, in the direct calculation of volume, the perimetric representation, the area representation, and the mixed representation that we had already observed in the first experiment. This was even the case with the block item; many young students made drawings in which perimetric and area models appeared clearly (Schema 5.28).

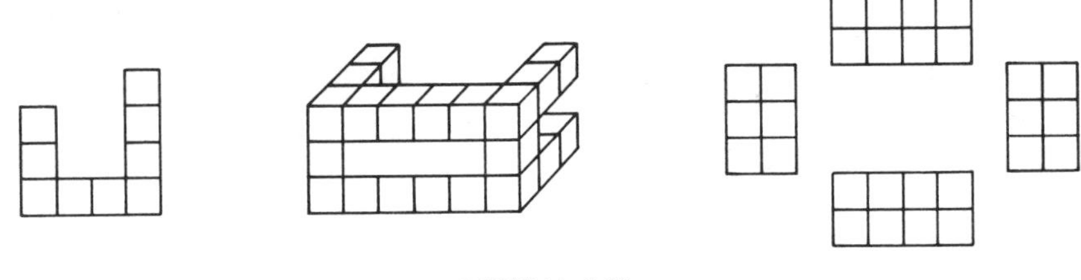

SCHEMA 5.28

The perimetric representation disappeared completely in the last grade, but this was not the case with the area representation: confusions between area and volume seem to be long lasting.

The stability of the procedures used for the calculation of the volume of the aquarium and of the classroom was quite good (75%). An expected finding was that some students could not express their results in cubic meters or centimeters, but would express them as meters or centimeters, or even centiliters.

The definitions of volume were varied. It is difficult to translate into English the exact wording of students. Table 5.3 is an attempt at such a translation. It shows a very interesting variety of definitions and also a good evolution from perimetric to volumic representations.

The two-aquarium item produced interesting results: out of 39 students who responded to the item, one finds 17 correct responses: 10 students gave the answer, $2 \times 3 \times 2 = 12$; 4 students paved the big aquarium mentally with the little one; and 3 students attributed hypothetic dimensions to the little one, calculated the dimensions of the big one, the volume of both, and then the ratio. This last "detour" shows how conceptually difficult it is to compose ratios in this structure.

Of the 22 incorrect responses, 9 students gave the additive answer, $2 + 3 + 2 = 7$; 3 students gave the "average" answer, "between 2 and 3;" and 1 student gave the modal answer, 2. Other students either tried a good procedure but got mixed up, or repeated the information given by the experimenter, or decided that it was impossible to say, because "one did not know the measure of the little aquarium." This very last answer confirms the conceptual difficulty of composing ratios in the absence of any information on the associated magnitudes.

As we expected, the sphere item was very difficult. One interesting fact was that students who gave the correct answer, after having hesitated for a while, used the gestural metaphor of putting together little spheres to fit the space occupied by the big sphere. This is of course a pure metaphor, borrowed from the cube model, but it does help. The most frequent incorrect answers were 2 and 4, as expected. The answers for the weight were not any different from those given for the volume; the students were deeply surprised when they discovered that 8 little spheres were necessary to make the equilibrium. None of the 19 subjects

TABLE 5.3

Definitions of Volume

	Age				
	11–12	12–13	13–14	14–15	Total
The room occupied by			1	1	2
The space occupied by			2	2	4
Length by width by height			1		1
The three dimensions of which one can move			1	1	2
The inside of		1	3		4
The interior of	1	2	2	2	7
The quantity contained in (a bottle, a room, a box . . .)	1	4	1	2	8
The capacity of	1	2	2	2	7
What is contained		2		1	3
A quantity (pencils, sheets, air, water . . .)	3	3		1	7
Filling up something (with water, cubes . . .)	1	2			3
The weight	1		1		2
The mass in the air		1	1	2	4
The area	1			1	2
All the room (gestures indicating surface)			2		2
The surface	2				2
The total of square meters	1				1
All the length of the room	1				1
The whole outline, contourline	2	1		1	4
What the room measures altogether (gestures)	2				2
All the sides	1		1		2
All the dimensions		1	1		2
Others	1		1		2
No answer	1	1	—	3	5
Total	20	20	20	20	80

that had given a wrong answer was able to explain correctly this astonishing fact: they used ad hoc explanations (for instance: multiplication of D by d) or refused to make any comment.

For the L-item, most students tried to imagine mentally the Big L and to count the blocks; most of them failed. Only four students attempted to compose directly the similarity ratio; one of them failed.

In conclusion, most students do not master the simplest elementary trilinear properties of volume, in the case of the parallelpiped, until the ninth grade (14–15-year-olds). Only a few of them can understand more difficult cases (L-item and sphere). Even the direct calculation of volume or the inverse calculation of one dimension knowing the volume and the other dimensions are still difficult for 13- to 15-year-olds, although they have been taught it. The multiplicative three-dimensional model contradicts more "natural" models such as the perimetric model and the area model. It is important to help students overcome these models and differentiate them from one another and from volume ideas. The difficulty of the concept of volume is considerably underestimated by teachers and programs, certainly in France. Volume is a good example of what was said in the introduction to this paper about how concepts develop over a long period of time.

Didactic Experiments

In the short space available in this chapter, I can only illustrate with a few examples the kind of experiments that we have developed inside the classroom. We usually have two aims: (*a*) to improve students' knowledge and know-how; (*b*) to make reliable observations and discover didactic facts, that is, facts concerning transmission and acquisition of knowledge.

The first important part of our work consists of arranging a series of didactic situations and making as explicit as possible both our didactic objectives and our hypotheses about what might happen. All members of the research team (mathematics teachers, psychologists, and mathematicians) participate in the choice of conceptions and ideas, and in the choice of situations (context, values of the different situation variables, order of problems, allowed suggestions, and so on). For instance, in the didactic sequence on volume, described below, we explicitly meant to start from a unidimensional conception of volume, as a quantity that can be compared and measured directly, to arrive at a tridimensional conception of the volume of parallelepipeds and prisms (see Rogalski, 1979; 1981, pp. 120–125, for interesting results). We also considered two important intermediary steps: the paving of the parallelepiped (as a natural transition from the unidimensional to the tridimensional conception) and the differentiation between volume, lateral area, and edge–periphery of a building (see the architect's problem). We also organized deliberately different difficulties and jumps in the questions posed to students.

But, before describing any situations or results, I need to clarify a few methodological points.

All situations are carefully described and written down for the teachers and the observers, with the different phases of each 50-minute lesson, the formulation of questions, and suggestions authorized to help students. This is the sort of care taken by psychologists when they interview subjects; although it is impossible to copy that model, we find it necessary to get as close to it as possible.

Students are usually divided into groups of four. Most of the lesson time is devoted to small group work. At some previously determined moments, explanations to the whole class are delivered, at the blackboard, by students representing their group, or by the teacher. During these phases, the different group conclusions are summarized and new questions raised.

When a series of 50-minute lessons has been programmed, it is run (with minor changes) in different classes to permit comparisons and the discovery of recurrent facts. By *different classes* we mean different classes of the same grade, and eventually classes of different grades. In each class, one group is permanently video recorded whereas (most) other groups are watched by one "observer" each. The observer tries not to intervene, but still does help the group when necessary. After each lesson, observations are discussed. The whole-class work phases are also video recorded.

One of the aims of such complex experiments is to observe regularities and establish and interpret reliable didactic facts. Pretest–posttest evaluations can also be organized, but this is not essential to the methodology.

AN EXAMPLE: A DIDACTIC SERIES ON VOLUME FOR SEVENTH GRADERS

Volume is a geometric–physical magnitude that can be, in certain cases, directly compared and measured. For instance, bottles, glasses, cups, vases, and other kinds of containers can be compared. This is not a difficult job for children. More difficult is the comparison of *full volumes,* such as stones, pieces of plasticine, or complex block shapes like those used in this experiment, and the comparison of full volumes with *hollow volumes,* or containers. These comparisons must usually be achieved with the help of some liquid (water, for example), and involve indirect comparisons.

In the first lesson, each group had first to compare four containers and to order them, $A < B < C < D$. Because these containers had been cut from tops and bottoms of different plastic bottles, the order was not obviously perceptible and students had to fill them with water to decide. During this phase we observed that although the task was easily done by seventh graders, the transitivity axiom was not always used and some redundant comparisons were made.

Each group was then given two full volumes, a block shape, and a plasticine one and had to place them in an ordered series:

Either E and H such that $E < A,\ C < H < D$

or F and G such that $A < F < B < G < C$.

The group sometimes had difficulties comparing full volumes with each other, and had even greater difficulties in comparing them with hollow volumes. The reason for these difficulties is the fact that such comparisons require complementary volumes and reasoning on complements. If F is put into X, and then G into X (X being a large container filled with water), then from $X - G < X - F$ the correct conclusion is $F < G$, not $F > G$.

Still more difficult is the comparison of a hollow with a full volume, B and F for instance, because one may have to find a liquid equivalent of F by complementing $X - F$ to X (double complement), or by reasoning on levels of water in X and planning a sequence of actions and measurements that is sometimes too complicated for the group, especially if two students have different plans. The protocols cannot be described in detail here, but we were struck by the unexpected difficulty of the task. Some children did not even bother to keep F totally under water, and drew inadequate conclusions because of this faulty procedure. In one class, no group was able to give a correct ordered series of full and hollow volumes. In other classes it was difficult for most groups. The task was easily done in only one class. Clearly for seventh graders (and probably for older students), the results show the relevance of this kind of situation and type of task.

In the second lesson, children were asked to measure all volumes. There are several ways of associating a number with an object. The volume formulas are one way to do it. In this lesson, a more direct method was used: two different capacity units were used for containers, different groups using different units. A natural solid unit (the block unit) was used for the block shapes E and F. It is fairly easy to count the number of capacity units in a container or to count the number of blocks in a block shape. More difficult is the association of a number of capacity units to full volumes or a number of blocks to containers and to plasticine shapes. These tasks involve either reasoning on complements or solving rule-of-three problems. So there are three different units: two liquid units, u_1 and u_2,and the block unit, u_3, related as follows:

$$2u_1 = 3u_2$$

$$u_1 = 18u_3$$

These equivalencies were not given and had to be discovered by students during the phase of calculation of the value of each volume in each unit system. The only way to complete this task is to find a common reference point: same volume measured with different units. This is not difficult for the first equivalency and for containers A, B, C, and D because the containers are the same in all groups and can be easily measured with u_1 and u_2. It is much more difficult for full volumes and for u_3 measures.

Table 5.4 summarizes the situation; it is a table of proportional numbers in which students have to fill empty squares. This table is not given to the students; it is drawn at the end of the second lesson, or at the beginning of the third lesson to summarize the results. Some scalar and function procedures used by students to calculate u_1 and u_2-measures of A, B, C and D are also described: (a) scalar procedure: Suppose you know that $m_1(A) = 6$, $m_2(A) = 9$ and $m_1(D) = 12$; because D is twice as big as A, then $m_2(D) = 18$; (b) function procedure: because $m_2(A)$ is $1\frac{1}{2}$ times as much as $m_1(A)$, then $m_2(D) = 12 \times (1\frac{1}{2}) = 18$. The above procedures can also be expressed in terms of column-to-column operators or line-to-line operators in the table.

The aim of the next two lessons was to help students move from a unidimen-

TABLE 5.4

Measures of Recipients A, B, C, and D and Full Volumes E and F

	E	A	F	B	C	D
u_1	—	6	—	8	10	12
u_2	—	9	—	12	15	18
u_3	72	—	126	—	—	—

sional conception of volume to a tridimensional one, to coordinate both conceptions, to compare different ways of paving and different sorts of elementary units, and to analyze the meaning and the homogeneity of the formula.

Two parallelepipedic boxes, A ($180 \times 120 \times 60$) and B ($180 \times 90 \times 75$), were used. Large numbers of two sorts of elementary units were provided: small parallelepipeds ($30 \times 20 \times 15$) and small cylinders ($d = 15$; $h = 30$). Measures were expressed in millimeters. They were not given to students.

Students were asked to measure and compare A and B. The idea is that both kinds of elementary units may be used, but that cylinders do not occupy all empty space and that Box B cannot be regularly and completely paved in all directions whereas Box A can. Further, the complete paving of A and B takes a long time, which can be saved by counting the number of elements that can be put in the length, in the width, and in the height, then multiplying. Finally, cubic units are indifferent to the orientations and optimize the procedure. This last idea is then tested with blocks of 1 cc. Although a large number of such cubic blocks was provided to each group, this number is not large enough for the complete paving of Boxes A and B, and the use of the formula is then unavoidable. The complete paving would take too much time anyway.

In a later phase, questions were raised concerning enlargement of parallelepipedic volumes by a ratio of 2, 3, or 10 on lengths—2^3, 3^3, and 10^3 on volumes)—and concerning all different possible combinations to make a parallelepiped of 24 blocks, or 48 blocks.

We observed most interesting behaviors during the paving process. In starting to work, many groups did not even think that it would save time not to pave completely Box A and Box B (they probably enjoyed paving the boxes). But unexpectedly, we also found groups that did not even find it necessary to make their paving regular; they would pave one line in length, then one line in width, along the same dimension of elements (see Schema 5.29 for examples of non-canonic pavings); or they would try to fill the box as completely as possible by paving the box with several lines regularly displayed and one more line differently organized. This of course does not allow them to make any simple multiplicative calculation.

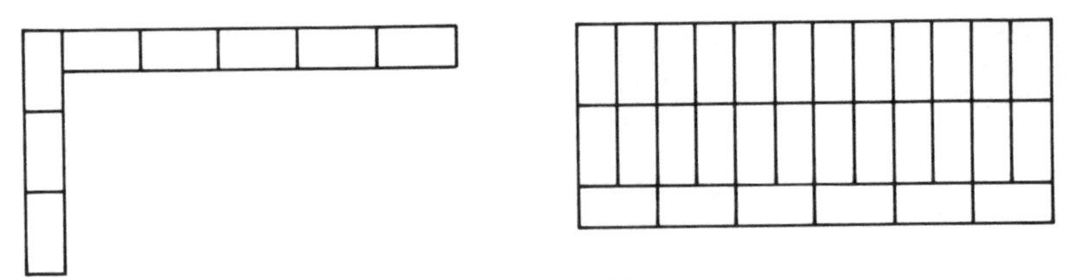

SCHEMA 5.29

Another interesting cognitive conflict occurred when the students tried to count the number of parallelepipeds in the length, in the width, and in the height. Beside the fact that they might arrange the elements along the length in all three dimensions, as mentioned above, some students were faced with the contradiction between the additive direct conception of volume and the multiplicative tridimensional conception. This contradiction was made especially obvious by a block-in-the-corner incident that took place in many groups. It usually appeared as a conflict between two members of the group, one of them saying that the block-in-the-corner should not be counted twice: "as it has been counted in the length, one should not count it in the width, or in the height," the other member of the group saying (without always being able to explain why) that it should be counted in the length, in the width, and in the height. The first position results from an additive direct conception of volume (in which it is actually true that the same partial volume cannot be counted twice), and the second position is associated with a multiplicative conception, in which blocks along the length, width, or height are not volumes but instruments to measure length, width, and height. The first conception is also reinforced by a perimetric view of volume as a composition of edges. This conflict, observed many times in different groups and different classes, is typically a didactic fact, due to the situation in which both unidimensional and tridimensional conceptions can be used. We would not have expected it to arise naturally in groups of 12–13-year-olds.

We also observed interesting behaviors mentioned before in the experiment on volume, concerning enlargement by a similarity ratio of 2, 3, or 10.

In the sixth lesson, the architect's problem was posed to students. In this problem, students were supposed to calculate different interesting magnitudes in a building. The base was either a rectangle (35×16) or a square (20×20) and the total area available for offices and apartments was known (5600 m^2). The height of each floor was given (3 m). The architect had to calculate the height of the building; the lateral area (that would be covered with glass); the total length of edges, except the base (aluminium devices were supposed to be fixed along the edges); and finally the volume (for the heating capacity). The aim of this problem was to oblige students to differentiate distinct kinds of perimetric, volumetric, and area magnitudes. The main obstacle for students was that the building's capacity (volume, in a way) was actually given through the intermediary of the total area available for apartments and offices. As a consequence, the number of floors was sometimes interpreted as the height of the building. The calculation of the lateral area was not very easy either, and some students tried to calculate it directly with the basic area or the total area available. The question about edges was, as expected, the easiest one. As for the volume, most groups did find it, but only one of them used the synthetic information: area available, 5600 m^2; height of each floor, 3 m. The other groups used the formula

$$L \times W \times H.$$

The last lessons were devoted to the study of triangular prisms, all having the same height and different base areas. In the first phase, students had to cut different bases from pieces of cardboard (8 cm or 16 cm wide) and to predict (before making them) the ranking order of the prisms built on these bases (six different bases, including equivalent-area bases). The prisms were then made, and sand was used to check the predictions. In order to explain the differences and the equivalencies, the concept of basic area was analyzed. To double the basic area, one can use twice the same area, or double the height of the triangle, or double the base of the triangle. A double-dependence table was drawn for the area of triangles and then for the volume of prisms (Schemas 5.30a and 5.30b).

(a) area of the triangle

		Height			
		4	8	12	16
Base	3				
	4				
	6				
	12				

(b) volume of the prism

			20	30	40
Basic area	16				
	24				
	32				

SCHEMA 5.30

Students worked on the first table and on the formula $A = 1/2\ BH$ by filling different cases. Their attention was drawn to the proportion between area and height (when base is held constant) and between area and base (when height is held constant). There are different possible procedures for the calculation of unknown cells: either the use of the formula or the application of adequate operators to previously known cells. Similar work was done on the second table.

Although all of these situations and behaviors deserve more detailed descriptions and explanations to be thoroughly understood, this chapter on multiplicative structures would have been very incomplete if this type of didactic experiment had not at least been briefly reported. Other didactic experiments are discussed in two previous papers (Rouchier, 1980; Ricco *et al.*, 1981; Vergnaud *et al.*, 1979).

Further Analysis and Experiments

As discussed in the first part of this chapter, problems met in ordinary economical and technical life involve different kinds of magnitudes and different categories of relationships. Asking students to solve such problems requires them to use logically distinct but psychologically interdependent topics in mathematics. Studying these topics in isolation is psychologically artificial. This is not specific to multiplicative structures: most didactic situations are conceptually pluridimensional. Three complementary approaches seem to be essential in this case: (*a*) fractions, ratios, and rational numbers, (*b*) linear and *n*-linear function with dimensional analysis, and (*c*) vector spaces.

Fractions, Ratios, and Rational Numbers

It is clear that multiplicative structures, because they imply multiplications and divisions, can be analyzed in a way that leads to fractions, ratios, and rational numbers. The main problem for students is that rational numbers are *numbers* and that entities involved in multiplicative structures are not pure numbers but measures and relationships.

The concept of rational number is defined, in mathematics, as an equivalence class of ordered pairs of whole numbers. This is a late construction in the history of mathematics. Fractions and ratios are not so well defined: the word *fraction* is sometimes used for a fractional part of a whole, sometimes for a fractional magnitude (that cannot be expressed by a whole number of units), sometimes for an ordered pair of symbols p/q, and sometimes for a relationship linking two magnitudes of the same kind. *Fraction* is rarely used for a relationship linking magnitudes of different kinds; the words *ratio* and *coefficient* are preferred. But *ratio* is also used for a relationship linking magnitudes of the same kind, and for an ordered pair p/q.

It would be futile to try to standardize the vocabulary. But as students' conceptions of rational numbers necessarily come from their conceptions of fractions and ratios, it is important to try to sort them out, in the light of our first analysis of multiplicative structures. Although decimals are very important in the development of rational-number concepts, I will not devote any special attention to them here. Brousseau (1980; 1981) and Douady (1980) have made very interesting contributions to the study of decimals.

THEORETICAL CONSIDERATIONS

Sharing a whole into parts is undoubtedly the very first experience with fractions. It involves a direct proportion between shares and the magnitude to be shared (isomorphism of measures). This magnitude can either be discrete (a set of sweets, a packet of playing cards) or continuous (a pie, a sheet of paper, a bottle of lemonade). An important difference is that one does not usually know the measure of the continuous magnitude to be shared (neither the weight nor the area of the pie) whereas a discrete magnitude can usually be counted. Consequently, the unit value (one person's share) is necessarily expressed as a fractional quantity (one-fourth, one-sixth) in the continuous case, whereas it may also be expressed as a number of elements (three sweets each) in the discrete case. (See Schema 5.31.)

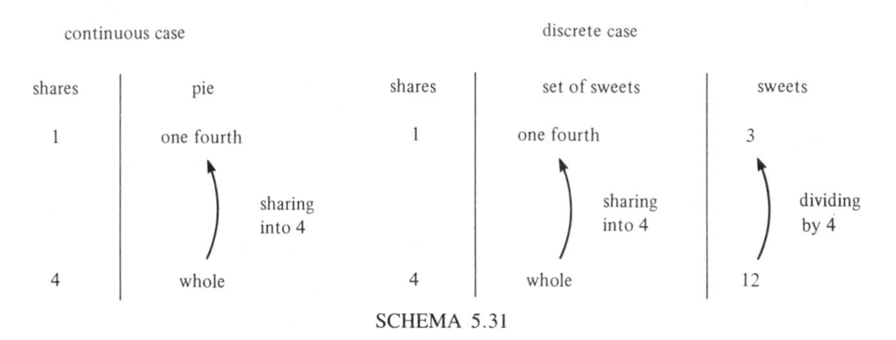

SCHEMA 5.31

The first type of division, mentioned in the description of the isomorphism of measures structure above, can easily be recognized in the discrete case (first and last columns) but it is not so easily recognized in the continuous case, or when one considers the discrete set as a whole (second column). In these latter cases, children have to recognize that sharing a whole into four parts, for example, requires dividing the unit 1 by 4. (See Schema 5.32.)

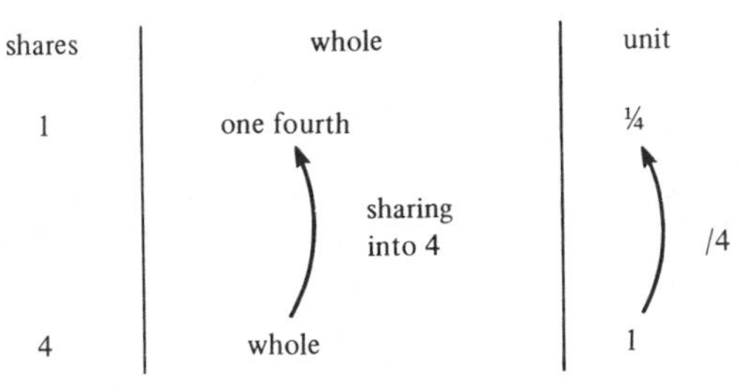

SCHEMA 5.32

This is a problem for elementary school children. Whereas whole numbers can be directly associated to quantities by counting, fractions (even the most elementary ones, $1/n$) cannot be associated directly to quantities; they are relationships between two quantities. This conceptual difficulty may be different for continuous and discrete quantities, and for different numerical values, as we will see later.

Once unit fractions have gained meaningfulness for children through sharing operations, many further steps remain for thorough understanding. We describe these steps in terms of non-Archimedean fractions p/q, smaller or bigger than 1, and equivalence relationships, and operations. The main conceptual problem for students is that fractions can be quantities, or scalars, or functions, and that these different concepts have to be integrated into one synthetic mathematical concept: the rational-number concept.

Because fractional quantities and magnitudes cannot be conceived without the help of scalar operators, I will start with this concept. As a scalar operator, a fraction links two quantities of the same kind. Being a quotient of two quantities of the same dimension, expressed in the same unit, it has no dimension and no unit. The problem of linking two quantities is raised either in comparisons (which quantity is bigger and how much bigger is one than the other?) or in proportional reasoning, as seen in the preliminary analysis above. Actually, comparisons do not necessarily involve fractions and ratios; differences (obtained by subtraction) are also appropriate in many comparison situations. In proportional reasoning, students must move from an additive method of comparing to a multiplicative one. Incidentally, this probably explains the well-known additive error in proportion tasks (Karplus & Peterson, 1970; Lybeck, 1978; Noelting, 1980a,b; Piaget, Grize, Szeminska, & Bang, 1968) and some other sophisticated errors (see error S′ in the rule-of-three experiment described above).

Schema 5.33 shows examples of scalar operators (whole and fractionary) from the simple multiplication and division cases to more complex cases (reducible and nonreducible ones).

There are two categories of ratios in comparison and proportion problems: either one quantity is part of the other (inclusive case) or there is no obvious inclusion relationship (exclusive case).

Scalar operations (whole and fractionary)

multiplication

division

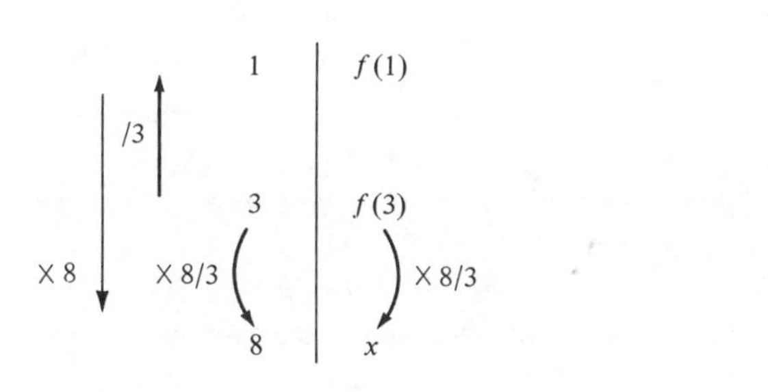

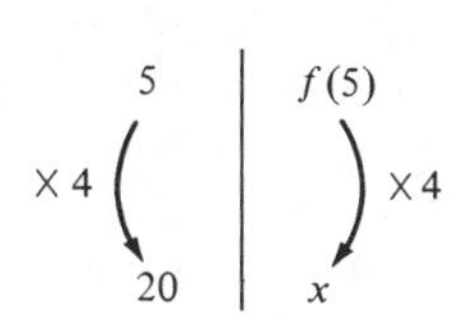

rule-of-three (nonreduceable)
fractionary operator

rule-of-three (reduceable)
whole operator

SCHEMA 5.33

1. Inclusive examples: p out of q. Peter ate two-fifths of the sweets. Three-fourths of the marbles are blue.
2. Exclusive examples: p to q. Peter's collection of miniature cars is three-fourths of John's collection. The distance covered in 5 hours is five-thirds of the distance covered in 3 hours.

There are some important differences between these categories:

1. Inclusive fractions or ratios are always smaller than 1 (except when the magnitude or the set is compared to itself), whereas exclusive fractions or ratios can be greater than, less than, or equal to 1.
2. Inclusive fractions are never reducible to whole numbers (except to 1 in the trivial case), whereas exclusive fractions can be either reducible or non-reducible (see examples in the previous schema.
3. Inclusive fractions have no inverse (for young students) because it is meaningless to consider the whole as a fraction of the part, whereas exclusive fractions have "natural" inverses: if Peter's collection is three-fourths of John's, John's collection is four-thirds of Peter's.
4. Inclusive fractions can be made meaningful to young students either by sharing operations, or more generally by subset–set proportions, whereas exclusive fractions necessarily involve comparisons.

Most children, at the end of elementary school, are unable to conceive exclusive fractions as fractions. Their model is the inclusive fraction model. This is a problem because comparisons and ratios between any two quantities of the same kind are a more powerful model than inclusive fractions, providing a more general foundation for scalar operators or ratios.

Fractional measures result from the application of fractional operators to other measures considered as wholes or units, in the nonreducible case. For example:

1. 3/10 cm is a fractional measure that results from dividing 1 cm into 10 parts and taking 3 parts, or else taking three-tenths of 1 cm.
2. 5/3 of 205 kms (1025/3) is a fractional measure of the distance covered in 5 hours, that results from applying 5/3 to the distance covered in 3 hours.

Nonunit fractional measures (i.e., p/q), like unit fractional measures (i.e., $1/n$), are necessarily relationships to other measures, via scalar ratios. But fractional measures can be added and subtracted, whereas scalar operators can only be composed in a multiplicative fashion.

Fractional scalar operators p/q are themselves the *concatenation* of one division by q and one multiplication by p, and all fractional scalar operators can be concatenated and composed into one single fractional scalar operator:

$$(\times\ p/q) \circ (\times\ p'/q') = (\times\ pp'/qq')$$

Addition and subtraction of fractional scalar operators are nearly meaningless.

The situation is not quite symmetrical with fractional measures, because they can be, most obviously, added and subtracted, but they can also be multiplied (or divided) by one another, when the structure is a product: "find the area of a rectangle that is 8/3 cm long and 4/7 cm wide."

Next, let us consider function operators. They are quotients of dimensions, and they raise conceptual difficulties for students. Nevertheless they provide the most natural way to introduce the concept of an infinite class of ordered pairs, as can be seen in Schema 5.34 taken from the wheat–flour problem.

Kg of wheat	Kg of flour
1.2	1
12	10
18	15
24	20
30	25
$6n$	$5n$
6	5

SCHEMA 5.34

If 1.2 kg of wheat is used to make 1 kg flour, one can establish a correspondence table between 12 kg wheat and 10 kg flour, 18 and 15, 24 and 20, and so on. All ordered pairs ($6n$, $5n$) belong to this infinite class; the function from left to right can be expressed by any operator $\times$ 10/12, $\times$ 15/18. . . . The simplest operator is $\times$ 5/6 and (5,6) is the simplest element of the class of ordered pairs. It is also possible to exhibit the equivalence of scalar operators, but the infinite character of the class is not exemplified.

In conclusion, it appears that the meaning of concatenation of division and multiplication comes from fractional scalar operators and ratios, the meaning of addition and subtraction comes from fractional quantities, and the infinite character of each rational-number class comes from fractional function operators and ratios. Multiplication of rational numbers can be made meaningful through composition of scalar operators, composition of function operators, and even product of measures. The synthesis of all three aspects can occur only if measures, scalar operators, and function operators lose their dimensional aspects and the distinction between element and relationship, and if the concept of rational numbers as pure numbers is built up. But this concept inherits all three aspects. Teachers should not expect this construction to be easy, fast, or immediately understood by students. Students cannot work on meaningless objects and one should not be surprised when they try to make pure numbers meaningful by interpreting them as quantities or operators.

The above analysis is convergent with Kieren's (1978–1979–1980) analysis, although it was developed quite independently. The main originality of this analysis is that it is strongly related to the general framework of multiplicative structures, in a way that clarifies differences between quantities and operators and between scalar and function operators.

SCALAR VERSUS FUNCTION

The distinction between scalar and function aspects has been mentioned by other authors, but it may be very ambiguous in some situations. Freudenthal (1978) and Noelting (1980a; 1980b) have used the distinction internal–external; Lybeck (1978), the distinction within–between; and Karplus, Pulos, and Stage (this volume) have discussed findings on the preference of students for scalar or internal aspects rather than function or external aspects. I would like to stress a few points in order to clarify the main theoretical issues discussed.

First, problems of comparing tastes, concentrations, or densities, as used in most studies on the development of ratio, (e.g., Noelting, 1980a; 1980b) are different from direct proportion problems (isomorphism of measures).

In direct proportion problems, there are only two variables and an invariant relationship (the function) between these two variables: the cost of goods, the speed, or the density is given as constant. The problem is to find $x = f(c)$, knowing a, $b = f(a)$, and c; it is not to compare two functions f and f'.

In most studies on ratio-concept development, the function is also a variable: for instance, in Noelting's excellent studies (1980a; 1980b) the experimental paradigm involves three variables: the number of glasses of water, the number of glasses of orange juice, and the taste or concentration, which is a quotient of the two other variables. The problem may be viewed in different complementary ways:

1. Direct comparison of two function-ratios
2. Decomposition of this problem into two other problems, relying upon two theorems:
   a. concentration is proportional to orange juice, provided water is held constant;
   b. concentration is inversely proportional to water, provided orange juice is held constant.

In this decomposition, one can easily recognize the structure of multiple proportion: the quantity of orange juice is a bilinear function of the quantity of water and of the concentration wished. Ratio-comparison situations cannot be analyzed as simple-proportion problems.

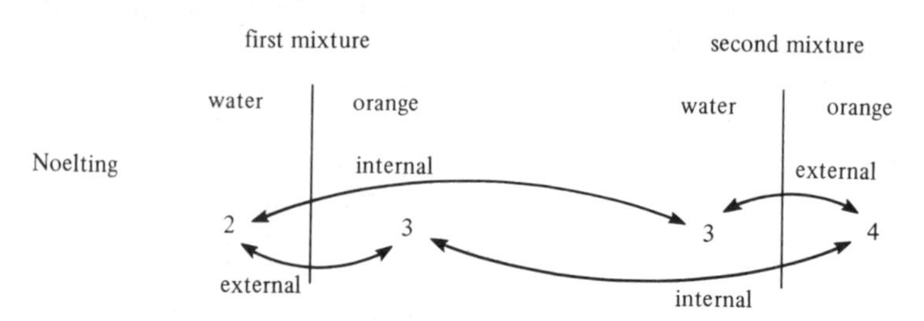

all ratios, internal and external, may vary.

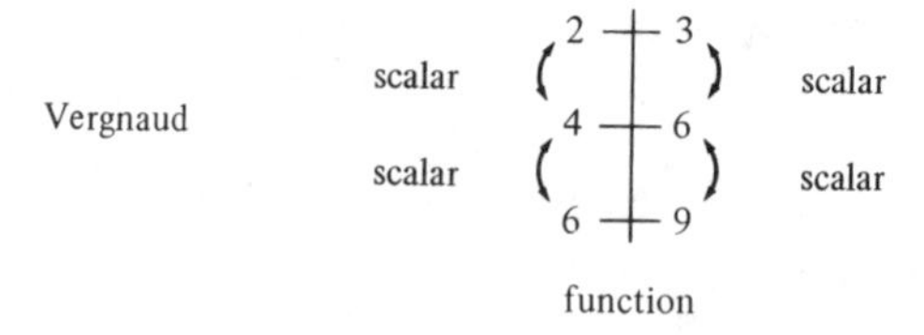

the function-ratio is invariant; scalar ratios may vary, but they are equivalent on both sides between two same lines.

SCHEMA 5.35

Second, the dimensional analysis of function ratios is clear when M_1 and M_2 have different dimensions, such as: time in hours and distance in kilometers for speed, volume in cc and mass in g for volumic mass, water in glasses and sugar in spoonfuls for sugared-water concentration, or even when M_1 and M_2 are the same magnitudes, expressed in different units, as is the case in problems of conversion from one unit system into another one. In many studies on ratio-concept development, because experimenters want the unit to be the same (glasses of the same size for water and orange juice), the quotient of two dimensions does not appear as clearly. Moreover, it is not invariant. The internal–external distinction made by Noelting is different from the scalar–function distinction, as can be seen in Schema 5.35. Although some results on ratio-concept development are well established (Hart, 1981; Karplus *et al.*, Chapter 3, present volume; Noelting, 1980a; 1980b; Suydam, 1978), more experiments are needed to clarify the dependence of this development on the different aspects of multiplicative structures.

FIRST NOTION OF FRACTION

I now report some results on the scalar operator concept that have been obtained with a different experimental paradigm. In her dissertation on the first ideas of children on fractions, Mariam Salim (1978) has used three different tasks:

1. Knowing the referent quantity and the fraction operator, find the compared quantity
2. Knowing the referent and the compared quantity, find the fraction operator
3. Knowing the fraction operator and the compared quantity, find the referent quantity.

The tasks involved different sorts of magnitudes:

1. Discs (continuous) on which lines had been drawn, to divide them into *n* parts (2, 3, 4, 5, 6)
2. Strips of paper (continuous) with no drawing
3. Sets of pearls (discrete)

and different numerical values:

1. $1/N$ ($N < 10$)
2. P/Q ($P < 4$ $Q < 10$).

Her most important finding is that understanding fractions depends heavily on their numerical value. Whereas all three tasks are easily achieved with continuous and discrete quantities for 1/2 and 1/4, they are still very difficult for other fractions.

There is a four year *decalage* on most tasks between 1/2 and other unit fractions, except 1/4 which is intermediary, and 1/3 which is a bit more difficult than other unit fractions (in French *un tiers* does not refer to *trois* as simply as *un cinquième* does to *cinq*).

In the first grade (6–7-year-olds), 1/2 fractions are fully mastered by 30% of the students and partly mastered by 50%. In the second grade, a few students deal fairly well with 1/4 fractions and start dealing with other unit fractions. In the third grade, nearly 40% of the students reach that stage, but this proportion increases rather slowly in the fourth and the fifth grades. In the fifth grade, all students master 1/2 fractions but only 40% master 3/4 and 2/5 fractions.

The difficulty of the three tasks was expected to be different and also to vary for discrete and continuous quantities. Salim did find some differences, but not as large as she expected, and very small compared with the differences due to the numerical values. It seems to be typical of unit fractions that only slight differences exist between the "find the compared quantity" task, the "find the operator" task, and the "find the referent quantity" task. Differences are bigger for P/Q fractions, as we see subsequently.

If these findings were confirmed, it would be an argument in favor of the relational character of fractional quantities: either the three-term relationship is mastered in all three tasks, and then the fraction is understood, or none of the tasks is solved.

These findings must be contrasted with another result obtained by Salim in a simpler task: as students were presented discs (with drawn division lines) and asked to show one-half, one-quarter, one-fifth, — three-quarters, two-fifths, 50% of the first-grade students were successful in the most difficult items. But this success is ambiguous because the procedure used by students was counting. One typical error of young students in the "find the compared quantity" task (discrete case) illustrates this ambiguity: when asked to find one-fifth of a set of pearls, many young students made a subset of 5 pearls.

Another error illustrates the inclusive character of the very first notion of fraction for children: when asked to find one-fifth of a set of pearls, or a strip of paper, many students were satisfied with just *a part*.

Another interesting result was found by Salim in comparison tasks: except for 1/2 and 1/4 fractions, students were not able to compare $1/n$ and $1/n'$ fractions until the third grade. But first graders were able to say that 3/5 was bigger than 2/5 by using a model that has nothing to do with fractions: 3 "something" is bigger than 2 "something." Third graders and fourth graders were not as successful on this item, because they tried to take the fractional character into account (U-shape curve). Most interesting were the explanations given by students: when asked to compare $1/n$ and $1/n'$ fractions, younger students referred to whole numbers n and n' and failed, whereas older students (third and fourth graders) referred to the number of shares. These results support the thesis that the first notion of fraction is inclusive, and refers to sharing operations.

In another experiment (Vergnaud, Errecalde, Benhadj, Dussouet, 1979) with older students (fifth and sixth graders), we systematically compared the inclusive case and the exclusive case. In three different contexts (customers in a restaurant, trees in a forest, load of a truck), students were faced with the three tasks: find the compared quantity, find the fraction operator, find the referent quantity. All fractions were P/Q fractions (where P and Q were different from 1).

None of the tasks was trivial, but we did observe that the "find the compared state" tasks were easier than the two others (60% success instead of 40%).

Then we compared the "find the fraction operator" tasks in the inclusive case (comparison of the part and the whole) and in the exclusive case (comparison of one part to another part). In the exclusive case, the success rate was about half the rate obtained in the inclusive case: from 10 to 20% instead of 40%.

This last result has convinced me that it is necessary to study the ratio-concept development in different contexts and in different frameworks. It would probably be fruitful to plan an extensive study on multiplicative structures, with different experimental paradigms involving different aspects of the ratio concept.

Linear Function and *n*-linear Function: Dimensional Analysis

The second approach for developing multiplicative structures with students includes the concepts of variable, function, linear function, n-linear function, and dimensional analysis. Although a linear function is a mapping from the set of real numbers into itself, and not from rational measures into rational measures, the linear model fits very well with the multiplicative structures. A formal derivation has been attempted by Kirsch (1969). I do not repeat here the analysis described above. What I would like to stress is the necessity of identifying very clearly for students the different variables, the different operators, and the different ways of solving the same problem.

One way to clarify these distinctions is to use symbolic representations that discriminate among different variables, different relationships, and different operations. For example, representing data and solutions in tables helps discriminate magnitudes of different dimensions (different columns or lines for different kinds of magnitudes) and relationships of different types (scalar relationships, function relationships, inverses, composed scalar and function relationships). We have used such representations in different situations (Rouchier, 1980) and I will report here only one example used with eighth graders.

THE FARM PROBLEM

"A farm, in the Beauce country, has an area of 254.5 ha. Half of it is devoted to growing wheat. The average crop is 6800 kg of wheat per ha. One needs 1.2

kg grain to make 1 kg flour; 1.5 kg flour to make 4 loaves. One loaf is, on the average, the daily consumption of two persons.''

Several tasks have been proposed to students:

1. Formulate and discuss a variety of questions. Are they well and completely formulated? Which ones are the same, under different formulations?
2. Make a table to represent the data and the relevant questions, and organize spatial correspondences.
3. Represent line-to-line and column-to-column operators.
4. Express the dimensional characteristics of these operators.
5. Explain the different solutions used by different groups of students for the same question and analyze them.
6. List the class of ordered pairs between any two columns (see the example of the wheat–flour function above) and the simplest fractional operator.
7. Identify $f(1)$ with the corresponding function operator.
8. Express and explain rules for composing column-to-column operators.

I can only summarize some experimental findings. More details can be found in Rouchier (1980). First, many questions are incompletely formulated and ambiguous. The equivalence of two different formulations is not immediately recognized. Second, the use of different columns and lines for different dimensions and for values that are not source and image of each other is not easily discovered by students. Once in use, however, the spatial organization of data and questions helps to clarify relevant relationships and calculations. The discrimination between simple proportion and multiple proportion can be more easily perceived and analyzed. Finally, function operations are difficult, except in the simple multiplication case.

Schema 5.36 provides examples of possible tables and questions.

area	wheat	flour	loaves	persons in one day
1	6800			
	b		1	2
	1.2	1		
		1.5	4	
127.5	*a*			*c*

Examples of questions

a What is the crop of the farm?

b How much wheat is needed to make one loaf?

c How many persons can one feed during one day with the crop?

d How much wheat is needed to feed 100,000 persons during 1 week?

		persons 100,000	*c*	
	1		*a*	
days	7	*d*		
	365			wheat

SCHEMA 5.36

One can easily imagine, in the first table, the succession of functions, with their dimensional meaning.

$\times$ 6800	$\times$ 1/1.2	$\times$ 4/1.5	$\times$ 2
wheat/area	flour/wheat	loaves/flour	persons/loaf

Interesting tasks can be organized around inversions and compositions of these functions.

Vector space

The vector-space model is again too strong for the sort of situations described here. It would be misleading to develop with seventh and eighth graders a formal presentation of vector-space theory. Yet the distinction between measure spaces, scalar operators, and function operators is directly related to vector-space theory: for students, measures behave as vectors, scalar operations as linear combinations, and functions as linear mappings. Actually, measure spaces are only semivector spaces (measures are positive), scalars are only rational numbers, and measures are only one-dimensional vectors.

It is nevertheless possible to make students confront less trivial vector spaces and linear mappings. For example, suppose a restaurant buys 4 different sorts of fruit every day: x_1 kg grapes, x_2 kg peaches, x_3 kg pears, x_4 kg apples. Two other important variables are the total weight y_1 and the total cost y_2. If, for a period (a week, for instance), the cost of each sort of fruit is constant (a_1, a_2, a_3, $a_4$per kg), there is a nice nontrivial linear mapping from (x_1, x_2, x_3,x_4) vectors into (y_1, y_2) vectors that make certain calculations easier. As

$$y_1 = x_1 + x_2 + x_3 + x_4$$
$$\text{and } y_2 = a_1x_1 + a_2x_2 + a_3x_3 + a_4x_4,$$

all isomorphic properties can be used and explained.

$$f(V + V') = f(V) + f(V')$$
$$f(\lambda V) = \lambda f(V)$$
$$f(\lambda V + \lambda' V') = \lambda f(V) + \lambda' f(V')$$

Conclusion

One might think that there is a major contradiction in this chapter between some experimental results showing the difficulty for students of multiplicative structures and some theoretical developments. Actually, these developments show that problems met by students, even at an early stage of their school curriculum, involve complex structures and concepts. This complexity is inescapable. How should we deal with it in mathematics education?

I have not referred to the Piagetian formal stage of development because I cannot see where one could trace the limit between a concrete stage and a formal stage in the development of such a diverse conceptual field. For instance, contrasting concrete numbers and pure numbers would be an oversimplification. It would be misleading to view measures as concrete numbers, or handling of operators as a concrete stage of understanding numerical operations; there are important formal ideas and theorems about measures and operators.

It is true that most general properties of rational numbers cannot be expressed and explained unless rational numbers are viewed as pure numbers. This is the case of cross-multiplying, for instance:

$$a/b = c/d, \qquad ad = bc, \qquad a/c = b/d$$

What would it mean to multiply a and d if a were an M_1-measure and d an M_2-measure? But it is true and important that understanding multiplicative structures does not rely upon rational numbers only, but upon linear and n-linear functions, and vector spaces too.

Many mathematics teachers have the illusion that teaching mathematics consists of presenting neat formal theories, and that when this job is well done, students should understand mathematics. In fact, concepts develop by problem solving, and this development is slow. Problem-solving situations that make concepts meaningful to students may be far removed from an advanced mathematical point of view. They are nevertheless essential and they must be carefully and completely analyzed so that the development of concepts may be traced and mastered.

Another tempting and inappropriate attitude is postponement: wait until students have reached a certain stage. This may be misleading too; there is no reason why students would develop complex concepts if they do not meet complex situations. The pedagogical illusion (teach it properly, they will know it) and the natural development fallacy (wait until they reach the stage) are Scylla and Charybdis obstacles in mathematics education. The framework of conceptual fields, which provides teachers with a variety of situations and different-level analyses, should help them to make students progress, slowly but operationally. Still, there is a long way to go before we fully understand the development of multiplicative structures.

References

Most references to American and English literature can be found in other chapters in this volume, especially in Behr *et al.*, Lesh *et al.*, and Karplus *et al.* One finds below only French references and a few specific references in English.

Brousseau, G. Problèmes de l'enseignement des décimaux. *Recherches en didactique des mathématiques,* 1980, *1,* 11–58.

Brousseau, G. Problèmes de didactique des décimaux. *Recherches en didactique des mathématiques,* 1981, *2,* 37–127.

Douady, R. Approche des nombres réels en situation d'apprentissage scolaire: Enfants de 6 à 11 ans. *Recherches en didactique des mathématiques,* 1980, *1,* 77–111.

Freudenthal, H. *Weeding and sowing. Preface to a science of mathematical education.* Dordrecht. Boston: Reidel, 1978.

Hart, K. *Children's understanding of mathematics* (Vol. 11–16). London: Murray, 1981.

Karplus, R., & Peterson, R. W. Intellectual development beyond elementary school: Ratio, a survey (Vol. 2). *School Science and Mathematics,* 1970, *70,* 813–820.

Kieren, T. E. The rational number construct—its elements and mechanisms. In T. E. Kieren (Ed.), *Recent Research on number learning.* Columbus, Ohio: ERIC/SMEAC, 1980.

Kieren, T. E., & Nelson, D. The operator construct of rational numbers in childhood and adolescence—An exploratory study. *The Alberta Journal of Educational Research,* 1978, *24,* 22–30.

Kieren, T. E., & Southwell, B. The development in children and adolescents of the construct of rational numbers as operators. *The Alberta Journal of Educational Research,* 1979, *25,* 234–247.

Kirsch, A. An analysis of commercial arithmetic. *Educational Studies in Mathematics,* 1969, *1,* 300–311.

Lybeck, L. *Studies of mathematics in teaching of science in Goteborg.* Goteborg, Institute of Education, 1978, No. 72.

Noelting, G. The development of proportional reasoning and the ratio concept: Differentiation of stages (Part 1). *Educational Studies in Mathematics,* 1980, *11,* 217–253. (a)

Noelting, G. The development of proportional reasoning and the ratio concept: Problem structure at successive stages; problem-solving strategies and the mechanism of adaptive restructuring (Part 2). *Educational studies in Mathematics,* 1980, *11,* 331–363. (b)

Piaget, J., Grize, J. B., Szeminska, A., & Bang, V. *Epistémologie et psychologie de la fonction.* Paris, Presses Universitaires de France, 1968.

Ricco, G. *Le développement de la notion de fonction linéaire chez l' enfant de 7 à 12 ans.* Thèse de 3ème cycle, Travaux du Centre d'Etude des Processus cognitifs et du Langage, 1978, No. 11.

Ricco, G., & Rouchier, A. Mesure du volume: Difficultés et enseignement dans les premières années de l'enseignement secondaire. *Proceedings of the Fifth Conference of the International Group for the Psychology of Mathematics Education, Grenoble,* 1981, 114–119.

Rogalski, J. Quantités physiques et structures numériques. Mesure et quantification: Les cardinaux finis, les longueurs, surfaces et volumes. *Bulletin de l'Association des Professeurs de Mathématiques de l 'Enseignement Public,* 1979, *320,* 563–586.

Rogalski, J. Acquisition de la notion de dimension des mesures spatiales de longueur et surface. *Proceedings of the Fifth Conference of the International Group for the Psychology of Mathematics Education, Grenoble,* 1981.

Rouchier, A. Situations et processus didactiques dans l'étude des nombres rationnels positifs. *Recherches en Didactique des Mathématiques,* 1980, *1,* 225–276.

Salim, M. *Etude des premièes connaissances sur les fractions chez l'enfant de 6 à 12 ans.* Thèse de 3ème cycle. Paris, 1978.

Suydam, M. N. Review of recent research related to the concepts of fraction and ratio. *Proceedings of the Second Conference of the International Group for the Psychology of Mathematics Education, Osnabruck,* 1978.

Vergnaud, G. Didactics and acquisition of ''multiplicative structures'' in secondary schools. In W. F. Archenbold, R. H. Driver, A. Orton, & C. Wood-Robinson (Eds.), *Cognitive development research in science and mathematics.* Leeds: University of Leeds Press, 1980.

Vergnaud, G. *L'enfant, la mathématique et la realite.* Berne: Lang 1981.

Vergnaud, G. A classification of cognitive tasks and operations of thought involved in addition and subtraction problems. In T. P. Carpenter, J. M. Moser, & T. A. Romberg, (Eds.), *Addition and subtraction: A cognitive perspective.* Hillsdale, New Jersey: Erlbaum, 1982.

Vergnaud, G. Cognitive and developmental psychology and research in mathematics education: Some theoretical and methodological issues. *For the Learning of Mathematics,* 1982, 3, *2,* 31–41.

Vergnaud, G., Ricco, G., Rouchier, A., Marthe, P., & Metregiste, R. Quelle connaissance les enfants de sixièe ont-ils des structures multiplicatives élémentaires? *Bulletin de l'Association des Professeurs de Mathématiques,* 1978, *313,* 331–357.

Vergnaud, G., Rouchier, A., Ricco, G., Marthe, P., Metregiste, R., & Giacobbe, J. *Acquisition des structures multiplicatives dans le premier cycle du second degré.* RO No. 2, I.R.E.M. d'Orléans, Centre D'Etude des Processus cognitifs et du Langage. 1979

Vergnaud, G., Errecalde, P. Benhadj, J., & Dussouet, A. La coordination de l'enseignement des mathématiques entre le cours moyen 2ème annèe et la classe de sixième. *Recherches Pèdagogiques,* 1979, No. 102.

CHAPTER 6

Space and Geometry

Alan J. Bishop

Geometry is the mathematics of space, and mathematicians approach space differently from artists, designers, geographers, or architects. They search for mathematical interpretations of space. Mathematics educators, therefore, are concerned with helping pupils gain knowledge and skills in the mathematical interpretations of space. Depending on many factors, such as one's philosophy of mathematics education, geometry education can range from learning well-established geometries, such as Euclidean or more modern transformation geometry, to developing the pupil's own geometrical ideas. For example, many secondary school geometry texts emphasize the fact that school geometry concerns learning theorems such as "the angle in a semicircle is a right angle" and "the line joining the midpoints of two sides of a triangle is half the length of the third side." On the other hand, the theme of a book like *Starting Points* by Banwell, Saunders, and Tahta (1972) is that any situation, including of course a geometric one, can be investigated and explored. This approach involves experimentation with materials and representations, classifying, defining, and analyzing why certain relationships occur. Perhaps it even involves building coherent systems of relationships based on selected assumptions and using that crucial mathematical ingredient, logical deduction. In short, one can emphasize either the results of other people's investigations into the mathematics of space, or the actual mathematizing of space by pupils. Moreover, as Robinson (1976) demonstrates, geometry itself is not a uniquely defined discipline, and therefore no reason exists to expect a consensus on what constitutes geometry education.

What then of mathematics education research that considers space and geometry? Compared to the other main focus of mathematics, *number*, there has been little research in this area. In a recent definitive volume on research in mathematics education (Shumway, 1980), space and geometry received minimal attention.

ISBN No. 0-12-444220-X

In the annual listing of research articles in the *Journal for Research in Mathematics Education,* July 1980, only 10 of the 161 entries concerned space and geometry in some way. (In the previous year, the figure was 8 out of 215.) Only two or three papers at the Fourth Annual Conference on the Psychology of Mathematics Education at Berkeley in 1980 were at all related to our concerns here. Whether this lack of attention reflects problems with geometry, with geometry education, or with research in geometry education is not clear at present, but the fact remains that mathematics educators do not have an extensive or comprehensive corpus of research from which they can draw ideas in tackling the issues surrounding the teaching of geometry.

However, there is more to space than geometry, and fortunately there is an extensive literature that pertains to spatial ability and that offers many interesting constructs and techniques, although not all viewed from the perspective of mathematics education. Therefore, I hope to do justice to the twin headings in the title by demonstrating that there is indeed a great deal of potential value in research on space and geometry.

I considered presenting two separate reviews of research, one on space and the other on geometry, or even one on developmental psychology and the other on factor analysis (because these have been the two general schools of research emphasis). As each has undoubtedly suffered because of an apparent unawareness of the other's work, an attempt was made, however, to synthesize the different research efforts. Two different types of constructs are therefore used to structure the review, one concerned with meanings and understandings, and the other with abilities and processes.

Meanings and Understandings

The focus of the first part of the research review is well conveyed by the titles of two classic books by the Geneva school, *The Child's Conception of Space* (Piaget & Inhelder, 1956) and *The Child's Conception of Geometry* (Piaget, Inhelder, & Szeminska, 1960). Research with this focus attempts to uncover the child's understanding of spatial and geometric phenomena and is potentially a most significant area for mathematics educators.

Within this area the influence of the Geneva school is all-pervasive. Whether one considers the topics studied, the research methods used, or the context for the interpretation of meaning, the towering presence of Piaget and his co-workers is felt. The two books listed above, together with a third, *Mental Imagery in the Child* (Piaget & Inhelder, 1971), have so defined the field that it is sometimes difficult to consider research in this area from any other perspective. There is no point in writing yet another review of the Piagetian literature (see Smock, 1976, for an excellent summary). The general construct of *meanings and understandings* was chosen to aid in interpretation and suggest further lines of research.

With respect to content, the mathematics educator is interested in the range and generalizability of the topics of recent studies. Because one cannot reasonably expect researchers to cover the entire geometry curriculum, it is important to know what has been considered, and the range, if not the amount, certainly is impressive. Rosskopf's (1975) book of six Piagetian studies includes one on bilateral symmetry, one on topological understandings, one on area, and one on limit. Martin's (1976) collection includes discussions that range from object permanence to spatial groups of transformations, from perimeter–area relationships to representations of Möbius strips. Fuson and Murray (1978) were concerned with recognition of geometric shapes, whereas Perham (1978) was interested in slides, flips, and turns—transformations in geometry. Vollrath (1977) investigated geometrical similarity, whereas Mitchelmore (19980a) looked at representations of three-dimensional objects.

In general one can find examples of studies that consider topological ideas, as defined by Piaget, and use young children as subjects, thereby reflecting the assumption that topological understanding occurs early. This does not imply that meanings and understandings of topology (in a mathematical sense) have been systematically explored with different ages or different types of test materials. Many studies exist that use Euclidean ideas as their topic, although more recently there seems to be an increase of interest in transformational rather than "formal" Euclidean geometry (see Lesh, 1976). This is related to a greater concern with spatial abilities, which will be discussed in the next section.

However, one might have expected more research on the *meanings* dimension of projective space. The majority of studies in that area seem to focus on orientation, cross sections, and surface development, which again seem to belong more to the *abilities and processes* section. Consider the classic three-mountain task used by Piaget and Inhelder to explore children's coordination of perspectives, which apparently matures only in the late concrete-operational stage. Several training studies have attempted to teach this ability, some successfully (e.g. Cox, 1977) and some not (for example Eliot, 1966). However, not much attention has been paid to meanings in the task. Hughes and Donaldson (1979), for example, report an important study that uses two policeman dolls, in which the young subject is required to hide a third doll from both. It is quite clear that many very young children are able to coordinate perspectives in this task. What we do not know is *why* they can in this situation when same-age children typically cannot coordinate perspectives in the three-mountain tasks. Complaints are often voiced that the varying stimuli and experimental conditions used by different experimenters make coordination of findings difficult to interpret. That objection may be valid concerning ability training. Concerning meanings, however, the complaint may be that the variation of stimuli and conditions has not been sufficient or systematic *enough*, so that we cannot interpret the effect of the child's *understanding* of the situation on task performance.

With regard to the example above, it seems intuitively obvious that a hiding game is likely to be more meaningful to a child than a puzzle involving model mountains, particularly if one were brought up in a flat landscape or an inner city. Further, one could say that systematic variation of task conditions may still only enlighten us with regard to the ability to coordinate viewpoints. If we want to know more about the child's meanings of space and its representation, we need research that focuses more on the relationship between real space and model space, real space and drawn space, real space and photographs of space, and so forth. One can characterize the experimental conditions of most geometry–space research in terms of using small-scale, modeled, drawn, or photographed versions of space—little is done with real, large-scale space. Perhaps we need to consider the research interests of geography educators. For example, Blaut and Stea (1974) demonstrate very well what real-world meanings young children are able to impose on objects, and indeed the whole field of cognitive mapping (see for example Tolman, 1948) may offer the mathematics educator much in the areas of meaning, understanding, and conception of space.

However, the framework generally used for the interpretation of the child's meanings is developmental; that is, a child's response to a particular task is evaluated in terms of "less" or "more" developed responses to that task. Typically, cross-sectional samples of different age groups are studied in order to relate individual performance to generalized stages of development, and typically, though not exclusively, it is the Piagetian ideas of development that are used as the interpretative framework. For example, in an interesting study by Wagman (1975), a sample of 8-, 10- and 11-year-olds was presented with several tasks in a one-to-one setting. The tasks covered various aspects of *area* as defined within a plane geometric context, including conservation of area, congruence of regions, unit area measure, and addition of area. Performance on these tasks was related to the Piagetian stages, with subdivisions, and four specific stages (Premeasurement, Beginning, Transition, and Attainment) were defined for the concept of area measure. Using Laurendeau and Pinard's (1970) criterion for attainment of stages, Wagman determined that the 8-year-olds had reached the Transitional stage. A few of the latter also had reached the Attainment stage.) She concluded that the 10- and 11-year-old children would be most likely to profit from appropriate instruction in area measure.

The use of stages and levels to aid interpretation of reponses and performances does not imply, however, that it is Piagetian theory that offers the best context. In particular, Piaget's topological, projective, Euclidean sequence of stages is most often rejected, if not by the data then by the researcher. Geeslin and Shar (1979), for instance, offer an interpretation of transition based on their ideas of progressive distortion. Other studies, such as those by Schultz (1978), Mitchelmore (1980a), or Küchemann (1980), suggest that the child's understandings of space and geometry change from being qualitative to being more quantitative and differentiated. The analyses of these researchers seem to agree with the general view of spatial development described by Werner (1964), who sees the progression to be from globality to increased differentiation. Geography educators appear to be more interested in Werner's work than are mathematics educators at present (see Hart & Moore, 1973).

Nussbaum and Novak's (1976) work on children's concept of the earth sensitizes us to other determinants of meaning. They focused on three aspects of the Earth concept: (*a*) that it is a spherical planet, (*b*) that it is surrounded by space, and (*c*) that objects fall toward the center. Among other data, they present an interesting transitional meaning stage in which the child's interpretation of gravity entails visualizing the earth as a sphere resting in an overall vertical gravitational field, so that objects "above" the earth fall toward it whereas objects on it tend to fall "downward" and off it! As a result of their work, they present four stages through which children progress until they reach stage five in which they "demonstrate a satisfactory and stable notion of the Earth concept" (p. 546). However, they prefer to relate their findings not to Piagetian stage theory but rather to Ausubel's (1968) ideas of meaningful learning. They also refer to some children's overseas travel experiences as significant "developers" of the Earth concept, instead of relating this specific acquisition to the child's overall cognitive development.

Expanding on this point, it is clear that different interpretations of children's meanings can be offered by different theoretical frameworks. Thus, although most of the reported research on geometrical and spatial meanings is interpreted within a developmental frame, there are others that may be more helpful to the mathematics educator. It is not useful to a mathematics educator just to know that particular meanings appear at specific stages in a child's overall development. An educator is more concerned with intervention—with placing learners in chosen contexts and confronting them with particular problems clothed in selected materials and sequenced in specific ways. The educator works within a given cultural framework and must make many decisions about content and methodology. The increasingly differentiated, quantitative, and hierarchical development that appears to take place does not occur in a vacuum. It must in some way be shaped by education and acculturation.

It may be unreasonable at this stage to argue for more research that would treat overall cognitive development as a dependent variable, but there is no reason to ignore other variables that could account for variations in children's understanding of a spatial phenomenon. Examples of such research are rare but some do exist. Vollrath (1977) looked at children's understanding of similarity and shape, with respect to interference from colloquial uses of words such as *similar*. Many ideas in mathematics and specifically in geometry have relationships with words used in everyday contexts, and quite clearly the colloquial meaning which a child associates with a word will affect the meaning associated with that word when it is used in a geometric context. Research on intuition (for example, by Fischbein, Tirosh, & Hess 1979) shows the same problem in a nonverbal form and this can affect nongeometric ideas in other areas of mathematics as well. For example, children develop the strong intuition that addition and multiplication make something larger, whereas subtraction and division make things smaller. These intuitions cause great difficulties when the child first meets negative numbers and fractions less than one.

Mitchelmore (1980b) identifies the difficulties that children have in representing parallels and perpendiculars, and reminds us that problems with visual symbolism are not trivial. Like children's problems in writing and speech, they tell us a great deal about their understanding. Goodnow's (1977) book on children's drawing also demonstrates difficulties in representation. In her study on bilateral symmetry, Genkins (1975) experimented with a paper-folding method and a mirror method, showing how, even with a study cast in a Piagetian frame, one can uncover relationships that inform us about other variables' effects on children's understandings. Prigge (1978) investigated the effects of children's use of manipulative apparatus on their understanding of geometric concepts. This study echoes an earlier study by Bishop (1973) that explored the relationship between the use of manipulatives in primary schools and scores on subsequently administered tests of spatial ability. Fuson and Murray (1978) divided their sample by what they termed *background* with "Background A: upper middle class/half day school/Montessori preschool and Background B: middle class/day care/traditional preschool" (p. 54). They reported that on their recognition-of-shapes task, "where there were differences, Background A scores were consistently higher" (p. 76). However, they also maintained that "The Background variable . . . reflects a combination of factors. . . . Although significant differences in construction and drawing performance were found, it was not the aim of the authors to encourage further research concerning the exact locus of this effect" (p. 80). This seems a strangely restricted response to an interesting finding.

As well as words, apparatus, and activities, much learning of geometric meanings involves the use of diagrams. Several Soviet researchers have been particularly interested in the effect that drawings can have. Zykova (1969) refers to the rigidity associated with geometric drawings in textbooks and the inability of pupils to generalize from these drawings. One problem in geometry teaching is that it is impossible to draw a generalized diagram. For instance, one cannot draw a general triangle—once it is drawn it is specific. It is therefore necessary to present many diagrammatic examples of a geometric concept if the learners are not to be restricted by the specificity of the diagram. It would seem valuable to use film to clarify the meaning of some generalized geometric relationships (such as, "vertical angles are equal").

In a more active mode, pupils may generate their own drawings to explore a geometric concept. In Fielker's excellent article (1979), examples of such geometric activities are given:

> How do you construct a square, given each of the following, *either* on a geoboard, *or* on plain paper, with restrictions to various choices from compasses, rules, setsquares, protractors or just folding? (i) one side (ii) one diagonal (iii) midpoints of opposite sides (iv) midpoints of two adjacent sides (v) midpoint and centre (vi) one vertex and centre (p. 112).

As he says, "this helps to get at the *essence of the idea* of what a square is" (p. 112), which clearly offers more meaning to the child than does merely teaching the concept of a square by discriminating it from other shapes.

Zykova (1969) makes the point more strongly by asking, "How does one master concepts whose essential feature is their relationship to other concepts?" (p. 151). No doubt Fielker's methods can help in this, but many reasonable questions are still begged.

Abilities and Processes

With respect to abilities and processes, a strong influence is again exerted by one particular interpretative framework, in this case, factor analysis. However, research efforts over the past decade have widened in scope and the situation is much less rigid than that described in the previous section.

Spatial abilities have interested factor analysts for many years, with Spearman (1927) and Thurstone (1938) leading their respective "schools" of research. Not until the 1940s and 1950s, however, did mathematics educators take a serious interest in the field and begin to concern themselves with relationships between spatial and mathematical abilities. Murray (1949), Wrigley (1958), and Barakat (1951) found that ability on spatial tests correlated more highly with ability in geometry than in algebra, and MacFarlane Smith (1964) argued that spatial ability was essential for mathematical ability. The field was split, however, by a general controversy as to whether spatial ability was a unitary trait; MacFarlane Smith claimed that it was, whereas Michael and others (Michael, Guilford, Fruchter, & Zimmerman, 1957) claimed the opposite. In fact, problems of definition, and of what constitutes a spatial test abound and are still being debated (see Clements, 1981). It is clear (to a mathematics educator, at least) that there can never be a "true" definition of spatial ability; we must seek definitions and descriptions of abilities and processes that help us to solve our own particular problems.

What abilities, therefore, have been identified as spatial? McGee's (1979) analysis follows Michael's and results in a description of two types of spatial ability:

1. Spatial visualization (Vz), which involves "the ability to mentally manipulate, rotate, twist or invert a pictorially presented stimulus object" (p. 893), and
2. Spatial orientation (SR-O), which involves "the comprehension of the arrangement of elements within a visual stimulus pattern and the aptitude to remain unconfused by the changing orientations in which a spatial configuration may be presented" (p. 893).

Although the Geneva school and the factor analysts have worked independently, the findings of the former do shed light on this area. As children develop, they move from a position of total egocentrism in the perception of space, to a stage in which multiple viewpoints can be considered, to a position of being able to conceptulaize and operate mentally on hypothetical space. Furthermore, growth in the ability to imagine and to internalize actions enables children to detach themselves from the constraints of the real world. Piaget and Inhelder (1956) make distinctions between perceptual and representational thinking, and between figurative and operative thinking. *Perceptual* thinking is related to sensorimotor actions, whereas *representational* thinking is necessary for the internal manipulation of imagery. *Figurative* thinking concerns static patterns and images, whereas *operative* thinking deals with patterns in the movement of objects and the manipulation of visual images. Furthermore, several researchers such as Fishbein, Lewis and Keiffer (1972) and Smothergill, Hughes, Timmons and Hutko (1975) work on spatial problems within a developmental framework, although they are clearly concerned with spatial *abilities*. There is therefore much in common between the developmentalists and the factor analysts regarding the kinds of abilities and processes impiled by a "loose" label like *spatial ability*.

Another approach to this research area is from the *individual-difference* perspective. Age or cognitive level could be considered as a significant individual difference. However, an equally important difference relating to the mathematics educator's concerns is that between more- and less-able pupils. Hamza (1952), for example, tested two groups of grammar-school boys in depth, one group retarded specifically in mathematics and the other a normal group (meaning above average at grammar school). Comparing the factor patterns for the two groups, he found a strong spatial factor for the normal boys which was missing for the retarded group. Krutetskii (1976), in a very different type of study, also noted a difference between more- and less-able students in their ability to grasp the problem as a whole. The capable pupils could integrate the separate elements into a significant, ordered structure that enabled them to recognize problems of a similar or opposite type. In contrast, less-able pupils were tied to the detail of the problem, seeing only disconnected facts and treating them all with equal significance. Later, however, Krutetskii did not include spatial ability as one of the components of mathematical giftedness, preferring to suggest that it characterizes a particular type of giftedness.

Krutetskii's work is encouraging to those interested in individual differences in preferred modes of processing. He finds one distinction to be important, between what he calls the *analytic* type of person and the *geometric* type. The former shows

> an obvious predominance of a very well developed verbal–logical component over a weak visual–pictorial one. They operate easily with abstract schemes; they have no need for visual supports for visualizing objects or patterns in problem-solving, even when the mathematical relations given in the problem "suggest" visual concepts (p. 317).

The geometric type, however, shows thinking that

> is characterized by a very well developed visual–pictorial component. . . . These pupils feel a need to interpret visually an expression of an abstract mathematical relationship and demonstrate great ingenuity in this regard: In this sense, relatively speaking, figurativeness often replaces logic for them (p. 321).

Others have also identified this distinction, including Hadamard (1945) and Richardson (1977). Recently, Moses (1979) and Lean and Clements (1981) have picked up this theme by defining a visual–verbal processing continuum and relating an indvidual's location on that continuum to ability on mathematical and spatial tests.

Another relevant individual difference research area is the difference in spatial abilities between boys and girls. Generally, boys seem to perform better than girls on spatial tests, but there are great fluctuations in the data. In addition, most of the research has been done with group tests, emphasizing mean scores, rather than using individual clinical procedures to ascertain what processes are actually being used. For example, Werdelin (1961) in a factorial study, found that a spatial factor emerged from the boys' scores but not from the girls'. MacFarlane Smith (1964) reports a study by Taylor (1960) in which a memory-for-designs test was marked in three ways after being given to 100 boys and 100 girls approximately 14 years old. The method of marking that considered accuracy of details produced results in favor of girls, whereas marking for correctness of proportion favored the boys. This is reminiscent of Krutetskii's difference between more- and less-able pupils, and is echoed by Wood (1976) who notes the particular difficulties experienced by girls in solving problems involving proportions, fractions, and probability. Harris (1978), in perhaps the most comprehensive review of sex differences in spatial ability, raises possible explanations in terms of brain laterality, genetic and hormonal influences, and various external factors.

Though these relationships may be interesting, the mathematics educator may still see little value in this type of research for guiding decisions about issues and procedures. The research method is, after all, predominantly correlational and causal inferences are at best speculative. Fortunately, in the past decade, there has been an increase of interest in the trainability and teachability of various spatial abilities. For example, Bishop's (1973) study, as referred to previously, showed that differences in spatial ability preformance could be related to a child's previous experience with manipulative apparatus, but studies and findings generally have been rather haphazard and inconclusive.

Part of the difficulty derives from inadequate definitions of spatial abilities and the lack of guidance on how best to test for them. Spatial tests almost invariably use figural stimuli. Though this may seem a curious objection, the use of figures has important ramifications. As MacFarlane Smith (1964) says of Michael's two principal spatial abilities: "The high correlation between SR-O and Vz tests is due to the fact that both types of test involve the same broad factor" (p. 96). The problem with considering an ability like *visualization* in the context of mathematics is that it is not necessary to have figural stimuli; that is, visualization can occur in arithmetic and algebra as well as in geometry. If one wants to test for visualization in mathematics one should use tests that include both figural and nonfigural stimuli.

Another ramification is that the figural stimuli on spatial tests have their own conventions, their own visual vocabulary so to speak. This is clearly a potential source of problem for some subjects. Third, the usual group-administered, timed format precludes discovering how the subjects reached their answers. Recent research by Guay, McDaniel, and Angelo (1978) and others has shown that subjects actually use idiosyncratic methods to solve spatial ability tasks—some use analytic procedures and some Gestaltist, for instance. Questions are therefore raised concerning what precisely it is that spatial ability tests are measuring.

To eliminate some of this confusion, and to help mathematics educators focus on relevant training and teaching research, Bishop (1980) proposed two different types of ability constructs as follows:

1. The ability for interpreting figural information (IFI). This ability involves understanding the visual representations and spatial vocabulary used in geometric work, graphs, charts, and diagrams of all types. Mathematics abounds with such forms and IFI concerns the reading, understanding, and interpreting of such information. It is an ability of content and of context, and relates particularly to the form of the stimulus material.

2. The ability for visual processing (VP). This ability involves visualization and the translation of abstract relationships and nonfigural information into visual terms. It also includes the manipulation and transformation of visual representations and visual imagery. It is an ability of process, and does not relate to the form of the stimulus material presented.

This IFI–VP dichotomy clearly relates to the SR-O and Vz dichotomy of Michael and McGee, but the IFI–VP dichotomy is more extensive and refined. The description of IFI extends SR-O by including geometric and graphical conventions not generally employed in SR-O tests and sharpens it by emphasizing the interpretation demanded by those representations. This ability is probably responsible for much of the relationship between spatial and geometric abilities found in the literature. It also links with the "meanings and understanding"section earlier in terms of the figural representations used to describe geometric and spatial concepts. Clearly such concepts will not have much meaning for a child who cannot interpret the language used to represent them. Bishop (1974, pp. 165–189) drew attention to the many figural forms used in mathematics, showing again that geometry is not the only branch of mathematics to use such diagrams.

VP has much in common with Vz but it extends and sharpens it by emphasizing the process aspect rather than the form of the stimulus. Interestingly there is support from Piagetian theory for this construct because the growth of imagery in the child appears to depend on the internalization of *action,* not form.

Another feature of this IFI–VP dichotomy concerns testing. Whereas it is possible to test the ability for IFI in a group mode, it seems much more necessary to use individual clinical procedures to test the ability for VP, and Bishop's (1980) paper contained many suggestions for suitable tasks.

However, it is when we consider training and teaching studies that the IFI–VP dichotomy could have its main value. Intuitively, IFI ability seems much easier to train and develop than VP, if only because of the public, communicable nature of IFI in comparison with the private, personal nature of VP.

Lean (1981) has summarized the literature on spatial training according to the IFI–VP distinction and produces some interesting results. Looking first at research that seemed to be concerned with training IFI ability, he reviews studies on "training the perception of pictorial materials" (for example, Deregowski, 1974; Leach, 1975), on "training in graphics" (for example, Blade & Watson, 1955; McCloskey, 1979), the studies of Vladimirskii (1971), and the effect of practice (for example, Connor, Serbin, & Freeman, 1978; Drauden, 1980). As a result he concludes:

> The evidence . . . indicates that these various skills are trainable given the appropriate experiences. Brief training with pictorial materials is sufficient to induce pictorial depth perception; relatively brief practice is sufficient to improve subjects' performances on spatial test items; the teaching of various spatial conventions and exercises with diagrams helps to improve geometry performance; and sufficiently long experience with technical drawing appears to improve performance on spatial tasks similar to those done in the drawing course (p. 10).

Turning to studies that are more concerned with the trainability of VP ability, Lean shows that the evidence is much less convincing. He first considers studies that have used mathematics courses to try to develop spatial ability. In comparing unsuccessful studies (like Cohen, 1959) with more successful ones (like Bishop, 1973) Lean comments that "these results may also indicate an age factor; the three geometry courses were all given to high school seniors where Bishop's study involved primary and seventh grade children" (p. 13). Relating to developmental theory, it may well be that there is an optimal time for such exercises in developing VP ability, and the hypothesis would be that the optimum is when the child is passing through Stage III (between approximately 7 to 12 years) and beginning to grasp projective space ideas.

Lean then considers general spatial training programs such as those studied by Wolfe (1970), Brinkmann (1966), Rowe (reported in Clements, 1981), and Moses (1979). Finally he concludes: "The majority of the studies discussed in this section have not been concerned with either retention or transfer and thus it is not possible to judge whether changes in subjects' VP abilities have occurred" (p. 16). However, he infers from the research that a wide range of spatial activities seems more likely to develop VP ability than is a more limited range. Again, the more successful studies were carried out with younger children (5-year-olds, Grade 5, and Grade 7), whereas studies with older subjects have had little success. What are clearly needed now are more training studies using clinical testing procedures, and involving retention and transfer tasks.

Spatial Understanding and Processes of Students in Papua New Guinea

An example of research relating to the literature analysis presented in the two previous sections is some recent work done in Papua New Guinea. The details of this research are given in Bishop (1978b) but this is the first time that the data have been analyzed using the frameworks of meanings and abilities.

Papua New Guinea is a country of rapid change, with the phrase "from stone age to twentieth century in one lifetime" being very apt. The majority of the population live in small villages and, apart from those living in the few small towns, they have little contact with the technological culture and society that we know so well. However, a school system is developing and there are two Universities, one at the capital, Port Moresby, and one at Lae.

At Lae there is a thriving Mathematics Education Centre, attached to the Mathematics Department, which undertakes both basic and applied research and development. Under the auspices of the Indigenous Mathematics Project (see Lancy, 1978) I spent 3 months working there on the problems surrounding the students' difficulties with geometric and spatial ideas in mathematics and science.

A group test was administered to all 250 first-year students at the University of Technology. The test, developed by the author and validated in the United Kingdom, contained many problems involving diagrams and other visual representations in mathematics, particularly using two-dimensional pictures of three-dimensional situations. These used the conventions of dotted lines for hidden edges, of perspective, of cross-sections, and of nets of solids. In order to deal successfully with them, the student had to read and understand the words used, read and understand the diagram, imagine the situation thus described, imagine what changes (if any) needed to take place, and represent the result either in symbols or in a diagram. The actual mathematical knowledge needed was very limited.

Generally, the test proved to be difficult, with some items, like those below, being very difficult (facility values expressed in percentage within parentheses following each problem). The following is a sample from that test.

> This is a diagram of a cuboid made of small cubes. If the OUTSIDE of the cuboid is painted red, how many cubes will have only one face painted? (14% of the students were successful) (See Schema 6.1.)

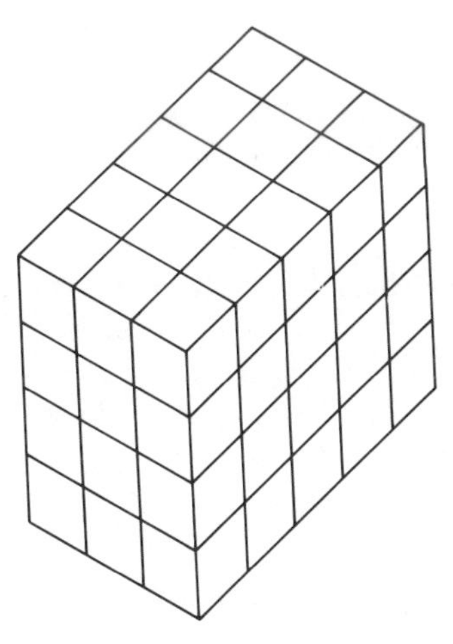

SCHEMA 6.1

> Draw the shape which you could cut out of paper, and fold, to make the six-faced object shown here. (25%) (See Schema 6.2.)

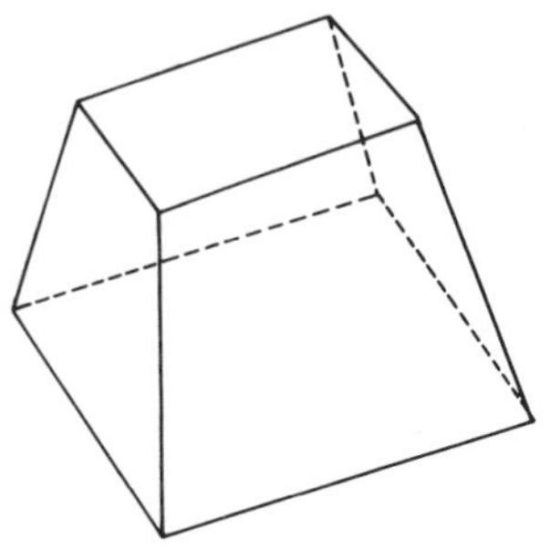

SCHEMA 6.2

Using the scores on this test, and other information supplied by colleagues at the University, a sample of 12 students was selected for follow-up interviews and individual testing. Three groups of four students were chosen, from Port Moresby (Urban culture), from the Enga province (Highland culture) and from Manus (Island culture), and a variation in performance on the initial test was obtained within each subgroup. The students were all male. The interviews took place in English in six or seven 1-hour sessions.

Many tasks were used, with the objective being to locate relative strengths and weaknesses among the students. The research was essentially exploratory and hypothesis generating. Some of the more significant tasks and the results are described and the conclusions summarized:

1. Map drawing.

   a. *The student was asked to sketch an outline map of Papua New Guinea and to indicate on it his home, the main towns, and the cardinal directions. Questions were also asked about directions of towns from each other.* Generally the shapes were accurate, with the "home" area better represented than the others. However cardinal directions were not always accurate. Manus students were better than the others at this task.

   b. *The student was asked to sketch a map of the university campus showing the route from the student's room to the testing office.* Generally this task was well done, with Port Moresby students drawing more accurate maps. Two of the Islanders did not include any roads on their maps and the buildings looked a little like islands!

2. Copy drawings. *The student was asked to copy the drawings from a specimen set, taken from Plate 1 of Bender (1938). The drawings involved straight and curved lines, dots, closed and open shapes, geometric and irregular shapes.* This task revealed two types of difficulty. First there was an obvious lack of expertise at drawing, with the Enga students showing this most. The other difficulty was with the criteria to be satisfied, that is, how accurate does a copy need to be? The idea of an *exact* copy did not appear to be familiar to the students.

TABLE 6.1

Number of Omits

English word	Omits	English word	Omits	English word	Omits	English word	Omits
Above	0	Side	1	Edge	1	Fat	3
Below	0	Turn	2	Surface	6	Narrow	2
Right	2	Follow	5	Bottom	1	Thin	1
Left	2	Lead	4	Top	1	Long	0
Far	0	First	1	Over	1	Short	0
Near	0	Second	5	Under	1	Wide	1
Furthest	2	Last	0	Size	6	There	1
Nearest	3	Straight	1	Beyond	5	Regular	5
In front	0	Inside	0	Nowhere	5	Irregular	7
Behind	0	Outside	0	Shadow	1	Shape	10
Opposite	7	Round	6	All around	6	Picture	6
Forwards	6	Sharp	1	Up	2	Pattern	7
Backwards	4	Jagged	8	Down	1	Spiral	9
Zig zag	7	Smooth	6	Here	1	Slope	7
Line	7	Valley	4	Corner	4	Direction	7
Crossroads	5	Hill	0	Deep	0	Horizontal	8
Between	0	Steep	6	High	1	Vertical	7
Middle	0	Flat	3	Tall	0		

3. Spatial vocabulary. *The student was asked to translate a list of 71 English spatial words into his home language.* The student was encouraged to omit a translation that involved several words. Generally only 15 out of the 71 words could be translated by all 12 students. Table 6.1 shows the number of *omits* (omissions) for each word—thus *shape* was the word for which only two of the students had an equivalent word in their own language. Interestingly, the Enga students only had 33 omits in total, fewer than Port Moresby students (76) and Manus (109).

The relative poverty of the Manus languages in this area was indicated by numerous "overlaps," in which one local word was the translation of several English words. For example in one Manus student's language all the following translated into one word: *above, surface, top, over,* and *up.*

4. Models. *The student was asked to make models from drawings using plasticene "corners" and toothpicks. The drawings were similar to those used by Deregowski (1974).* This task revealed the difficulty experienced by the Enga students, particularly of reading depth into a two-dimensional drawing of a three-dimensional object. For example, two Enga students produced two-dimensional (flat) objects when shown the diagrams in Schema 6.3. The oblique convention used to represent part of a cube was not generally known, nor was the foreshortening effect.

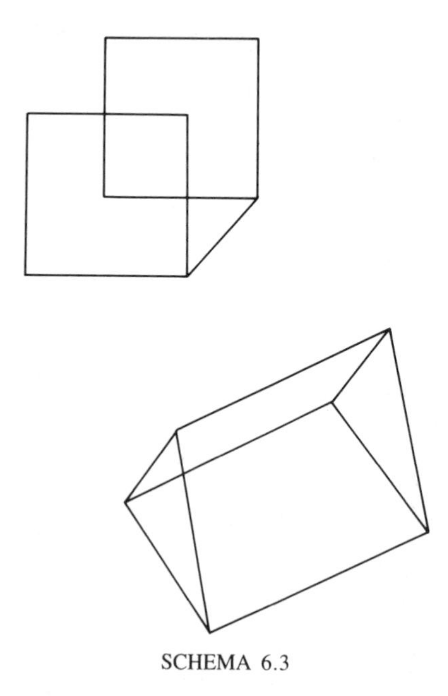

SCHEMA 6.3

5. Imagination. *The student was asked to imagine a situation and then was asked questions about it. For example: think of the number 458. Which digit is on the left? On the right? What would it read backwards? Another example: think of a triangle, label the vertices A, B, and C. Which is the longest side? Which is the shortest side? Largest angle? Now draw it. Other examples: which capital letters could you write in one movement? Which number is opposite 10 on a clock face?* These tasks were generally well done; difficulties occurred not with the imagining, but with the language. For example the most frequent answer to the last question above was "2." Also *edges* was confused with *corners* when a cube and a pyramid were being described in another imagining task.

6. Location cards. *Location cards were based on the diagrams used by Asso and Wyke (1973). They had two or three intersecting lines and a small circle, and the student was asked to say where on the card the circle was drawn.* For example, the diagram in Schema 6.4 produced a few "bottom right hand corner" codings whereas one student who had no real system in his repertoire described it as "diagonally below from the bottom angle to the right." Some drawings caused particular difficulty (see Schema 6.5). One Enga student said, "On top of the diagonal line (pause) right in the middle on top of the vertical line which slopes to the right—no left" (I had previously discovered that for this student *slopes* means *leans*). Another student (from Manus) described the same position as "On the top line drawn horizontal on the northwest direction." Generally this was an extremely revealing task, showing the extent to which some sort of coordinated reference system had been mastered.

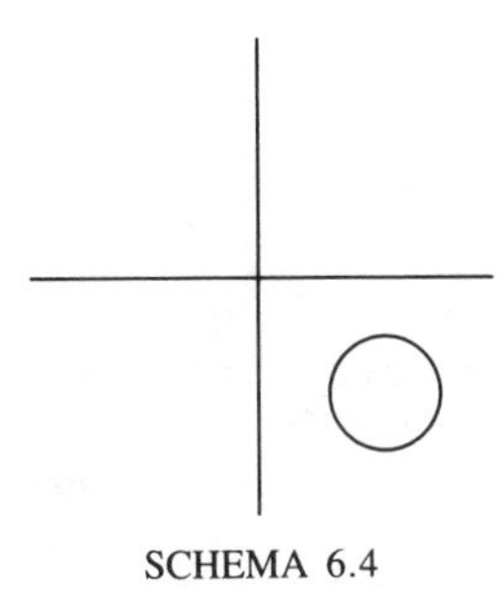

SCHEMA 6.4

7. Square mazes. *The students were asked to trace a route out of six square mazes. These were similar to those in the Porteus maze test.* The general performance on this task was very good. No student failed to solve any maze, although a few false starts were made on the more difficult ones. The task was easily understood and enjoyed by the students.

8. Camera location. *The Eliot–Price test (Professor J. Eliot, University of Maryland) consists of a series of photographs taken of a set of small abstract objects placed on a shaped mat. The student was asked to identify the place from which the camera took the photograph.* This was one of the hardest tasks for the students and it created a great deal of anxiety for many of them. With five students the task was terminated early, which produced much relief. The difficulties centered on the students' unfamiliarity with reading photographs and their inadequate knowledge about cameras. Out of possible total scores of 72 (not all sets were used), and the Manus students scored 51, the Enga 34, and the Port Moresby students 26.

9. Visual memory. *The visual memory task was developed from one used by Kearins (1976). Twelve small objects were presented in a 3 × 4 array for 45 seconds, after which the pieces were removed. The student was asked to replace them in their correct positions. There was no time limit for replacement.* In contrast with the previous task, this one seemed to tap a strength rather than a weakness. Only one student made any mistake (and he was suffering from malaria at the time). Other versions of this task using playing cards, shells, and feathers were more difficult. The Enga and Manus students were generally better at this tasks than were those from Port Moresby.

SCHEMA 6.5

There was rapt attention given to this task by all the students, who in most cases worked slowly and deliberately (except for the easy objects) sometimes taking 15 minutes to put back all the pieces. The typical Westerner seems to prefer to work quickly, in fear of losing the memory. What was therefore interesting to follow up was retention. Certain students were asked to replace the small objects again one day later, and they all succeeded; some succeeded again 1 week later, and one student was successful again 2 weeks later, at which time he corrected a mistake he had made 1 week earlier! This delayed-memory task was clearly not perceived to be unreasonable, and most of the students attempted it as if they were confident of success. Only the urban students, from Port Moresby, had difficulties with retention.

10. Block drawing. *The student was presented with a small wooden block made of 1-cm cubes. Twenty-one cubes were used and the student viewed the block in this orientation presented in Schema 6.6. The student was first asked to sketch the block as it appeared to him. Then he was asked to sketch how it would look to the experimenter sitting on the opposite side of the table.* This task caused difficulty for most of the students though most completed the first part successfully. The second part is more difficult and produced several errors. Of great interest was the fact that in this reverse position all four Enga students drew the configuration in its correct orientation, though some of the drawings did contain inaccuracies.

11. Embedded figures. *The subtest "Form Recognition" from Spatial Test 1 produced by the National Foundation for Educational Research. United Kingdom, was used. Fifteen items (untimed here) required the student to identify which of four simple figures were embedded in the more complex ones.* This was another very difficult task for these students. Those from Port Moresby per-

SCHEMA 6.6

formed much better than the others, and many items were omitted by students who could not see any of the four simple embedded figures. There was much anxiety expressed about this task.

12. Matchbox corners. *The matchbox corners was a subtest from Spatial Test 2 produced by the National Foundation for Educational Research, United Kingdom. A matchbox is drawn, with dotted hidden lines, and a large dot is placed at one corner. Four matchboxes are then shown in different orientations and the student must place the dot on the correct corner of each. Five different sets were used and the task was untimed.* The students enjoyed this task and performed well; eight students made fewer than two errors. The matchbox drawings were easily understood.

13. Word completion. *The word-completion task was a subtest from the Multi Aptitude Test, Psychological Corporation. Eighteen typed English words were presented in varying degrees of obliteration and the student was asked what the original word was.* Despite the fact that English was their third or sometimes fourth language, many words were indentified correctly. According to the validation data (from a sample of graduate students in the United States) all the Port Moresby students, two Enga and one Manus student scored less than 1 SD below the U.S.A mean.

14. Picture completion. *The picture-completion task paralleled the previous one, but used outline drawings from the study by Kennedy and Ross (1975). The drawings showed people, pigs, houses, and so forth with 40–80% obliteration. The student was asked to identify the object represented.* In contrast to the previous task, identification was almost impossible with 80% obliteration. With only 40% obliteration most students were able to recognize the objects. The most difficult objects to identify were two birds and a Papua New Guinea warrior. Clearly, even when the objects were well known, their representations were not.

Analysis of the Results

In terms of the constucts *abilities* and *meanings,* the most obvious difficulty for the students on these tasks was related to their IFI ability, which affected performance on several of the tasks, particularly Task 4 (models), Task 8, and Task 11 (embedded figures). With the last two of these, the anxiety caused by this disability was so acute that the tasks were terminated early. In Task 2 (copy drawings) and Task 10 (block drawing) it was clear that a general lack of experience with drawing "reality" was underlying much of the IFI weakness. The performance on Task 14 (picture completion) showed the symbolic and general nature of the supposedly realistic drawings we use, which makes them not instantly recognizable to someone who has not had a great deal of experience with many types of visual conventions and representations.

Two tasks that involved particular conventions were done sucessfully. Task 7 (square mazes) and Task 12 (matchbox corners). The first has much in common with mapping, and Task 1 (map drawing) was done generally well. Perhaps there is something very basic about the notion of maps, recalling Blaut and Stea's (1974) research. Regarding Task 12 (matchbox corners), the representation was a very clear one of a familiar object; the task seemed really to be testing their VP ability.

These data highlight the need to focus on the IFI ability tested by spatial and goemetric tasks. Although it is often reported that students from non-Western cultures have poor spatial skills, it is likely that unfamiliarity with Western conventions and a weakness at IFI ability generally are to blame. Fortunately, we know from various training studies, reported earlier, that particular conventions are relatively easy to teach; it is also likely that IFI ability can be improved by concentration on the act of representing and on the variety of representations used (see Mitchelmore [1976] for a similar argument.)

The students' performance on tasks that seemed to tap visual processing ability was generally very good. Task 5 (imagination) specifically tested VP ability, and it posed no real problem. However the most impressive aspect of the students' visualizing ability was shown by Task 9 (visual memory). Clearly memory plays a part in visualization, although it is not certain how. For these students, this task was obviously solved *visually*. In other words, they were not using any verbal or symbolic coding to assist their memory. They did not know the names of some shells, and there was no easy way to code the feathers verbally. They said nothing and there were no lip movements. They seemed instead to be "looking" at the remembered image, placing an object in position and then comparing the reality with the memory. Sometimes they would be satisfied, but at other times they would shake their heads, pick up the object, and "look" at their image again.

Other tasks involving VP ability were Task 1 (map drawing), Task 2 (mapping part of the campus), Task 10 (block drawing b) and Task 12 (matchbox corners). Task 7 (square mazes) should also perhaps be included here, particularly because the students used a very visual, Gestalt-like, approach to it rather than an analytic, stepwise method. Also, Task 13 (word completion) tapped much VP ability with these students. They imagined drawing letters, sometimes trying to "see" the whole word and sometimes concentrating on individual letters.

VP ability, therefore, appeared to be a relative strength for these Papua New Guinea students. It was also interesting to see the personal nature of visual processing. With no time pressure, it was possible for students to use whatever method they liked to solve the tasks and all of the tasks in this section could have been solved with analytic methods and verbal or symbolic coding. They chose instead to use visual processing.

Although none of these tasks demanded any great understanding of geometry, with respect to meaning it is reasonable to ask: What was learned about the student's understanding of space and its representation? The map-drawing task has already been discussed but is worth further comment. The students clearly felt at ease with mapping, and later in the interviews, when discussing their village gardens or fishing areas, they would enthusiastically and often spontaneously draw a sketch map with many details. Among the Enga (highland) students, village territorial limits were well known and carefully drawn; the Manus and Port Moresby students, who fished a great deal, drew excellent navigational maps with "lines of sight" clearly shown. The particualr convention used in the maps varied, suggesting that the drawings of maps was not a schooled ability but somehow culturally learned. Their concepts of physical space seemed well developed.

Performance on Task 6 (location cards), on the other hand, showed a distinct lack of a concept of coordinate system and relative position location among some of the students. Others who used phrases such as "top hand side" clearly did have a useful concept, though an unconventional one. Language is, however, a confounding variable. In a place like Papua New Guinea, it is sometimes difficult to separate conceptual limitations from linguistic confusion. As Task 3 (spatial vocabulary) showed, the students' home languages are not particalarly rich in the kinds of spatial words used in Western-type education. There was no systematic attempt, however, to discover the richness of the local languages for dealing with their own spatial ideas. Evidence from one Enga student suggested, for example that his people had several ways to describe slopes of hills.

With over 750 different languages identified in Papua New Guinea (and also upward of 250 different counting systems), the country seems an excellent place to study linguistic–cultural–spatial interactions. Certainly as Jones (1974)has shown, there seem to be considerable gaps in understanding various significant (for mathematics) spatial concepts, and language gaps are surely supporting evidence for this. Berry (1971) suggests also that cultures with strong spatial skills have rich spatial languages, but the evidence from this study does not specifically support that view. In comparing the three groups of students, for example, the Enga students seemed to have the richest spatial language but they were by no means the best at the tasks. They did, however, seem better at visual processing, which could support Berry's hypothesis, although the small numbers and the exploratory nature of the tasks make the testing of any hypothesis impossible.

Jones' (1974) data show how different spatial concepts and understanding can be in an indigenous culture such as that in Papua New Guinea. A series of questions were devised covering many basic topics in mathematics (measuring, classifications, ordering, and counting systems). These comprised a questionnaire that was sent to all the field workers of the Summer Institute of Linguistics in different parts of Papua New Guinea. Respondents were asked to assess how difficult or easy it was to express the sense of a particular (English) sentence in the vernacular where they were working. Regarding measuring, statements concerning length and weight were made relatively easily in the vernaculars, but those regarding height, distance, and speed were more difficult; those about area and volume caused considerable difficulty. Some of the particular responses (translated) illustrate well the differences in concepts:

> [Describing a length measurement:] Put your arm three times and get me a length of wood as long as that.
>
> Only relative weights exist—heavy, very heavy, heavy little bit, not heavy.
>
> The local unit [of distance] is "a day's travel" or "a stage" which is not very precise.
>
> One can only say [in relation to area] "my garden is bigger than yours"—just by looking at it—but not using units of area.
>
> The people would not think of [the volume of] a solid rock and liquid water using the same words. A rock's volume is in a different class from the water's volume.
>
> There is no comparison construction. '*A* runs fast. *B* runs slow."
>
> Height and distance are not abstracts in the language. Neither is heard.
>
> It could be said [that two gardens are equal in area] but it would always be debated.
>
> [Regarding volume of rock and water:] This kind of comparison does not exist, there being no reason for it.

Perhaps the most striking illustration I had of different meanings of spatial concepts was in a conversation with one of the students at the University of Technology:

> *A.J.B.:* How do you find that area of this [rectangular] sheet of paper?
> *Student:* Multiply the length by the width.
> *A.J.B.:* You have gardens in your village. How do your people judge the area of the gardens?
> *Student:* By *adding* the length and width.
> *A.J.B.:* Is that difficult to understand?
> *Student:* No, at home I add, at school I multiply.
> *A.J.B.* But they both refer to area.
> *Student:* Yes, but one is about the area of a piece of paper and the other is about a garden [A.J.B. then drew two different rectangles on paper.]
> *A.J.B.* If these were two gardens, which would you rather have?
> *Student:* It depends on many things, I cannot say. The soil, the shade.

Clearly this student's concern was with two different problems. Size of gardens was a problem embedded in one context, rich in tradition, folklore, and the skills of survival. The other problem, area of rectangular pieces of paper, was embedded in a totally different cultural context. How could they involve the same concept?

The problems of teaching mathematics in less technologically oriented cultures are immense, and it is not the purpose of this paper to tackle them. However, research like this can provide some suggestions for helping with spatial and geometric development, and Bishop (1978a) and Mitchelmore (1976) have both concluded that there is value in creating general spatial training programs. What this review has shown perhaps is that such programs need to be wider than the label "Spatial Training" suggests, and that the meaning and conceptual aspects of space and geometry also need more attention. Perhaps more studies like Jones (1974), like Harris' (1980) similar study with the Australian Aborigines, and like Lewis' (1972) work on Polynesian navigators can increase our general awareness of the environmental and cultural influences on spatial and geometric concepts.

Guidelines for Future Research

This review has been written by someone working in the field of mathematics education, someone who sees empirical research as a way of shedding light on curricula and pedagogical issues by collecting data systematically, by interpreting those data, and by reflecting on their implications in relation to practice and to other research. The search is not necessarily for consistent theory, but for alternative ways of construing phenomena (Kelly, 1955), and for ideas that will enable us to derive better tests, task materials, teaching materials and procedures, and educational practices to enable more children to feel successful and confident in mathematics.

There are many research avenues that could be explored, some suggested by research, some by theory, some by practice, and some indeed by hunch. The focus, however, will be provided by the two main constructs of the review: meanings and abilities.

Meanings

A major theme of this chapter has been the limited value, for mathematics education, of developmental theory for interpreting children's meanings and conceptual understanding in geometry. It would seem essential, therefore, to consider this aspect first in a section devoted to future research directions.

There are many potential influences on meanings and it would certainly be of value to know more about which are actually operative. Cultural influences, which were much of the theme in the section on research in Papua New Guinea, suggest not only that more cross-cultural studies would be of value, but that differences within cultures would also be interesting to explore. We know little of the effects of family, home, and social background on the child's understanding and meaning of spatial ideas. Vollrath (1977) has shown us one effect of colloquial speech and it is very likely that there will be other aspects of "colloquial learning" that will influence children's school learning. Although we may feel that environmental influences are more the province of geographers, it is clear that for many of the concepts associated with measurement (e.g., length, area, time), the child's environment is likely to have an important effect. One feature of environment is specific experiences involving travel, for example, and the role of such events in a child's understanding is relatively unknown.

If the child's local knowledge (in map terms) is quick to develop, as seems to be the case, that local environment can offer many opportunities to the teacher for experiences in mapping, surveying, drawing, and measuring. Perhaps, therefore, too much of our attention in geometry has been focused on small-scale objects and drawings and we have ignored the importance of the larger space surrounding the child and its effect on the child's spatial development.

As well as activities within the larger environment, it is important to consider the place of activities that scale down that environment. Modeling, photographing, and drawing are all examples of scaling down, and relationships among all these can be very revealing for children; for example, where was the model cameraman standing on this model in order to take this photograph? Not only is this likely to offer meanings to lines, angles, and projections, it may well also assist with developing children's VP ability.

Considering children's interaction with their environment, it is still important for research to explore features such as the extent of active manipulation necessary, and the value of discussion with other children. How children's conceptual understanding in geometry is affected by experimentation and discussion is still not known. Studies of small-group interactions in geometry experiments would surely be very revealing, particularly where limited apparatus was provided so that some children were able to carry out manipulations while others merely watched. Also, ideas of spatial orientation would certainly be enhanced by putting children in small groups around certain objects.

Discussion has, of course, at least two features—the reception by learners of "viewpoints" other than their own, and the opportunity for learners to "talk themselves into understanding." Both aspects are in need of some analysis and both could shed important light on issues surrounding the "individualization of instruction," for example.

Abilities

It would be fruitful to explore further the dichotomy suggested here between the ability for interpreting figural information and the ability for visual processing. The first certainly seems to bear a strong relationship to the knowledge of visual conventions, but the suggestion here is that there is a more general ability for IFI.

The best analogy is with reading. One learns a certain orthography but through increasing experience with reading one is able to read in orthographies that one has not been taught specifically. It is of course interesting that reading and writing are often taught simultaneously—the argument being that by learning the process of writing one is also learning about written forms. Likewise one would expect that practice with drawing and using different visual forms would develop IFI ability, which would enable one to interpret other conventions than those one had been specifically taught. Research on the development of IFI ability, on generalizability, and on transfer would be extremely interesting. The many studies that have shown that individual conventions can be taught form an excellent basis for more extended work.

It would be particularly valuable if some of the standard spatial tests were used in these investigations because it is likely that in previous spatial training studies that have shown increases in spatial test scores, either IFI ability has been improved or the particular figural convention used in that spatial test has been taught.

The major question about VP ability is whether it is teachable. As was indicated earlier, the research is not conclusive and it is likely that the difficulty lies with the assessment of VP as much as with its development. As has been pointed out, because it is a preferred mode of processing it is something that may only be testable individually using a variety of clinical tasks. Krutetskii (1976) has given a lead and Lean and Clements (1981) indicate another approach. What is clear is that figural *and* nonfigural stimuli need to be used.

It would be valuable to reassess some of the factor analytic work involving spatial ability to clarify the nature of VP. For example, one could readminister the same set of tests as in a previous factor study, and include some nonfigural but visually provoking tasks as well. Changes in factor structure might be revealing.

Training studies are also needed here, particularly using individualized techniques, and it is important to know when in a child's development might be the optimal time for training. It would also be useful to explore the specificity or generality of VP ability. Does VP training within a mathematical context have a greater mathematics achievement payoff than it does within a more general context? If it is done within geometry does it transfer to arithmetic and algebra? What individual differences among pupils interact with the training? For example, it is more likely to be of benefit to less-able than to more-able pupils as the latter, by definition, have presumably developed successful processing abilities. Can VP training experiences help to redress the reported imbalance between boys and girls in mathematics achievement?

Conclusion

It is apparent that there is no shortage of research avenues to explore in this field. Despite the many ideas being experimented with in schools and the many stimulating articles being written about geometry teaching (see, for example, Fielker, 1979), we are still relatively ignorant about the learning of spatial and geometric ideas. The hope is that by concentrating research efforts on the interpretations of children's meanings and on the development of IFI and VP abilities we can generate ideas that will help to make geometry teaching more imaginative and more successful than it is at present. The prospects look favorable.

References

Asso, D., & Wyke, N. Verbal descriptions of spatial relations in line drawings by young children. *British Journal of Psychology,* 1973, *64,* 233–240.

Ausubel, D. P. *Educational psychology: A cognitive view.* New York: Holt, Rinehart and Winston, 1968.

Banwell, C. S., Saunders, K. D., & Tahta, D. G. *Starting points.* Oxford: University Press, 1972.

Barakat, M. K. A factorial study of mathematical abilities. *British Journal of Psychology: Statistics Section,* 1951, *4,* 137–156.

Bender, L. *A visual motor Gestalt test and its clinical use.* New York: American Orthopsychiatric Association, 1938.

Berry, J. W. Ecological and cultural factors in spatial perceptual development. *Canadian Journal of Behavioural Science,* 1971, *3,* 324–336.

Bishop, A. J. Use of structural apparatus and spatial ability: A possible relationship. *Research in Education.* 1973, *9,* 43–49.

Bishop, A. J. Visual mathematics. *Proceedings of the ICMI–IDM Regional Conference on the Teaching of Geometry, IDM* West Germany: University of Bielefeld. 1974.

Bishop, A. J. *Developing spatial abilities* (Mathematics Education Centre Report No 4). Mathematics Education Centre, University of Technology, Lae, Papua New Guinea, 1978. (a)

Bishop, A. J. *Spatial abilities in a Papua New Guinea context* (Mathematics Education Centre Report No 2). Mathematics Education Centre, University of Technology, Lae, Papua New Guinea, 1978. (b)

Bishop, A. J. *Spatial and mathematical abilities: A reconciliation.* Paper presented at the Conference on Mathematical Abilities at the University of Georgia, Athens, June 12–14, 1980.

Blade, M. F., & Watson, W. S. Increase in spatial visualization test scores during engineering study. *Psychological Monographs,* 1955, *69,* (1, Whole No. 397).

Blaut, J. M., & Stea, D. Mapping at the age of three. *The Journal of Geography,* 1974, *73,* 5–9.

Brinkmann, E. H. Programmed instruction as a technique for improving spatial visualization. *Journal of Applied Psychology,* 1966, *50,* 179–184.

Clements, M. A. *Spatial ability, visual imagery, and mathematical learning.* Paper presented at the Annual Meeting of the American Educational Research Association, Los Angeles, 1981.

Cohen, L. *Evaluation of a technique to improve space perception through construction of models by students in a course in solid geometry.* Unpublished doctoral dissertation, Yeshiva University, New York, 1959.

Connor, J. M., Serbin, L. A., & Freeman, M. Training visual–spatial ability in EMR children. *American Journal of Mental Deficiency,* 1978, *83,* 116–121.

Cox, M. V. Teaching perspective ability to five-year-olds. *British Journal of Educational Psychology,* 1977, *47,* 312–321.

Deregowski, J. B. Teaching African children pictorial depth perception: In search of a method. *Perception,* 1974, *3,* 309–312.

Drauden, G. M. *Training in spatial ability.* Unpublished doctoral dissertation, University of Minnesota, 1980.

Eliot, J. *The effects of age and training upon children's conceptualization of space.* Unpublished doctoral dissertation, Stanford University, 1966.

Fielker, D. S. Strategies for teaching geometry to younger children. *Educational Studies in Mathematics,* 1979, *10,* 85–133.

Fischbein, E., Tirosh, D., & Hess, P. The intuition of infinity. *Educational Studies in Mathematics,* 1979, *10,* 3–40.

Fishbein, H. D., Lewis, S., & Keiffer, K. Children's understanding of spatial relations: Coordination of perspectives. *Developmental Psychology,* 1972, *7,* 21–33.

Fuson, K., & Murray, C. The haptic-visual perception, construction and drawing of geometric shapes by children aged two to five: A Piagetian extension. In R. Lesh & D. Mierkiewicz (Eds.), *Recent research concerning the development of spatial and geometric concepts.* Columbus, Ohio: ERIC/SMEAC, 1978.

Geeslin, W. E., & Shar, A. O. An alternative model describing children's spatial preferences. *Journal for Research in Mathematics Education,* 1979, *10,* 57–68.

Genkins, E. F. The concept of bilateral symmetry in young children. In M. F. Rosskopf (Ed.), *Children's mathematical concepts.* New York: Teachers' College, Columbia University, 1975.

Goodnow, J. *Children's drawing.* London: Fontana, 1977.

Guay, R. B., McDaniel, E. D., & Angelo, S. Analytic factors confounding spatial ability measurement. In R. B. Guay & E. D. McDaniel (Eds.), *Correlates of performance on spatial aptitude testes.* Purdue University, U.S. Army Research Institute for the Behavioral and Social Sciences, 1978.

Hadamard, J. *The psychology of invention in the Mathematical field.* London: Dover, 1945.

Hamza, M. Retardation in mathematics among grammar school pupils. *British Journal of Educational Psychology,* 1952, *22,* 189–195.

Harris, L. T. Sex-differences in spatial ability: Possible environmental, genetic and neurological factors. In M. Kinsbourne (Ed.), *Assymmetrical functions of the brain.* Cambridge: Cambridge University Press, 1978.

Harris, P. *Measurement in tribal aboriginal communities,* Australia: Northern Territory Department of Education, 1980.

Hart, R. A., & Moore, G. T. The development of spatial cognition: A review. In R. M. Downs & D. Stea (Eds.), *Image and environment: Cognitive mapping and spatial behaviour.* London: Arnold, 1973.

Hughes, M., & Donaldson, M. Hiding games and coordination of viewpoint. *Educational Review,* 1979, *31,* 133–140.

Jones, J. *Quantitative concepts, vernaculars, and education in Papua New Guinea* (E.R.U. Report 12) Educational Research Unit, The University, Port Moresby, Papua New Guinea, 1974.

Kearins, J. Skills of desert children. In G. E. Kearney & D. W. McElwain (Eds.), *Aboriginal cognition: Retrospect and prospect.* New Jersey: Humanities Press Inc., 1976.

Kelly, G. A. *The psychology of personal constructs* (Vols. 1 and 2). New York: Norton, 1955.
Kennedy, J. M., & Ross, A. S. Outline picture perception by the Songe of Papua. *Perception,* 1975, *4,* 391–406.
Krutetskii, V. A. *The psychology of mathematical abilities in schoolchildren.* Chicago: University of Chicago Press, 1976.
Küchemann, D. E. Children's difficulties with single reflections and rotations. *Mathematics in School,* 1980, *9*(2), 12–13.
Lancy, D. F. (Ed.). The Indigenous Mathematics Project. *Papua New Guinea Journal of Education* (Special Issue), 1978, *14.*
Laurendeau, M., & Pinard, A. *The development of the concept of space in the child.* New York: International Universities Press, 1970.
Leach, M. L. The effect of training on the pictorial depth perception of Shona children. *Journal of Cross Cultural Psychology,* 1975, *6,* 457–470.
Lean, G. A. *Spatial training studies and mathematics education: A review.* Unpublished paper, Department of Education, University of Cambridge, 1981.
Lean, G. A., & Clements, M. A. Spatial ability, visual imagery, and mathematical performance. *Educational Studies in Mathematics,* 1981, *12,* 267–299.
Lesh, R. Transformation geometry in elementary school: some research issues. In J. L. Martin (Ed.), *Space and geometry.* Columbus, Ohio: ERIC/SMEAC, 1976.
Lewis, D. *We, the navigators.* Honolulu: The University Press of Hawaii, 1972.
MacFarlane, Smith, I. *Spatial ability: Its educational and social significance.* London: University of London Press, 1964.
Martin, J. L. *Space and geometry: Papers from a research workshop.* Columbus, Ohio: ERIC/-SMEAC, 1976.
McCloskey, P. The facilitation of spatial ability and problem-solving in adolescent pupils through learning in design. *Educational Review,* 1979, *31,* 259–267.
McGee, M. G. *Human spatial abilities.* New York: Praeger, 1979.
Michael, W. P., Guilford, J. P., Fruchter, B., & Zimmerman, W. S. Description of spatial visualization abilities. *Educational and Psychological Measurement,* 1957, *17,* 185–199.
Mitchelmore, M. C. Cross-cultural research on concepts of space and geometry. In J. L. Martin (Ed.), *Space and geometry.* Columbus, Ohio: ERIC/SMEAC, 1976.
Mitchelmore, M. C. Prediction of developmental stages in the representation of regular space figures. *Journal for Research in Mathematics Education,* 1980, *11,* 83–93. (a)
Mitchelmore, M. C. *The representation of parallels and perpendiculars in children's drawing.* Paper presented at the Fourth International Congress on Mathematics Education. Berkeley, California, August, 1980. (b)
Moses, B. E. *The effects of spatial instruction on mathematical problem-solving performance.* Paper presented at the Annual Meeting of the American Educational Research Association, San Francisco, April, 1979.
Murray, J. E. Analysis of geometric ability. *Journal of Educational Psychology,* 1949, *40,* 118–124.
Nussbaum, J., & Novak, J. D. An assessment of children's concept of the earth utilizing structured interviews. *Science Education,* 1976, *60,* 535–550.
Perham, F. An investigation into the effects of instruction on the acquisition of transformation geometry concepts in first grade children and subsequent transfer to general spatial ability. In R. Lesh and D. Mierkewicz (Eds.), *Recent research concerning the development of spatial and geometric concepts.* Columbus, Ohio: ERIC/SMEAC, 1978.
Piaget, J., & Inhelder, B. *The child's conception of space.* London: Routledge and Paul, 1956.
Piaget, J., & Inhelder, B. *Mental imagery in the child.* London: Routledge and Paul, 1971.
Piaget, J., Inhelder, B., & Szeminska, A. *The child's conception of geometry.* London: Routledge and Paul, 1960.

Prigge, G. E. The differential effects of the use of manipulative aids on the learning of geometric concepts by elementary school children. *Journal for Research in Mathematics Education*, 1978, *9*, 361–367.

Richardson, A. Verbalizer, visualizer, a cognitive style dimension. *Journal of Mental Imagery*, 1977, *1*, 109–126.

Robinson, E. Mathematical foundations of the development of spatial and geometrical concepts. In J. L. Martin (Ed.) *Space and geometry*. Columbus, Ohio: ERIC/SMEAC, 1976.

Rosskopf, M. F. (Ed.), *Children's mathematical concepts: Six Piagetian studies in mathematics education*. New York: Teachers College, Columbia University, 1975.

Schultz, K. Variables influencing the difficulty of rigid transformations during the transition between the concrete and formal operational stages of cognitive development. In R. Lesh & D. Mierkiewicz (Eds.), *Recent research concerning the development of spatial and geometric concepts*. Columbus, Ohio: ERIC/SMEAC, 1978.

Shumway, R. J. (Ed.). *Research in mathematics education*. Reston, Virginia: National Council of Teachers of Mathematics, 1980.

Smock, C. D. Piaget's thinking about the development of space concepts and geometry. In J. L. Martin (Ed.), *Space and geometry*. Columbus, Ohio: ERIC/SMEAC, 1976.

Smothergill, D. W., Hughes, F. P., Timmons, S. A., & Hutko, P. Spatial visualizing in children. *Developmental Psychology*, 1975, *11*, 4–13.

Spearman, C. E. The abilities of man: Their nature and measurement. *London: Macmillan, 1927.*

Taylor, C. C. A study of the nature of spatial ability and its relationship to attainment in geography. Unpublished Ph.D. Thesis, University of Durham, Durham, U.K., 1960.

Thurstone, L. L. Primary mental abilities. *Psychometric Monographs*, 1938, *1*.

Tolman, E. C. Cognitive maps in rats and men. *Psychological Review*, 1948, *55*, 189–208.

Vladimirskii, G. A. An experimental verification of a method and system of exercises for developing spatial imagination. In J. Kilpatrick & I. Wirszup (Eds.), *Soviet studies in the psychology of learning and teaching mathematics* (Vol. 5). Stanford, California: School Mathematics Study Group, 1971.

Vollrath, H. J. The understanding of similarity and shape in classifying tasks. *Educational Studies in Mathematics*, 1977, *8*, 211–224.

Wagman, H. G. The child's conception of area measure. In M. F. Rosskopf (Ed.), *Children's mathematical concepts*. New York: Teachers' College, Columbia University, 1975.

Werdelin, I. *Geometrical ability and the space factors in boys and girls*. Lund, Sweden: Gleerups, 1961.

Werner, H. *Comparative psychology of mental development*. New York: International Universities Press, 1964.

Wolfe, L. R. *The effects of space visualization training on spatial ability and arithmetic achievement of junior high school students*. Unpublished doctoral dissertation, State University of New York at Albany, 1970.

Wood, R. Sex differences in mathematics attainment at G.C.E. Ordinary level. *Educational Studies*, 1976, *2*, 141–160.

Wrigley, J. The factorial nature of ability in elementary mathematics. *British Journal of Educational Psychology*, 1958, *1*, 61–78.

Zykova, V. I. The psychology of sixth-grade pupils' mastery of geometric concepts. In J. Kilpatrick & I. Wirszup (Eds.), *Soviet studies in the psychology of learning and teaching mathematics* (Vol. 1). Stanford, California: School Mathematics Study Group, 1969.

CHAPTER 7

Van Hiele–Based Research*

Alan Hoffer

Introduction

Van Hiele–based research and development deals with the interaction between teaching and learning, as well as with an analysis of each. Past publications (e.g., Wirszup, 1976) have communicated the work of the van Hieles in terms of a stratification of human thought as a sequence of levels of thinking into which people may be classified. Thought levels comprise only one of the three main components of the van Hiele model. The other components are insight and phases of learning.

Insight

The van Hieles were interested at the outset in finding ways to develop insight in their students. They define *insight* as follows (e.g., van Hiele, 1973): A person shows insight if the person (*a*) is able to perform in a possibly unfamiliar situation; (*b*) performs competently (correctly and adequately) the acts required by the situation; and (*c*) performs intentionally (deliberately and consciously) a method that resolves the situation. To have insight, students understand what they are doing, why they are doing it, and when to do it. They can apply their knowledge in order to solve problems.

*This research was supported in part by the National Science Foundation under grant number SED-7920568. Any opinions, findings, and conclusions expressed in this report are those of the authors and do not necessarily reflect the views of the National Science Foundation.

ISBN No. 0-12-444220-X

Thought Levels

The *thought levels* attached to the learning of a particular topic are inductive in nature. (*a*) At level $n - 1$ certain limited versions of objects are studied. Some relationships are stated explicitly about the objects; however, there are other relationships, possibly quite accessible, that are not stated explicitly. (*b*) At level n the objects studied are now the statements that were explicitly made at level $n - 1$ as well as explicit statements that were only implicit at level $n - 1$. In effect, the objects at level n consist of extensions of the objects at level $n - 1$.

One major purpose for distinguishing the levels is to recognize obstacles that are presented to students. If students who are thinking at level $n - 1$ confront a problem that requires vocabulary, concepts, or thinking at level n, they are unable to make progress on the problem, with expected consequences such as frustration, anxiety, and even anger.

Phases of Learning

To move students from one thought level to the next within a subject, the van Hieles proposed a sequence of five *phases of learning,* a prescription for organizing instruction. The van Hiele phases may be compared to what Polya refers to as the *principle of consecutive phases,* consisting of exploration, formalization, and assimilation (Polya, 1965). In a later section we compare the van Hiele phases with the learning cycle of Dienes (1963). A common element in these and other formulations (see also Greenwood & Anderson, Note 1) is the continuous manner in which ideas are generated, refined, and extended by the students.

The van Hiele phases are entitled *inquiry, directed orientation, expliciting, free orientation,* and *integration.* These are described in the next section.

Work of the Van Hieles

As high school teachers in the Netherlands, Pierre van Hiele and Dina van Hiele-Geldof were troubled by the way their students performed in geometry. It was while studying some of the works of Piaget that Pierre van Hiele formulated his system of thought levels in geometry. He noticed that, as is apparent in some of Piaget's interviews, the problems or tasks that are presented to children often require a knowledge of vocabulary or properties beyond the children's level of thinking. If teaching occurs at a level above that of the student, the material is not assimilated properly into long-term memory. In the words of Freudenthal (1973, p. 130):

> As long as the child is not able to reflect on its own activity, the higher level remains inaccessible. The higher level of operation can then, of course, be taught as algorithm though with little lasting consequence. This has been proved by the failure of teaching fractions.

In 1957, the van Hieles, at the University of Utrecht, completed companion doctoral dissertations to evolve a structure for and experiment with thought levels for the purpose of helping students develop insight into geometry. P. van Hiele (1957) formulated the scheme and psychological principles. D. van Hiele-Geldof (1957) focused on the didactics experiment to raise students' thought levels. Nearly half of D. van Hiele-Geldof's dissertation consists of a detailed and exciting log of the teaching experiments. (For brief descriptions in English, see van Hiele & van Hiele-Geldof, 1958, or Freudenthal, 1973.)

Questions surrounding the teaching of geometry were popular in the Netherlands in the 1950s (Freudenthal, 1958), and the van Hieles were actively involved. P. van Hiele's (1959) article delineating the thoughts levels and the phases of learning attracted the immediate attention of Soviet psychologists.

Here is a capsule version of the thought levels as they apply to geometry:

Level 0 Students recognize figures by their global appearance. They can say *triangle, square, cube,* and so forth, but they do not explicitly identify properties of figures.

Level 1 Students analyze properties of figures: "rectangles have equal diagonals" and "a rhombus has all sides equal," but they do not explicitly interrelate figures or properties.

Level 2 Students relate figures and their properties: "every square is a rectangle," but they do not organize sequences of statements to justify observations.

Level 3 Students develop sequences of statements to deduce one statement from another, such as showing how the parallel postulate implies that the angle sum of a triangle is equal to 180°. However, they do not recognize the need for rigor nor do they understand relationships between other deductive systems.

Level 4 Students analyze various deductive systems with a high degree of rigor comparable to Hilbert's approach to the foundations of geometry. They understand such properties of a deductive system as consistency, independence, and completeness of the postulates.

To help students raise their thought levels, the van Hieles specified a sequence of phases that moves from very direct instruction to the students' independence of the teacher. The phases of learning in this plan are as follows:

Phase 1: inquiry The teacher engages the students in (two-way!) conversations about the objects of study. The teacher learns how the students interpret the words and gives the students some understanding of the topic to be studied. Questions are raised and observations made that use the vocabulary and objects of the topic and set the stage for further study.

Phase 2: directed orientation The teacher carefully sequences activities for student exploration by which students begin to realize what direction the study is taking, and they become familiar with the characteristic structures. Many of the activities in this phase are one-step tasks that elicit specific responses.

Phase 3: expliciting The students, building from previous experiences, with minimal prompting by the teacher, refine their use of the vocabulary and express their opinions about the inherent structures of the study. During this phase, the students begin to form the system of relations of the study.

Phase 3 has been incorrectly translated as *explanation* by other writers. It is essential here that students make the observations explicitly rather than receive lectures (explanations) from the teacher.

Phase 4: free orientation The students now encounter multistep tasks, or tasks that can be completed in different ways. They gain experience in finding their own way or resolving the tasks. By orienting themselves in the field of investigation, many of the relations between the objects of the study become explicit to the students.

Phase 5: integration The students now review the methods at their disposal and form an overview. The objects and relations are unified and internalized into a new domain of thought. The teacher aids this process by providing global surveys of what the students already know, being careful not to present new or discordant ideas.

At the close of the fifth phase, the new level of thought is attained.

After his wife's untimely death, P. van Hiele continued to teach at the secondary level in the Netherlands and to develop textual materials (see van Hiele, 1979). He published several articles in English, French, German, and Nederlands. The contents of several of these articles were organized in the book *Begrip en Inzicht* (van Hiele, 1973), which is the best single reference for van Hiele's ideas.

More recently, van Hiele's focus has been on *structure*, especially as it pertains to didactics. He gave a brief presentation (van Hiele, Note 2) of these ideas during the research reporting session prior to the 1980 Annual Meeting of the National Council of Teachers of Mathematics in Seattle, Washington. He has just completed a book on structure that is to appear soon in the Netherlands. The structures that van Hiele considers are not restricted to the algebraic or geometric structures that were the organizational themes of, for example, the School Mathematics Study Group (SMSG). Van Hiele considers structures that are deeply involved in learning and teaching and that form a basis for solving problems. For example, a piece of squared grid paper provides an underlying structure for discussing plane and solid geometric figures as well as for working with fractions, areas, and so forth.

Soviet Studies

Soviet psychologists and educators have been active for nearly a half century in their study of how to improve their curriculum and of how students learn (see Kilpatrick & Wirszup, 1969–1977). The Soviets were attracted immediately to the van Hiele ideas. Van Hiele's 1959 paper is referred to extensively in a 1963 report by A. M. Pyshkalo, which was revised in 1968. Student difficulty with geometry was as common in the Soviet Union as it was and is in the United States. Pyshkalo (1968, p. 4) asks, "Why is it that many children who successfully master most subjects in the school curriculum 'get nowhere' in the study of geometry?" He gives a partial answer:

> It is well known that geometry instruction in the schools begins late, and then takes up measuring right away, thus passing over the qualitative phase of transforming spatial operations into logical ones. It is also known that the children's development of geometric operations proceeds in the opposite direction—that is, their first geometric operations are not quantitative but qualitiative. Therefore, in the system which we propose, the pupils' familiarization with geometric objects begins with the formation of qualitative geometric operations (the study of shape, mutual position, relations, etc.); only somewhat later are quantitative operations (measurements) gradually developed. Such an approach—as is verified by experimental data—insures the possibility of an earlier beginning of the study of geometry [Pyshkalo, 1968, p. 23].

As a basis for developing a new 8-year program in geometry, Pyshkalo refers to and quotes extensively from van Hiele's 1959 paper. (Pyshkalo numbers the levels 1 through 5 instead of 0 through 4.) The Soviets used the levels to analyze the student materials in Grades 1 through 8 (ages 7–15) that were being used in the Soviet Union in 1960. They also used the levels in research studies. The major findings were as follows (Pyshkalo, 1968):

1. In the first five grades, a significant number of children were perceiving figures as "wholes."
2. Pupils stayed at the first level of geometric development for a considerable time. By the end of Grade 5, only 10–15% reached the second level (van Hiele Level 1), which is needed as a basis for further study of geometry.
3. With respect to pupil understanding of solids, the delay was even longer, that is, until Grade 7.
4. Pupils' development was altered by purposeful study of geometric material. Familiarizing second graders with solids enabled them to reach the

> second level (van Hiele Level 1), surpassing the progress of seventh graders in the traditional school.

The preliminary studies (1960–1964) led the Soviets to the development of "a single continuous line of geometric development for the pupils" (Pyshkalo, 1968, p. 23). Pyshkalo (1968, p. 34) describes the results of the experimental course as follows:

> Our data enable us to establish criteria for choosing geometry material such that by the end of grade 3, *all* pupils could reach the second level of geometric development, having established a sufficiently stable stock of geometric notions. This permits them to advance to a higher stage in the study of geometry by grades 4–5.

In apparent reference to the higher van Hiele level, Pyshkalo (1968, p. 336) states: "In experiments in recent years it has been shown that the capacity in children of primary school age for ordered thinking is significantly higher than has usually been believed and higher than the level at which traditional instruction begins."

Preliminary Work in the United States

Professor Isaak Wirszup at the University of Chicago formally introduced the van Hiele ideas to American audiences in 1974 at the Annual NCTM Meeting in Atlantic City. Wirszup described breakthroughs in the teaching of geometry and reported on the Soviet studies and the work of the van Hieles as published in their 1959 paper. At about the same time, Professor Hans Freudenthal's (1973) comprehensive work, *Mathematics as an Educational Task,* provided an example of the thought levels in the learning of mathematical induction. Indeed, he asserts that mathematical induction actually developed along the levels. He states, "History moved according to these levels" (Freudenthal, 1973, p. 123).

In spite of the earlier appearance of Freudenthal's book, it was Wirszup's presentation and subsequent publication (1976) of his talk that attracted the attention of American educators to the van Hiele's work, several of whom were captured by the simplicity of the ideas.

It is quite natural when working with the level structure to form a collection of questions and problems and to attempt to identify the minimum thought level that is required to answer each question or solve each problem. In working with children, it is discouraging how often one overestimates the level; that is, students are usually at a much lower thought level than one expects, considering their ages. It is interesting that college students, even upper division mathematics majors, often respond at Level 1. It is not uncommon for someone to ask, "Is a square also a rectangle?'' Few students understand the difference between a definition, a postulate, and a theorem.

Even the most cursory analysis of textual material provides some glaring surprises. One is shocked by the inconsistency of level assumptions that are made within a textbook series from the same publisher. One popular American series, for example, had several Level 1 questions in the Grade 4 book, and almost entirely Level 0 questions in the Grade 5 book. To compound the situation, one can find in informal surveys that a large percentage of elementary school teachers skip the chapters on geometry because they do not consider geometry a ''basic skill.'' The situation is not much better at the junior high school grades because the geometry experiences that are offered there seem inadequate to prepare students for the thought levels that are demanded in the high school geometry course.

About the same time Wirszup introduced the van Hiele ideas in America, a project was funded by the National Science Foundation to develop resource materials for middle school mathematics teachers. One of the resources was devoted to geometry and visualization (Mathematics Resource Project, 1978a), and a companion book discussed appropriate didactic principles (Mathematics Resource Project, 1978b). The intent was to present a collection of activities from which teachers could select in order to help students round out the thought level on which they were operating and to prepare them for the next higher level.

At the secondary level, the textbook, *Geometry, A Model of the Universe* (Hoffer, 1979), was written with the van Hiele structure in mind, but with the limitation of not departing too radically from the traditional 1-year geometry course. The intent of this text was to devote the first semester of the ''geometry year'' to activities and investigations that would prepare students to work with a deductive system (van Hiele Level 3). Formal proofs were reserved for the second semester of the course. The materials were tested during the 1976–1977 school year, and at the end of the year the teachers involved composed a proof test that was administered to students in the experimental and traditional classes. No pretest was given, nor were comparisons made among students in the various classes. The teachers who participated in this informal study were pleased that students in the experimental sections could write original proofs at least as well as students in the regular classes. The teachers believed that the experimental classes ''learned more geometry,'' especially related to the topics of area, volume, and transformations. One teacher who had been hesitant about postponing formal proof in this way became quite enthusiastic with the approach. When the textbook was published, she used it a second year and reported that, on a problem-solving test, students using the approach outperformed students in the honors class who used one of the well-established textbooks that devote nearly a full year to proof (Wheeler, Note 3).

At this time there is very little published in English that describes the results of research related to the van Hiele model. Information undoubtedly will be published as a result of current research projects described in the next section.

Recent Studies in the United States

There have been three major research studies that deal with the van Hiele model in geometry. These studies address three areas in which research is needed to better understand structures in children (Coxford, 1978). Coxford calls for (*a*) carefully documented longitudinal case studies of children; (*b*) the gathering of data by age sampling to compare cognitive structures and developmental stages; and (*c*) an analysis of the effects of instruction on cognitive structures.

Following brief descriptions of the three projects are some of their preliminary results and observations from informal studies.

The Oregon Project: Assessing Children's Development in Geometry

The project was sponsored by the National Science Foundation from September, 1979 through February, 1982. The staff consisted of William Burger (Director), Oregon State University; Alan Hoffer, University of Oregon; Bruce Mitchell, Michigan State University; and Michael Shaughnessy, Oregon State University. The purpose of the study was to investigate the extent to which the van Hiele levels serve as a model to access student understanding of geometry (Burger, Note 4).

The project developed two sequences of tasks and companion scripts that the four researchers administered to students in Grades 1 through 12. The interviewer audio-taped the student's responses in two sessions of approximately 45 minutes each and continued through the sequences as long as the student was able to respond. The beginning tasks in a sequence were open and exploratory in nature whereas the later tasks were more specific and required a higher thought level. For example, in the sequence on quadrilaterals students were first asked to "draw several four-sided figures that are different in some way" so the researchers could identify what attributes the children would vary. Near the end of the sequence the children were asked if they could explain why (i.e., prove) the opposite sides of a parallelogram are equal in length and, if they had heard the terms, to explain the difference between a theorem, postulate, and definition. There is a similar sequence for triangles.

After trials with several students, the tasks and scripts were revised and administered to over 70 students in Oregon, Ohio, and Michigan. The interviewers and outside evaluators analyzed student responses and written materials looking for anecdotal descriptors of student perception and comprehension as well as placement of the student's responses within the level structure. The evaluations were then analyzed statistically.

The Brooklyn Project: Geometric Thinking among Adolescents in Inner City Schools

The Brooklyn project was sponsored by the National Science Foundation for the duration November, 1979 through January, 1982. The staff consisted of David Fuys, Dorothy Geddes (Director), C. James Lovette, and Rosamond Tischler, all of Brooklyn College.

The purposes of the study were to determine whether the van Hiele model describes how students learn geometry and how the model can be interpreted in the context of an American curriculum and environment—in particular, in the context of teachers and minority students in Grades 6 and 9 in a large urban area (Geddes, Note 5).

The project evaluated the geometric content of several school textbook series according to the level of the activities in the van Hiele model. The main part of the project consisted of the development and implementation of four instructional modules that were based on the van Hiele levels and phases. The modules, patterned after the experiments in Dina van Hiele's thesis, were used to obtain data on the effects of instruction on students' performances.

A videotape was made of each interviewer (not necessarily one of the investigators on the project) working through the modules with the students on an individual basis. Several sequences in the modules utilize concrete objects that appear to offer definite instructional advantages and aid a reviewer of the tape in identifying the student's thought processes.

The Chicago Project: Cognitive Development and Achievement in Secondary School Geometry

The project was sponsored by the National Institute of Education from July 1979 through June 1982. The staff consisted of Roberta Dees, Sharon Senk, and Zalman Usiskin (Director), all of the University of Chicago.

The purpose of the study was to determine the effects of the student's stage of cognitive development and performance on a test of mathematics prerequisites on student achievement in standard geometry concepts and proof (Usiskin, Note

6). Approximately 2900 students in high school geometry courses (honors through redmedial) in California, Florida, Massachusetts, Michigan, Oregon, and Illinois were included in the study.

The project used four tests, three of which were constructed by the staff. One multiple-choice test (P) deals with knowledge that the staff identified to be prerequisite to the study of high school geometry. A second multiple-choice test (V) deals directly with the van Hiele levels in which each question is associated with the minimum van Hiele level that is required to answer the question correctly. A third test (Pr) deals with proof-writing ability; each student was given two problems involving completion of parts of proofs and four traditional proofs. The project devised a rating scale by which three people rate the six problems on each student's paper. To measure geometry achievement, the project used a standardized multiple-choice test (A) that is available commercially.

Tests P and V were administered near the beginning of the school year (September, 1980). Near the end of the year (May, 1981) tests P and V were again administered, together with tests A and Pr. The project staff then analyzed the data and established various correlations among the test results.

Observations

More information about the van Hiele model will be forthcoming as the three projects publish their final results. The projects respond to each of the needs identified by Coxford:

1. The sequences of tasks and scripts of the Oregon project relate to Coxford's call for longitudinal case studies. The Oregon project found that the same script could be used with children of all age levels. Hence, the project essentially produced a prototype instrument to be used for a longitudinal study.
2. The identification of students' thought levels and performances in geometry for nearly 3000 high school geometry students contributes to Coxford's call for gathering data to compare cognitive structures and developmental stages.
3. The modules that were developed and tested by the Brooklyn project address Coxford's call for an analysis of the effects of instruction on cognitive structures.

One easily envisions a single long-term study that unites the aspects of the three projects by following a group of students over several years, periodically using the Oregon interviews, interspersed with the Brooklyn instructional modules, evaluated at the high school geometry age via the Chicago battery of tests, followed by posttests and interviews near the end of the students' high school work.

The van Hiele model provides us with a peephole through which we can use our mathematical eye to view children's interaction with mathematics. The observations that follow are simply glimpses of real-life phenomena, outgrowths of the three U.S. projects, and the author's experiences.

Language

The Oregon and Brooklyn projects found that starting in the elementary grades, students are familiar with several geometric words, such as *square, triangle, rectangle,* and *circle*. Middle school children are familiar also with such things as *right angles* and *parallel lines*. A large number of students at the middle school grades though high school geometry, although familiar with the words, interchange *rhombus* and *isosceles trapezoid*. The Chicago project found students entering the high school geometry course who cannot identify simple figures such as *triangles, squares, rectangles,* and *parallelograms*.

Even though they are familiar with the words, students have gross misconceptions or totally incorrect ideas regarding their meanings, at least compared with the way adults use the words. For example, many students process *triangle* to mean an equilateral triangle with a side parallel to a line through the student's eyes. Students actually will turn the page or physical object to view the triangle in what we might call *the textbook position*. Other students, through junior high school, allow figures to be called *triangles* even when the sides are curved outward or inward. They seem to focus on figures with "three points" and not three "straight" sides. Figures whose sides curve outward are more readily called triangles as long as they do not depart too far from equilateral. Many students, through junior high school, hesitate to (or definitely do not) allow scalene triangles to be called *triangles,* especially ones in which a side length is considerably greater than the corresponding altitude, or vice versa. Students consider these to be "lopsided" or "too narrow" to be triangles. There are similar misconceptions about quadrilaterals.

Students at all ages use physical properties to describe figures. They will say, "That is a perfect triangle," when they draw an equilateral triangle. Other triangles are "too tall," "flat," "like an arrowhead" (a nonconvex quadrilateral identified as a triangle), "nearly isosceles," and so forth. Also, students may understand an idea and yet describe the idea with nonstandard vocabulary. For example, students refer to parallel lines as "both are straight," and perpendicular lines as "going straight up."

Perception

The students' language indicates how they use visual cues to perceive geometric shapes. That they exclude relevant attributes while identifying shapes was noted above; for example, straight sides are not considered a necessary condition for a figure to be a triangle. Also, students include irrelevant attributes as necessary conditions when identifying shapes by requiring, for example, that all sides be congruent. Several students who have not taken high school geometry, irrespective of their age, exclude all triangles except the equilateral ones from the set of triangles. A ninth-grade algebra student, when asked how many different types of triangles there are, insisted that there are four types: "the ones that point up, the ones that point down, the ones that point to the left, and to the right." All the triangles he drew were equilateral.

A large proportion of students are unduly influenced by the orientation of figures. Triangles, for example, are not recognized as such unless they are in the textbook position. Opposite sides of a quadrilateral appear to be parallel to students in one orientation but not in other orientations. In working with kites, many students, even high school geometry students, believe that the opposite sides are parallel and the opposite angles are congruent. It is quite common for students and adults to refuse to admit that a rectangle could also be a parallelogram or that a square could also be a rectangle.

Among the more surprising perceptions of children, including some algebra students, is the claim that a rectangle has just one right angle, namely, the one at the lower left when the rectangle is oriented in the textbook position (that angle "opens to the right," like alligator jaws).

Reasoning

By the manner in which students answer quetions or respond to tasks, we can anayze their thought processes using the level structure. The previous examples of vocabulary use and perception indicate a preponderance of Level 0 thought. Even when students analyze properties of figures (Level 1) and are able to verbalize their perceptions, they allow a visual image of a shape to influence their descriptions, as in this example: "A parallelogram has both opposite sides equal and parallel. Two sides are longer than the other, and it doesn't have a right angle because then it would be a rectangle." A student near the end of the geometry course (taught at Level 3) defined a parallelogram as "a quadrilateral with opposite sides parallel and has no right angles." The student said this was a "bad definition" because it was different from the definition in the book, and yet it was clear during the interview that the student excluded rectangles from the set of parallelograms.

Students at all ages use converse reasoning in the sense that they convert necessary conditions of a figure into sufficient conditions. The reasoning pattern seems to be: with those properties the figure must be Type X because figures of Type X have those properties. For example, in the Oregon project an activity is called "What's my shape?" Students receive a sequence of clues and continue to ask for more clues until they are confident that they can correctly identify the target shape. In one sequence, the first clue was, "Has four sides." Several students immediately responded, "It is a square." They reasoned as follows: "A square has four sides. I am looking for a figure with four sides. It must be a square." In other cases, as soon as the students saw the clue that two sides of the figure were parallel, they decided that it must be a parallelogram; given that a four-sided figure has a right angle, they decided it must be a rectangle, and so forth.

Explicit understanding and application of various rules of logic escape many students. For example, a student who was described by the teacher as one of the best students in a high school geometry class incorrectly identified the trapezoids among figures on a sheet of paper by omitting those that were not isosceles. One trapezoid had two right angles, and the student insisted that it was not a trapezoid. The dialogue follows (I = investigator; S = student):

I Is this figure a trapezoid [pointing to the one with two right angles]?
S No.
I Why isn't it a trapezoid?
S Because it isn't.
I What is a trapezoid?
S It is a convex quadrilateral with only two sides parallel.
I What can you say about this figure [pointing to the same figure]?
S Well, it's a quadrilateral and has two parallel sides. The other sides aren't parallel.
I Is it a trapezoid?
S No.

The student was functioning in a Level 3 environment and yet was in effect thinking at a visual level.

Students are, for the most part, unable to contrast definitions, postulates, and theorems. They say, "I never really knew the difference between a theorem and a postulate;" "Postulates you prove from defintions, I think;" and "I know I read it a long time ago. I always forget it though."

Similar responses come from postcalculus students in a college geometry course prior to studying deductive systems. They say, "A postulate is something defined according to the definitions;" "A theorem is the proof of a postulate;" "Theorems are proved postulates which are statements of definition;" and "A postulate is a rule that might not be true."

When asked if they remember an axiom or postulate, even students who have studied high school geometry will more often offer a field postulate from algebra, such as an associative or commutative property, or an often quoted theorem, again from algebra, such as $(a + b)^2 = a^2 + 2ab + b^2$.

It is quite rare for students throughout high school grades and even juniors in college to have much of an understanding of what a deductive system is, that is, a well-rounded Level 3.

Texts and Educational Background

The Brooklyn project's initial evaluation of textbook series (K–8) indicates that most of the qualitative geometry material can be classed as Level 0, although some language may be at Level 2. Little material is classed at Level 1 except where properties of figures are stated in the book for children. The work largely requires identifying and naming shapes, parts of shapes, and some relationships, such as parallelism. Children are seldom asked to reason with the figures, so answers can usually be given using recall alone. This is consistent with preliminary results from the Chicago project, which confirm that there are students who enter the high school geometry course at low van Hiele levels and are not adequately prepared to work at the Level 3 geometry course.

Students seem to believe what they see in a textbook but not what they read. We found examples of this earlier with students whose concept formation differs considerably from the textbook presentation. In one example, the student was aware that his definition of a parallelogram was "bad" because it differed from the text definition. He did not accept the text defintion, which violated his visual image of the parallelogram. In another example, the student correctly parroted the text definition of a trapezoid and yet she refused to permit a trapezoid to have a right angle, again a violation of a visual image.

The claim by Pyshkalo that the educational experiences extended to children can increase their thought level is substantiated by the three United States projects. The instructional modules developed by the Brooklyn project did contribute to student movement through the lower levels on certain topics. The modules were intended to facilitate such movement. Also, during the interviews by the Oregon project, some students actually started to think at a higher level as they reflected on their own responses to the interview tasks. For example, one student refused to include squares in the set of rectangles; when asked to describe the shapes she did so in sufficient generality that, after reflecting on her descriptions, she decided that squares must be rectangles and rectangles must be parallelograms. This is a fine example of expliciting. Evidence from the Chicago project indicates that students who are identified at Level 2 or above in the fall of their high school geometry course do perform well on proofs by the end of the course.

Levels of Communication

The van Hiele model reveals an alarming lack of harmony in the teaching and learning of mathematics. Children do think on different levels, from each other and from the teacher. Words and objects are used by children in ways that differ from the way their teachers and textbooks intend. We have seen this, for example, in geometry, with the fundamental concept of a triangle.

There are numerous other occurrences of this disharmony. For example, concerning fractions, a middle school class was asked to imagine cutting a watermelon in half and then cutting one of the "big pieces" in half. The class was then asked what part of the whole was one of the small pieces. The students were puzzled until one student said, "If you cut the other piece in half, then the small piece is one-fourth." The students needed to have four pieces in order to talk about $\frac{1}{4}$. In algebra, the square root of 2 is understood to be 1.4 or 1.414; to the teacher it means a positive number whose square is equal to 2.

There seem to be harmful diversions placed in the path of student learning. We recognize these from such student misconceptions as the view of fractions described above. Other examples are so called error patterns, such as students who borrow when they should not, as in (a) below; students in algebra or calculus who cancel when they should not, as in (b); or trigonometry students who "simplify" expressions, as in (c).

(a) $\begin{array}{r} 147 \\ -23 \\ \hline 114 \end{array}$ (b) $\dfrac{x+7}{y+7} = \dfrac{x}{y}$ (c) $\dfrac{\sin 2x}{\sin x} = 2$

The lack of understanding may be caused by limited pictures and examples in textbooks, inadequate teacher explanations, improper sequencing of the learning material, and so forth. Misconceptions, once learned, seem to persist, as exemplified by adults who firmly believe that a parallelogram, as displayed in most textbooks, cannot have a right angle. Increasing chronological age does not automatically produce an increase in thought levels.

Other Aspects of the Van Hiele Model

Thought Levels as Categories

The level structure may be interpreted in terms of a modern mathematical concept; that of a category. Indeed, category theory is a study of structure. A *category* consists of a collection of elements, called *objects,* and a collection of relations between the objects, called *morphisms,* that satisfy a list of postulates (see MacLane, 1971). As examples, there is a category in which the objects are sets and the morphisms are functions between sets; there is a category in which the objects are groups and the morphisms are homomorphisms.

We propose to view each of the van Hiele levels in terms of a category and apply this structure to topics other than geometry. The objects for each of the levels may be described as follows:

Level 0: Objects are the base elements of the study.
Level 1: Objects are properties that analyze the base elements.
Level 2: Objects are statements that relate the properties.
Level 3: Objects are partial orderings (sequences) of the statements.
Level 4: Objects are properties that analyze the partial orderings.

Note the similarity and implied recursive nature of the system in the descriptions for Level 1 and Level 4.

By specifying only the objects, we do not completely define a category. It is necessary to identify the morphisms as well. We reserve comment on this for the moment.

Note that the levels imply a scheme for organizing material for students to learn. Goals and mathematical expectations for learning are inherent in the level structure. For example, the levels for geometry as they were described earlier aim to develop thinking about figures and proofs of propositions as organized in various deductive systems (Level 4). Although thought becomes increasingly abstract as one moves up the levels, there is a certain pattern being followed that is dictated by our present understanding of mathematics or any other structured discipline. In the Netherlands, the levels have been used to structure courses in chemistry and economics.

Hence, with the goals clearly established, we can use the above topic-free descriptions of the levels to organize at least the objects of various topics. Here are three examples:

Logic

Level 0: The objects are sentences and their symbolic representation.
Level 1: The objects are properties of the sentences, such as their truth value and their decomposition using conjunction and negation.
Level 2: The objects are statements that relate the properties, such as the chain rule, modus ponens, modus tollens, and logical equivalence.
Level 3: The objects are sequences of statements, such as proofs that have derivations in a set of postulates.
Level 4: The objects are properties that analyze the logical system by comparing it with different logics.

Geometric Transformations

Level 0: The objects are changes in figures, such as tilting, stretching, binding, and turning.
Level 1: The objects are properties of the changes in terms of what they do to figures, such as preserve lengths, reverse orientation, or distort shape.
Level 2: The objects are statements that relate the properties, such as the composition of two reflections being equal to a rotation or translation.

Level 3: The objects are sequences of statements, such as proofs that reflections generate the isometries, or reflections and dilations generate the similarities.

Level 4: The objects are properties that analyze groups of transformations relative to different geometries in the spirit of Klein's Erlanger program (see, for example, Martin, 1976, or Weinzweig, 1978.)

Real Numbers

Level 0: The objects are symbols that represent, for example, directed segments on a "number line."

Level 1: The objects are properties of the symbols, such as ordering, approximation by rationals, and representation as infinite sequences of integers.

Level 2: The objects are statements that relate the properties, such as the use of exponents and radicals, and elementary consequences of the field postulates.

Level 3: The objects are sequences of statements within a system, such as viewing real numbers as Dedekind cuts or limits of rational sequences, and deducing statements such as $(-a)(-b) = ab$.

Level 4: The objects are properties of a complete ordered field as compared with fields in general.

Learning Phases as Functor

In category theory a *functor* puts the objects and morphisms of one category into a correspondence with the objects and morphisms of another category to satisfy certain postulates (see MacLane, 1971). In a sense, the van Hiele phases of learning form a functor. The intent is to move the student from Category (Level) L_{n-1} to category L_n in such a way that the objects and morphisms (relations) at level $n - 1$ are, after Phase 5, interpreted and understood as the objects and morphisms at Level n. Hence, the naive interpretations that students give to words and relations at Level $n - 1$ become more sophisticated at the higher level. Recall that an essential feature that distinguishes Level $n - 1$ from Level n is that what is implicit at Level $n - 1$ becomes explicit to the student at Level n.

The van Hiele functor can be compared to what Dienes, building on Piaget's work, refers to as the *learning cycle* (e.g., Dienes & Golding, 1971). Van Hiele's inquiry phase is similar to Dienes' free play stage; the directed orientation phase is akin to Dienes's structured play stage. For the expliciting phase, Dienes describes stages of abstraction, representation, and description. Dienes asserts that in order to abstract, we need to gather common features of a large variety of experiences and reject the irrelevant features of these experiences. As the abstraction (at the student's level) begins to take on more structure, students, according to Dienes, tend to externalize or represent their thoughts in some way and then describe or make explicit the abstraction. Van Hiele's free orientation phase may be called a *prediction stage* in Dienes's cycle. Whereas van Hiele describes the phase in terms of students' being able to find their own way toward the resolution of a problem, Dienes describes this as a testing of the description stage to predict outcomes in possibly unfamiliar situations. Van Hiele's fifth phase, integration, is referred to at the end of Dienes's learning cycle as an abstract mathematical system in which the descriptions are interpreted as a system of axioms, and the predictions as the theorems.

Relations as Morphisms

If we view each van Hiele level as a category with the objects as described above, we are obliged to specify the morphisms. There is some choice in the matter. By specifying different sets of morphisms, one technically defines different categories that may be related by a functor. In general, the morphisms are chosen in a "natural" way. (For an explanation of "natural," see MacLane, 1971, or Manes, 1976.)

Consider the objects of the van Hiele levels for geometry that were described above. In the spirit of Klein's Erlanger program, we choose transformations to be the morphisms for each of the levels. Note that transformations as morphisms in the geometry categories are objects in the transformation categories. Also, transformations are used at Level 0 differently than they are used at higher levels. Three examples are given below to illustrate this point.

IDEALIZATION

The student interprets a word or object in what we would consider a restricted manner by transforming it to a special case. For example, at Level 0 each triangle is related to an ideal figure, namely an equilateral triangle in the textbook position. Students consider the ideal quadrilateral to be a square. When asked to draw different four-sided figures, some students draw only squares of different sizes. On the "Guess My Shape" activity in the Oregon project, when given the clue, "a four-sided figure," many students immediately said "square."

INVARIANCE

This may be a convoluted form of conservation that should be distinguished from conservation. Students will keep the same name for a figure even though the original figure is transformed into another. For example, a rhombus appears to students as a square that is transformed by "tilting" or "pushing it over," but to the student the rhombus is still a square! Rectangles appear as "stretched out squares," and so forth. Many students are convinced that the area of a quadrilateral can be obtained by transforming it into a rectangle with the same perimeter as the original figure. This notion of invariance may be a cause, in part, for the error patterns that we alluded to above.

PSEUDOEQUIVALENCE

The student's early use of equivalence does not agree with what is expected mathematically. For example, with fractions, students may transform a pie into three parts with equally spaced parallel lines, as in Schema 7.1 below, rather than use regions with equal areas. Errors then result because of an incorrect or inadequate representation. In geometry, a student may totally miss the point in proving a theorem such as: If the bisector of an angle of a triangle is perpendicular to the base, the triangle is isosceles. If the student's representation of *triangle* is equilateral, then the theorem is obviously true, so the student is confused by the need to prove such an obvious fact.

Analogous to Klein's description of a geometry in terms of a group of transformations, we consider a topic level in terms of the set of morphisms. That different morphisms are used at the various levels is a cause for friction between students and mathematical ideas.

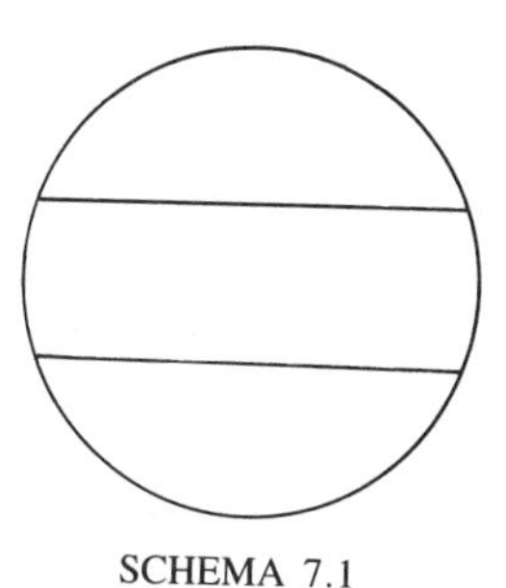

SCHEMA 7.1

Future Work

It is claimed that the results of the Soviet studies over more than 30 years have produced a curriculum in the Soviet Union that is far superior to the one in the United States. For example, M. Adler (1981, p. 73) states that, "the Soviet Union is demanding that all children take mathematics up through calculus." I.

Wirszup (Siddon, 1981, April 28, 1981) goes even further in saying that "a dangerous gap exists between our educational standards and those of the Soviet Union." In testimony before congressional subcommittees, Wirszup (Glocker, 1981, May 1, 1981 p. 5) claims that "the weaknesses of the American educational system have become a national malady that gnaws at our economic strength, our competitive edge in technology and production, and our ability to defend ourselves."

Amidst these ominous tones, we return to the urgent question of developing insight in our students. It is folly to ask whether thought levels as proposed by van Hiele actually exist. That they are being used as descriptive frames of reference guarantees their existence. On the other hand, we should avoid labeling people by the levels. The value of the van Hiele model resides not so much in a stratification of student thought as in a prescription for instruction, not only in geometry but in most structured disciplines.

The model provides us with a blueprint for future work. The main task is to interpret each topic that we want students to learn in terms of the model. We need to identify explicitly the objects and morphisms of each of the levels as perceived by students as well as by the goals of the mathematical topic (e.g., Hoffer, 1981). Also, we need to elaborate on each phase in the van Hiele functor. The model is, after all, a skeleton structure, a quotient space of sorts, that needs to be filled in. With this in mind, there are four critical questions that must be addressed. These are posed below in terms of geometry, but analogous questions apply to other topics.

First, just what are the objects? This needs expliciting at each of the levels. The Oregon and Brooklyn studies confirm that children interpret words differently from what adults might expect. More investigations of this sort are needed to provide, as Coxford advised, more documentation over a substantial period of time. An instrument like the one developed by the Oregon project, that can be used with students in Grades 1 through 12, might be most suitable in this regard. Some information can be obtained by administering a battery of tests like the ones used in the Chicago project with high school geometry students. However, standardized test results might be misleading when used to determine how people think. The Chicago project found, for example, that some children correctly answered certain Level 4 type questions and yet incorrectly answer lower level questions. This may be caused by poor questions, student guessing, or rote responding. Another possible explanation is what van Hiele refers to as "reducing the level" (van Hiele, 1973, Note 2); that is, some students are successful in removing the obstacle of a higher level question by using cues to answer the question at a lower level.

We need a well-coordinated network of clinical interviews and case studies to determine explicitly how children perceive the objects at different levels and how the children are influenced toward these perceptions. Only then can we plan effective learning sequences for children.

Second, what are the morphisms? What is the system of relations at each of the levels? Lesh notes, "especially in geometry, children make many mathematical judgments using qualitiatively different methods than those typically used by adults. Yet the nature of these differences is not clearly understood" (1976, p. 186). We emphasize that these different methods occur in topics other than geometry, as evidenced by the error patterns referred to above. In geometry, we observed the phenomena of idealization, invariance, and pseudoequivalence relations that children use. There are likely many other such maladies that we need to identify and understand before they can be treated. It is also likely that such information must be obtained from clinical interviews.

Reasoning patterns of children are quite closely connected with the system of relations that children form. Hence, a study of children's cognitive behavior assumes a knowledge of their logic, and this involves refinement of the concrete operational–formal operational partition. It was noted that children reason through a physical–visual mode for several years. Even after having experiences at higher levels, children and adults will resort to physical traits (e.g., "tall triangles"). It is pedantic to describe this thinking in terms of topological invariants. The children actually have images of physical objects. We need to determine whether reasoning at different levels is cumulative or whether certain patterns appear and then disappear. If the latter occurs, what are the causes?

The third question deals with the van Hiele functor, the phases of learning that move a student from one thought level to the next. There are prototype materials, such as D. van Hiele's thesis, P. van Hiele's books for technical schools, and Pyshkalo's textual material for Grades 1 through 4. Dienes (Dienes & Golding, 1971) gives examples of the steps in the learning cycle and the Brooklyn project has developed modules based on the levels. Quite clearly the ability to think at higher levels is not acquired from written materials alone and, at least for a while, not from computer materials alone. The phases suggest an interaction between student and teacher similar to one in which a vastly wise, knowledgeable, perceptive, and loving parent provides the child with the necessary and sufficient amount of help to enable the child to mature.

There are those who say that learning is continuous. In reality, learning is quite discontinuous. It is teaching that should be continuous. The third question then is how to put into effect the phases of learning to raise students' thought levels. Once we know how students perceive and reason with the objects and relations, what learning experiences can we provide to help the students gain insight into the subject? This takes insight on our part.

The van Hiele phases, currently explicated in outline form, provide a more complete teaching and learning plan than one finds in many existing programs. For example, so-called direct instruction models, such as the Montessori system or Distar, are based almost entirely on Phase 2: directed orientation. A perverse form of Phase 3, expliciting, is called the lecture method. Phase 4, free orientation, is prominent in so called problem-solving curricula in which students are expected to find their own ways in diverse topics without adequately, according to the van Hiele model, setting the stage and building a system of relations from which students can operate effectively.

By incorporating the best aspects of these programs and by involving students more in the learning process, using (for example) problem posing (Brown & Walter, 1982) and student self-evaluation (see Greenwood & Anderson, Note 1), we can round out the van Hiele functor. More descriptive experiments are needed.

The fourth question deals with the nebulous environment of the child. What does bring about concept formation and the development of insight? We need to know more about the positive and negative effects of textbooks and teachers.

In summary, the van Hiele model makes transparent for us the interface, often misaligned, between perceptions of children and the mathematical expectations dealing with objects, relations, and reasoning. These misalignments cause learning difficulties. As we view the original van Hiele model for geometry in terms of category theory, we are able to generalize it and apply it to other structured topics. This categorical view also provides a direction for future work on the model as we raise our own level of understanding about how people learn and develop insight.

Reference Notes

1. Greenwood, J., & Anderson, B. Some thoughts on teaching and learning math. *Arithmetic Teacher,* to be published.
2. van Hiele, P. M. *Levels of thinking how to meet them, how to avoid them.* Presented to the research presession prior to the Annual NCTM Meeting, Seattle, 1980.
3. Wheeler, N. Correspondence to Alan Hoffer, May 17, 1980.
4. Burger, W. F. *Thought levels in geometry.* Prepared for a Project Directors Meeting of the National Science Foundation. Washington, D.C., 1981.
5. Geddes, D. *Geometric thinking among adolescents in inner city schools.* Prepared for a Project Directors Meeting of the National Science Foundation, Washington, D.C., 1981.
6. Usiskin, Z. *Cognitive development and achievement in secondary school geometry.* Prepared for a Meeting of the National Institute of Education, Washington, D.C., 1981.

References

Adler, M. J. Education: We've grown weaker and Europe stronger. *U.S. News & World Report,* July 13, 1981.

Brown, S. I., & Walter, M. *The art of problem posing.* Philadelphia: Franklin Institute Press, 1983.

Coxford, A. Research directions in geometry. In R. Lesh & D. Mierkiewicz (Eds.), *Recent research concerning the development of spatial and geometric concepts*. Columbus, Ohio: ERIC/SMEAC, 1978.

Dienes, Z. P. *An experimental study of mathematics learning*. London: Hutchinson, 1963.

Dienes, Z. P., & Golding, E. B. *Approach to modern mathematics*. New York: Herder & Herder, 1971.

Freudenthal, H. (Ed.). *Report on methods of initiation into geometry*. Groningen: Wolters, 1958.

Freudenthal, H. *Mathematics as an educational task*. Dordrecht, Holland; Reidel, 1973.

Glocker, D. Wirszup testifies against education cutbacks. *The Chicago Maroon,* May 1, 1981.

Hoffer, A. *Geometry, a model of the universe*. Menlo Park: Addison–Wesley, 1979.

Hoffer, A. Geometry is more than proof. *Mathematics Teacher,* NCTM, 1981, *74,* 11–18.

Kilpatrick, J., & Wirszup, I. (Eds.). *Soviet studies in the psychology fo learning and teaching mathematics*. Stanford, California: School Mathematics Study Group, 1969–1977.

Lesh, R. Transformation geometry in elementary school: Some research issues. In J. L. Martin (Ed.), *Space and geometry: Papers from a research workshop*. Columbus, Ohio: ERIC/SMEAC, 1976.

MacLane, S. *Categories for the working mathematician*. New York: Springer-Verlag, 1971.

Manes, E. *Algebraic theories*. New York: Springer-Verlag, 1976.

Martin, J. L. The Erlanger Program as a model of the child's construction of space. In A. R. Osborne (Ed.), *Models for learning mathematics: Papers from a research workshop*. Columbus, Ohio: ERIC/SMEAC, 1976.

Mathematics Resource Project. In A. Hoffer (Dir.), *Geometry and visualization*. Palo Alto: Creative Publications, 1978. (a)

Mathematics Resource Project. Planning instruction in geometry. In A. Hoffer (Dir.), *Didactics and mathematics*. Palo Alto: Creative Publications, 1978. (b)

Polya, G. *Mathematical discovery* (Vol. 2). New York: Wiley, 1965.

Pyshkalo, A. M. Geometriya v I–IV klassakh. In *Problemy formirovaniya geometricheskikh predstavienii u mladshikh shkol' nikov*. [Geometry in grades I–IV.] In A. Hoffer (Ed.) and I. Wirszup, University of Chicago (Trans.), [*Problems in the formation of geometric ideas in primary-grade school children*] (2nd ed.). Moscow:Prosveschchenle, 1968.

Siddon, A. Science cuts may cripple U.S., says Chicago Prof. *Chicago Tribune,* April 28, 1981.

van Hiele, P. M. De problematiek van het inzicht. (Thesis). Utrecht, 1957.

van Hiele, P. M. La pensée de l'enfant et la geometrie. *Bulletin de l'Association des Professeurs Mathematiques de l'Enseignement Public,* 1959, *198,* 199–205.

van Hiele, P. M. *Begrip en inzicht*. Netherlands: Muusses Purmerend, 1973.

van Hiele, P. M. *van a tot z*. Netherlands: Muusses Purmerend, 1979.

van Hiele, P. M., & van Hiele-Geldof, D. A method of initiation into geometry. In H. Freudenthal (Ed.), *Report on methods of initiation into geometry*. Groningen: Wolters, 1958.

van Hiele-Geldof, D. De didaktiek van de meetkunke in de eerste klas van het V.H.M.O. (Thesis). Utretcht, 1957.

Weinzweig, A. I. Mathematical foundations for the development of spatial concepts in children. In R. Lesh & D. Mierkiewicz (Eds.), *Recent research concerning the development of spatial and geometric concepts*. Columbus, Ohio: ERIC/SMEAC, 1978.

Wirszup, I. Breakthroughs in the psychology of learning and teaching geometry. In J. L. Martin (Ed.), *Space and geometry: Papers from a research workshop*. Columbus, Ohio: ERIC/SMEAC, 1976.

CHAPTER 8

Trends and Issues in Mathematical Problem-Solving Research

Frank K. Lester, Jr.

Introduction

One day a few years ago, a close friend phoned to say she had decided to learn, at long last, how to ride a bicycle and to ask for my help. Because she was then in her mid-30s and had never shown any interest in any form of athletic endeavor, I was temporarily taken aback by this bold announcement. I agreed to help in any way I could. Today, in spite of my coaching, my friend rides as well as most of us who learned in our childhood.

I see several parallels between riding a bicycle and solving mathematical problems. First, it is difficult to explain either activity in words. In describing bike riding to a nonrider, one might say, ''Get on the bike, push off with one foot, and then pedal while balancing yourself.'' An explanation of this sort describes the essential components of bicycle riding but does not capture the essence of coordinating the acts of pedaling, braking, turning, watching where one is going, and maintaining one's balance. Also, showing my friend how to ride a bike seemed to be of little value to her. My advice, ''Watch me and then do what I did,'' caused more anxiety than help. Problem solving is similar insofar as no description, no matter how carefully worded to include all the key components, can express its complexity, subtlety, and highly personal nature. Also, good problem-solving behavior usually is not fostered by having students imitate how teachers solve problems. Because teachers typically demonstrate only correct moves, students often come to view problem solving as the act of delving into a mysterious bag of tricks to which only a select few are privy.

ISBN No. 0-12-444220-X

Because of the elusive and intractable nature of mathematical problem-solving, researchers in this area have made very slow progress in developing a stable and useful body of knowledge about this acquisition. Yet, reviews of mathematical problem-solving research indicate problem solving is one of the most popular areas of research in mathematics education (Kilpatrick, 1969; Lester, 1980b; Suydam, 1976). Indeed, it is a common belief that the development of students' abilities to solve problems is the most important goal of the study of mathematics. Evidence of this belief is found in conference reports and position statements by national organizations. For example, the Snowmass Conference report pointed to the need for "basic research into the nature and meaning of problem solving and mathematical thinking" (Report of the Conference, Note 1, p. ii). Four years later, the National Council of Supervisors of Mathematics (NCSM) stated that: "Learning to solve problems is the principal reason for studying mathematics" (National Council of Supervisors of Mathematics, 1977, p. 20). Although it may be argued that conference reports and position papers merely reflect the mood of a community, it appears that the emphasis on problem solving of the Snowmass Conference was instrumental in making funding agencies such as the National Science Foundation more receptive to research proposals in this area. At the same time, the identification of problem solving by the NCSM as a *basic skill* in mathematics has served to make curriculum developers, school-textbook writers, teacher educators, and classroom teachers more conscious of the fundamental importance of problem solving to the school mathematics program. The pervasiveness of agreement on the importance of problem solving in mathematics makes the slow pace of advancement of our understanding of problem solving particularly worrisome.

Past problem-solving research has not been conducted systematically within the mathematics education community. Due to lack of agreement on (*a*) what constitutes problem solving, (*b*) how performance should be measured, (*c*) what tasks are appropriate for research purposes, and (*d*) what the key variables are that influence behavior, it is extremely difficult to organize and synthesize the results of the myriad studies that have been conducted.

Within the past 5 or 6 years, however, there has been a discernible trend toward more systematic, focused, and coordinated efforts. This trend was given considerable impetus by the 1975 Mathematical Problem Solving Research Workshop sponsored by the University of Georgia Center for the Study of Learning and Teaching Mathematics under a grant from the National Science Foundation (Hatfield & Bradbard, 1978). Four working groups of researchers with related interests resulted, and two very thoughtful monographs have been produced (Goldin & McClintock, 1979; Lesh, Mierkiewicz, & Kantowski, 1979).

Despite the fact that problem-solving research in mathematics education is becoming less chaotic, the research remains largely atheoretical in nature. This condition appears to mirror the situation that exists in the psychological problem-solving literature. Greeno has observed: "we have not yet developed analyses of learning related to problem solving that achieve the degree of conceptual clarity and rigor that is characteristic of modern psychological theory" (Greeno, 1978, p. 266). Nevertheless, just as recent developments in cognitive psychology indicate an awareness of the need for careful development of theories of problem solving, there is growing interest in theory-based research among mathematics educators.

Along with the increased concern for theory-based research has come growing recognition of the inadequacy of strictly experimental, almost exclusively quantitative methods to study problem-solving processes. For example, several mathematical-problem-solving researchers have used "think aloud" techniques to collect problem solvers' protocols (e.g., Kantowski, 1977; Kilpatrick, 1968; Lucus, 1974; Webb, 1975). In addition, the series of Soviet studies edited by Kilpatrick and Wirszup (1969–1975), and especially the research of Krutetskii (1976), made American mathematics educators more aware of the potential value of clinical investigations of mental processes. The enthusiastic acceptance of nonexperimental procedures employing qualitative analysis has underscored the difficulties in observing, recording, and analyzing problem-solving behavior. In order to develop valid and reliable techniques to handle these difficulties, researchers have been forced to give heed to conceptual issues associated with the nature of mathematical problem solving.

The Nature of Mathematical Problem-Solving and Mathematical Problem-Solving Research

The terms *problem* and *problem solving* may be the terms defined most often in the entire literature on cognition and thinking. The many descriptions of problem solving in the psychological literature indicate several very different views of its nature. For example, Asher (1963) regarded problem solving as the inverse of learning because it involves the process of disrupting established concepts. On the other hand, Duncan (1959) and Davis (1966) regarded problem solving as a particular type of learning. Whereas Goldin (1982) argues for adopting a broad view of problem solving for research purposes, calling a task "a problem when steps or processes are detected between the posing of the task and the answer" (p. 97), the following definition seems to be reasonably descriptive of and suited to mathematics problems and will be used in this discussion.
A *problem* is a task for which:

1. the individual or group confronting it wants or needs to find a solution;
2. there is not a readily accessible procedure that guarantees or completely determines the solution; and
3. the individual or group must make an attempt to find a solution.

These three components have special implications for mathematics instruction. Posing the cleverest problems is not productive if students are not interested or willing to attempt to solve them. The second component requires initial failure on the part of the student, at least in the sense that mere recall of facts or application of a previously learned algorithm will not give a solution. Unfortunately, many, if not most, of our teachers learn in their teacher training courses that failure is harmful to their students' self-confidence. As Brownell (1942) emphasized over 40 years ago, it is important that teachers realize that some failure is not only a good thing, but also a necessary part of problem solving. The student must not be able to get the solution immediately, yet a solution must be within grasp. Students should not be led to believe that if a task cannot be done easily it cannot be done at all.

Considering all three components described above, a great deal of information is needed about each student to create real problem-solving experiences. Teachers must bear in mind that a problem for one person may not be a problem for another, due to difference in knowledge, experience, ability, or interest, in addition to other factors. It is also important to be aware of the influence of environmental factors (e.g., mode of instruction, external pressure) on problem-solving performance.

Because of the multitude of factors that affect problem solving, especially in mathematics, adequate analyses often seem nearly impossible to perform. However, there are some clearly identifiable categories of factors that serve to classify the preponderance of mathematical problem-solving research.

1. *Task Factors.* Task factors are those associated with the nature of the problem, for example, its mathematical content and structure, its context, and the syntax of the problem statement (Caldwell, 1977; Caldwell & Goldin, 1979; Goldin & Luger, 1975; Harik, 1979, 1981).

2. *Problem-solver Factors.* What the individual brings to a problem falls into the category of problem-solver factors. Characteristics of the problem solver, such as previous experience, mathematical background, sex, age, reaction under stress, degree of field independence, and spatial ability are included (Krutetskii, 1976; Meyer, 1978; Silver, 1979, 1982).

3. *Process Factors.* The overt and covert behavior of the problem solver involves numerous factors, many of which are tied closely to the Task and Problem-solver categories. The fact that such a wide range of activities, both mental and physical, are involved in problem solving warrants making a separate

Process category to include them (Goldin & Luger, 1975; Kantowski, 1977; Krutetskii, 1976; Schoenfeld, Chapter 10, present volume; Webb, 1975).

4. *Environment Factors.* Included in the Environment category are those features of the environment external to the problem and the problem solver. An extremely important class of factors within this category are those related to instruction (Kantowski, 1977; Post & Brennan, 1976; Putt, 1978; Schoenfeld, 1979, 1980; Vos, 1976).

In order to classify properly the majority of the mathematical problem-solving research literature, a fifth category is needed:

5. *Instrumentation and Research Methodology.* A significant number of recent investigations have been concerned with the development of valid and reliable instruments for measuring problem-solving processes and for evaluating performance. In addition, there is a controversy surrounding the best way to study instruction variables; for example, whether to employ teaching experiments or to conduct classical methods comparisons (Kantowski, 1977; Kulm, Campbell, Frank, & Talsma, 1981; Lester, 1980a; Lucas, Branca, Goldberg, Kantowski, Kellogg, & Smith, 1979; Proudfit, Note 2; Romberg & Wearne, Note 3; Schoen & Oehmke, Note 4).

These five categories are not the exclusive domain of mathematical problem-solving. It is probable that any type of problem solving involves factors from each of these categories. They are presented to provide a scheme for classifying problem-solving research within mathematics education.

Overview of Mathematical Problem-Solving Research

Kilpatrick (1969) has characterized the body of mathematical problem-solving research as atheoretical, unsystematic and uncoordinated, dealing primarily with standard textbook word problems (e.g., problems involving one-step translations from words to mathematical sentences), and interested exclusively in quantitative measures of behavior.

Much of the research has lacked sophistication in design and has been guilty of a variety of methodological errors. For example, Gorman (1968), in his analysis of elementary school mathematical problem-solving research, found that when criteria involving the control of variables affecting internal validity were applied to the nearly 300 studies he analyzed, only slightly more than 12% could be considered "acceptable."

Research Trends

Although the five categories of problem-solving factors described in the preceding section are useful to highlight trends in the research, they do not partition the set of contemporary studies. Many studies cut across categories.

TASK FACTORS

In the early 1970s, a primary focus of mathematical problem-solving research was to isolate key determinants of problem difficulty. Multiple regression was used to identify sets of variables that predict problem-solving performance. Exemplary studies of this type were conducted by Jerman and his associates, who sought to determine linguistic and computational variables that predict students' error rates on verbal arithmetic tasks (Jerman, 1973, 1974; Jerman & Rees, 1972). However, as Jerman noted, this approach was limited by the absence of theory to determine optimal weights to give variables or to explain the accuracy of the predictor variables.

Recently, multiple regression analyses have been replaced by state–space analysis as a primary approach to studying task variables. For a large class of problems, state–space analysis offers a complete representation of the algorithmic and logical structure of the task and also provides a means for studying the effect of problem structure on problem-solving behavior. The work of Goldin and his students is representative of this relatively new approach, which is adapted from artificial-intelligence research (Goldin, 1979; Goldin & Luger, 1974, 1975; Luger, 1979; Waters, 1979).

Interest in task variables is natural in view of the unquestioned influence they have on problem-solving performance. However only recently have attempts been made to design classification schemes for task variables. The staff of the Mathematical Problem Solving Project (MPSP) devised a scheme for classifying problems after observing and interviewing a large number of intermediate-grade children over the course of a year (Lester, 1978). The MPSP scheme consisted of four dimensions: setting, structure and mathematical complexity, mathematical content, and strategies applicable to the problem's solution. The dimension of complexity included: complexity of the problem statement, complexity of the focusing process, complexity of the solution process, and complexity of evaluation and generalization.

Kulm (1979) has proposed a somewhat different system for categorizing mathematical problems. He has identified syntax, mathematical content and non-mathematical context, structure, and heuristic processes variables as the four major categories of task variables. Detailed discussions of each of Kulm's categories have been given by Barnett (1979), Webb (1979), Goldin (1979), and McClintock (1979). Kulm believes that such a scheme is vital to the study of problem solving because it helps the researcher understand how task variables interact with the total task environment. Goldin (1982) adds another reason for identifying important task variables: "if the problem-solving researcher is unaware of important properties of the task itself, or does not describe it well enough for other researchers to construct identical instruments, the observations made will be of limited use" (p. 88).

The work on task variables begun by Jerman and continued by Goldin, Kulm, the MPSP, and others represents an honest and determined attempt to build a foundation for studying mathematical problem-solving in a systematic and scientific manner.

PROBLEM-SOLVER FACTORS

Among the characteristics of the individual that play an important role in problem solving are: mathematical background, prior experience with similar problems, reading ability, perseverance, tolerance for ambiguity, spatial ability, age, and sex. Research in this area has typically looked for correlates of problem-solving success. For example, Dodson (1972) identified discriminating characteristics of successful problem-solvers using data from the National Longitudinal Study of Mathematical Ability (conducted by the School Mathematics Study Group [SMSG]). Among the prominent differences he found between good and poor problem-solvers were that good problem-solvers were superior with respect to mathematical achievement, verbal and general reasoning ability, spatial ability, positive attitudes, resistence to distraction, field independence, and divergent thinking.

Dodson used mulltivariate statistical techniques (in particular, discriminant analysis) and, in this respect at least, his research is similar to Jerman's and is subject to the same limitations, namely, the absence of theory to guide the selection of variables to study and to explain results. Another shortcoming of Dodson's research is that it largely ignored problem-solving behavior factors. That is, it provided no indication of what good problem-solvers do that distinguishes them from poor problem-solvers. To this end, a primary goal of the 12-year study by Krutetskii (1976) was "to clarify the features that characterize the mental activity of mathematically gifted pupils as they solve various mathematical problems" (p. 78).

Krutetskii distinguished three types of students according to characteristic ways they tended to solve problems: analytic, geometric, and harmonic (a combination of the other two types). Important abilities that discriminated between good and poor problem-solvers were distinguishing relevant from irrelevant information, quickly and accurately seeing the mathematical structure of a problem, generalizing across a wide range of similar problems, and remembering a problem's formal structure for a long time.

Krutetskii's work set the stage for a major shift away from attempting to identify characteristics that are essentially unrelated to cognitive behavior toward characterizing problem solvers according to the types of mental processes they use. Recently, Silver's (1979, 1981) research represents this trend. Silver supported Krutetskii's conclusion that good problem-solvers are superior to poor problem-solvers in their ability to perceive the mathematical structure of a problem and to generalize problems having similar structure, even when the problem is presented in the form of the problem's solution. Other noteworthy results were: (*a*) differences in memory for problem structure between good and poor problem-solvers cannot be attributed to general memory differences. (*b*) A significant amount of transfer of information occurred from a "target" problem to a structurally related problem. This result stands in direct opposition to the results of research conducted by various cognitive psychologists using puzzle problems (e.g., Reed, Ernst, & Banerji, 1974). (*c*) Neither good nor poor problem-solvers tended to acknowledge having used information from the "target" problem to solve a structurally related problem.

The construction of models of both skilled and unskilled problem-solving behavior has been quite popular in recent psychological research (e.g., Chase & Simon, 1973; Larkin, McDermott, Simon, & Simon, 1980; Simon & Simon, 1978; Chi, Feltovich, & Glaser, Note 5). Within mathematical problem-solving research, the work of Schoenfeld and his associates is most prominent (Schoenfeld, Chapter 10, present volume; Schoenfeld & Herrman, Note 6).

Schoenfeld has sought to characterize the *managerial* aspects of expert and novice problem-solving behavior. Schoenfeld believes undue attention had been devoted by psychologists to studying what he calls *tactical decisions,* that is, "things to implement" (Schoenfeld, Chapter 10, present volume). In this respect at least, Schoenfeld's research differs qualitatively from the related psychological research. This difference may be attributable in part to the fact that, as a mathematician and teacher, he is more keenly aware than most psychologists of the insufficiency for successful problem-solving of having mastered the use of a collection of algorithmic and heuristic procedures.

It appears that interest among mathematics educators in studying problem-solver variables in connection with cognitive (and metacognitive) processes will continue for some time.

PROCESS FACTORS

As the discussion of problem-solver factors suggests, the mental processes used during problem solving have been the object of a considerable amount of study. It may be that problem-solving research gets its reputation for being such a chaotic area of study because such a wide variety of mental activities is involved. Webb (Note 7) reviewed several mathematical problem-solving studies and found very little commonality in the types of processes considered. For example, he reported that such different behaviors as divergent thinking, blind guessing, identifying a pattern, analytical reasoning, using drawings, and looking carefully at details were being studied. In addition, until recently, whenever cognitive processes have been studied as dependent variables, only quantitative measures have been employed. Although the nature of the mental activity required to solve mathematical problems has been of interest to mathematics educators for many years, it was not until Kilpatrick's dissertation study (1968) that cognitive processes (particularly heuristic processes) were investigated in a systematic fashion. The protocol-coding scheme he devised has served as a model for several subsequent investigations (e.g., Goldberg, 1974; Kantowski, 1977; Lucas, 1974; Putt, 1978; Webb, 1975).

It should be apparent from the discussion of task factors and problem-solver factors that processes are typically studied in conjunction with other factors. In particular, researchers interested in task factors are often interested in the interaction between the intrinsic structure of a problem and cognitive processes used to solve the problem. With certain kinds of problems, state–space analysis appears to be an especially useful technique for this purpose. For example, Harik (1979, 1981) has modified the standard techniques of state–space analysis to study algebraic tasks. Her work is discussed in more detail in the next section.

Investigations of the interaction between process and problem-solver factors have been restricted primarily to studies of how experts and novices differ or to studies of the effect of instruction on processes individuals use. Analyses of developmental changes in the use of various processes are extremely rare. This is unfortunate because far too little is known about which processes individuals use naturally, when they acquire these capabilities (or the ability to learn particular processes), and which processes individuals can be taught to use at a given stage of their development. In addition, it would be valuable to know more about the types of misconceptions children have about problem solving and the types of defective or inappropriate procedures they use during problem solving. Brown and Burton's (1978) diagnostic model for students' ''bugs'' in their basic mathematics skills and Matz's (Note 8) error-analysis model for high school algebra have potential for yielding information of this sort.

A primary focus of current process factors has been the development of instruments to record and analyze data on problem-solving behavior. The most prominent work in this area has been that of the Heuristics Working Group organized as a result of the problem-solving conference at the University of Georgia in 1975. This group developed a process–sequence coding scheme that is designed to organize and record processes as they are observed to occur during a problem-solving episode (Lucas *et al.*, 1979). Kulm and his associates (Kulm *et al.*, 1981) have modified this coding scheme to make it suitable for use with a large number of protocols. These schemes are the result of the joint efforts of several researchers and they represent a major step in the development of a valid and reliable coding system. Thus, although the work to create effective methods for recording and analyzing problem-solving processes is a slow and painstaking task, progress has been made.

The potential value of protocol-coding schemes notwithstanding, it is likely that protocols based on any sort of verbal reports would contain only a proper subset of the information in the problem solver's short and long term memories (Ericsson & Simon, 1980). Also, it is possible, perhaps even likely for young children, that a problem solver may not be aware of some of the thought processes he or she is using. For these reasons, I have attempted to develop a different procedure for observing certain cognitive processes used during a specific kind of problem-solving task; that is, multiple classification (Lester, 1980a). The procedure utilizes tasks that force the individual to manifest mental processes through the physical manipulation of objects. Consequently, problem solvers do not need to relate their thoughts verbally or to remember accurately what they did.

ENVIRONMENT FACTORS

There are a multitude of factors external to the problem and problem solver that influence performance. However, within the mathematics education community, the overwhelming majority of environment factors research has been concerned with instruction. This is natural because the pursuit of a fuller understanding of problem solving leads to an interest in the conditions that most enhance success and the degree to which individuals can be taught to be better problem solvers.

Most instruction-related research can be classified into four categories:

1. instruction to develop master thinking-strategies;
2. instruction to teach the use of specific *tool skills;*
3. instruction to teach the use of specific heuristics;
4. instruction to teach the use of general heuristics.

Master thinking-strategies are general problem-solving strategies that can be used in any content area. Typically, instruction of this type aims to improve students' originality, level of divergent thinking, and attitudes toward problem solving. As might be expected, there is no evidence that such an approach improves mathematical problem solving.

Instruction in the use of *tool skills* is designed to equip students with a collection of skills that are useful in implementing various strategies. For example, "making a table" is a tool skill that often facilitates the "looking for a pattern" strategy. Some other tool skills that are commonly taught are drawing a diagram or picture, writing an equation, and organizing a list. This approach is popular in many contemporary school mathematics texts and has been the object of a relatively large amount of research. Research results indicate that tool skills can be taught, but their efficacy for improving overall mathematical problem-solving performance has not been demonstrated.

Instructional methods designed to develop either specific heuristics or general heuristics have been based largely on the writings of Polya (1957, 1962, 1965). Specific heuristics include such strategies as looking for a pattern, simplifying the problem, working backward, and solving an analogous problem. Such heuristics are applicable only to certain types of problems. On the other hand, general heuristics (e.g., means–end analysis, planning, and organizing information) are applicable to a wide range of problems. Research results regarding heuristics instruction is equivocal. It appears that no ''pure'' approach is better than any other; rather ''good'' problem-solving instruction probably involves some combination of instruction in the use of both specific and general heuristics, with training to develop tool skills. In addition, experience in solving a wide variety of problems over an extended period of time is probably essential.

In addition to the above four general instructional approaches, there is at least one other. It focuses on having students discuss, think about, and generally become more aware of the various processes they use to solve problems; that is, it focuses on a metacognitive dimension (Brown, 1978; Flavell, 1976). An instructional model based on this approach was developed by the MPSP at Indiana University and is discussed in some detail in the next section of this chapter. A variation and elaboration of the MPSP model has been implemented in a large-scale, longitudinal study in the Parkersburg, West Virginia schools (Charles, 1982).

A very similar, but independently developed, ''metacognitively oriented'' instructional model is that of Silver, Branca, and Adams (1980). Like the MPSP model, the models of both Charles (1982) and Silver, *et al.* were designed for use with children in the upper elementary or middle school grades. At the college level, Schoenfeld's (1979, 1980) interest in studying managerial decisions in mathematical problem solving has a very definite metacognitive flavor.

To conclude this discussion of instructional factors, there appears to be a trend away from studying the effectiveness of instruction that focuses on only a single aspect of problem-solving activity (e.g., use of tool skills, use of task-specific heuristics) and toward instruction that attempts to create an appropriate atmosphere for good problem-solving. This trend typically incorporates the use of clinical studies and teaching experiments (e.g., Kantowski, 1978; Putt, 1978), described subsequently.

INSTRUMENTATION AND RESEARCH METHODOLOGY

Three types of research and development have been conducted that can be characterized as focusing on instrumentation or research methodology. These are: (*a*) the development of protocol-coding schemes; (*b*) the design and conduct of clinical studies, especially teaching experiments; and (*c*) the development of paper-and-pencil problem-solving tests. Protocol-coding schemes have already been discussed.

Teaching experiments were brought to the attention of American mathematics educators with the publication by the School Mathematics Study Group of the Soviet Studies series (Kilpatrick & Wirszup, 1969–1975, 14 vols.). Kantowski (1978) provides a concise description of the nature of the teaching experiment. A teaching experiment has the following characteristics:

1. It is nonexperimental in design
2. It is often longitudinal in nature
3. It attempts to catch processes as they develop
4. The teacher is not a controlled variable but is rather a vital part of the environment
5. Subjective analysis of qualitative data is often of more concern than are quantitative results.

Thus, the teaching experiment is an ''ecological experiment'' (Bronfenbrenner, 1976). Although teaching experiments and other clinical studies of problem solving have the potential for providing a wealth of information that could not be obtained using standard research procedures, the conduct of such studies is very difficult. At the present time, there do not exist techniques for observing, recording, and analyzing problem-solving behavior, as it occurs in its natural habitat, that have the same level of validity and reliability as procedures used in experimental studies.

Calls for more emphasis on problem solving in the school mathematics curriculum have fallen on deaf ears at least partly because there are no satisfactory means to measure students' performance. Consequently, significant attempts have been made to create paper-and-pencil tests for group administration (Proudfit, Note 2; Romberg & Wearne, Note 3; Schoen & Oehmke, Note 4). Paper-and-pencil tests have been developed in response to the inadequacy of standardized problem solving tests to provide appropriate data about students' behavior. A common feature of paper-and-pencil tests is their attempt to get reasonably valid measures of the individual's behavior at each of four stages in the problem-solving process: understanding, planning, using a plan, and evaluating. These instruments have easily identifiable flaws but they are substantially better than other available problem-solving tests.

Recent Studies at Indiana University

An immediate consequence of the 1973 Snowmass Conference (Note 1) was the formation at Indiana University of the Mathematical Problem Solving Project (MPSP). The MPSP existed for 2 years (1974–1976) under the cosponsorship of

Indiana University and the National Council of Teachers of Mathematics with funding from the National Science Foundation. The activities of the MPSP are not described here but are detailed fully in the final report to the NSF (LeBlanc & Kerr, Note 9). The main goal of the MPSP was to develop instructional materials for improving the problem-solving performance of children in Grades 4, 5, and 6. In particular, the emphasis was on improving children's abilities to use certain heuristics (e.g., looking for a pattern), tool skills (e.g., making a table), and other problem-solving tactics (e.g., identifying relevant information) to solve a wide range of elementary-school mathematics problems.

Although the MPSP was primarily a curriculum development effort, as is true of many educational development projects, several small-scale research studies sprang from questions that arose as work progressed. A great deal of attention was given to observing individuals and small groups. In addition, many children were interviewed both during and after problem-solving sessions and their written work was closely scrutinized in order to gain insights into the problem-solving behavior of children in this age range. The interviews, observations, and analyses of written work led to the formation of several conjectures about the way intermediate grade children solve problems. Among the conjectures that have been investigated both during the operations of the MPSP and in subsequent years are the following:

1. Children without training use only a random trial-and-error strategy in solving process problems, if they are unable to decide on a computation to perform. (In the elementary-school problem-solving literature a *process* problem is a mathematical problem that lends itself to exemplifying the nonalgorithmic procedures inherent in problem solving.)
2. Children are unable to coordinate simultaneously the multiple conditions present in a problem. This inability takes the form of ignoring one or more conditions as they work on a problem or of recognizing the need to coordinate multiple conditions but not being able to do so.
3. Children can be taught certain skills and strategies (heuristics) and they can use them to solve related (i.e., structurally isomorphic) problems. The skills include making a table, organizing a list of information, and "guessing and testing." Heuristics include looking for a pattern and simplifying the problem.
4. The single most effective means for improving children's problem-solving performance is to have them try to solve a wide range of types of problems over a prolonged period of time without specific instructional intervention.
5. Initial problem representations formed by children typically are based on syntactic interpretations only.

Some elaboration on these five conjectures may help make them clearer. Regarding Conjectures 1 and 2, consider the following problem:

> Tom gets \$1.75 each week for allowance. This week his mother gave him 18 coins. The coins were nickels, dimes, and quarters. How many of each kind of coin did Tom get?

The most common approach was to ignore the fact that 18 coins were needed and to simply ''play around'' until some numbers of nickels, dimes, and quarters totaled $1.75. For example, a first guess might be 5 of each coin. When that was found not to work, instead of using the information available from that attempt (e.g., that makes a total of $2.00, or $0.25 too much) the children often chose an apparently unrelated combination of coins (e.g., 7 nickels, 3 dimes, and 4 quarters). Thus, they did not appear to incorporate information gleaned from incorrect efforts into their existing conceptualizations of the problem. At the same time, if the observer pointed out that a proposed solution (e.g., 5 nickels, 3 dimes, and 4 quarters) did not give a total of 18 coins, the child often simply altered the answer to yield 18 coins (e.g., 5 nickels, 5 dimes, and 8 quarters), overlooking the condition that there be $1.75.

During the project's 2 years of operation, three skills modules were developed: ''Using Guesses to Solve Problems,'' ''Using Tables to Solve Problems,'' and ''Organizing Lists.'' Perhaps not surprisingly, children who studied these modules made significant gains in their ability to perform these three skills. However, improvement in overall problem-solving performance was not significant, even though there was evidence that the problem-solving behavior of children who studied the modules had substantially improved for certain categories of problems. In addition, students learned to look for patterns and to simplify problem statements. Whether or not they attained the higher-level ability to judge the value of these two heuristics for a particular problem was not determined. It is possible that the children were trained to look automatically for a pattern or to begin work on all ''hard'' problems by simplifying them. If this is the case, no real improvement in problem-solving ability occurred.

Conjecture 4 resulted from the observation that the most important component of all the various instructional methods devised by the MPSP was for students to engage actively in solving many problems. A more extreme version of this conjecture would be to claim that problem-solving ability cannot be taught, but can only be developed by solving problems. Although there is no doubt that problem-solving ability is enhanced by solving problems, both the necessity and sufficiency of this condition for making students good problem-solvers is unclear.

The fifth conjecture is more difficult to document, but the written work and verbal comments of children suggest that incorrect solutions often stemmed from internal representations that incorporated syntactic information only. For example, consider the following problem:

> A caterpillar is put into the bottom of a jar that is 8 inches high and 6 inches across. Every day it crawls up 4 inches. Every night it slips back 2 inches. How many days will it take for the caterpillar to touch the top of the jar?

Students who proceeded syntactically usually did something like this:

> "The jar is 8 inches high and the caterpillar advances 2 inches each day. Dividing 8 by 2 gives 4, so it takes 4 days."

Of course this is not purely syntactic, but it does suggest a failure on the students' part to get a feel for what is happening in the problem. Behavior of this type was extremely common for problems in which one or more arithmetic operations were suggested. In these cases, students tended to shift into a "number crunching" mode.

Each of these five conjectures has been investigated more closely by members of the MPSP staff both during the project's operation and since that time. The two sections that follow describe these investigations.

Learning to Solve Problems by Solving Problems

Three members of the MPSP staff devoted the entire second year of the project's operation to an investigation of the hypothesis that children become better problem solvers simply be solving problems without direct instruction from the teacher (Conjecture 4). Although this effort cannot be considered experimental research, valuable information about all five conjectures was obtained. More specifically, the goals of this investigation were to

1. Reduce impulsivity on the part of the children (i.e., the tendency to "crunch numbers" and generally to fail to gain "good" understanding of the problem statement before beginning to try to solve the problem).
2. Make the children aware that most problems can be solved in more than one way.
3. Decrease the children's tendency toward premature closure (e.g., make them aware that many interesting problems take more than 1 or 2 minutes to solve).
4. Make children aware that some problems may have more than one correct answer whereas others may have no answer, due to insufficient information.
5. Help children to realize the importance of, and to gain skill in, organizing information.
6. Increase children's willingness to engage in problem solving.

This list of goals indicated that the overall aims of the study were affective and metacognitive in spirit. Students must bring appropriate willingness and motivation to a problem-solving situation. Positive attitudes toward problem solving are a necessary but not sufficient condition for developing good problem-solving behaviors. At the same time, making students more aware of and knowledgeable about their cognitive processes helps them manage and monitor their own problem-solving activities. Training in the use of a collection of skills and heuristics (*tactical decisions,* as Schoenfeld (Chapter 10, present volume) calls them) without attention to metacognitive aspects of problem solving is inadequate.

The investigation proceeded in three stages. The first stage involved work with two groups of six fifth-graders once or twice per week for 8 weeks. One group of six children worked individually or in pairs as they chose. Children in the other group were asked to work together, to share ideas, and to ask questions of one another. Students were asked to write down as much detail of their work as possible and each session was either audio- or videotaped. During each session, a problem was read to the group and questions concerning the problem were answered. As the children worked on the problem, the leader (an MPSP staff member) was available to answer questions of a general nature only. In a few sessions the leader experimented with giving hints, initiating group discussion, and posing questions to refocus student attention on relevant information in the problem. Two observers were present at each session and immediately following a session the observers and session leader discussed what took place and analyzed the students' papers. The purpose of this stage was to develop a scheme for engaging an entire classroom of children in problem solving that would involve relatively little teacher direction. Secondary purposes were to determine what sort of teacher input would be most appropriate given the goals of the study and to select a set of suitable problems. Details of the instructional scheme, the role of the teacher, and the problems chosen are given in the report by Stengel, LeBlanc, Jacobson, and Lester (Note 10).

Briefly, the format that was used in the next two stages contained three phases:

1. Problem Presentation;
2. Solution Effort;
3. Problem and Solution Discussion.

Desirable teacher and student behaviors at each phase plus the intended outcomes are listed in Table 8.1.

Stages 2 and 3 of the investigation consisted of two short-term teaching experiments. The first experiment involved one third-grade, four fourth-grade, and three fifth-grade classes of from 22 to 28 students each. Teachers of these classes represented a wide range of experience and styles. Over a 5-week period the students in these classes worked on 10 problems, using the instructional procedure outlined in the preceding paragraph. The purposes of this experiment were to gather data regarding the effectiveness of various teacher behaviors as well as to observe student behavior in large-group settings. Conclusions drawn from observations of problem-solving sessions were that the children were eager to work on the problems and they had little difficulty adjusting to a classroom environment that was less teacher-directed than usual. However, teachers experienced a great deal of uneasiness. Several found it difficult initially to keep from

TABLE 8.1

First Stage of the MPSP Investigation: Intended Outcomes and Desirable Teacher and Student Behaviors

Step	Intended outcomes	Teacher behaviors	Student behaviors
Problem presentation	Students develop appropriate understanding of the problem.	Presents problem to class. Answers questions about problem comprehension. Organizes students to begin work on the problem.	Reads and/or listens to problem statement. Asks questions for clarification. Records pertinent information.
Solution effort	Students develop and implement a plan for solving the problem.	Encourages students to share ideas with others. Poses questions to refocus students' attention on relevant information. Provides hints to help students make progress (only as a last resort).	Works alone or with peers on problem.
Problem and solution discussion	Students gain new insights into the problem and good problem-solving behavior.	Allows students (3–5) to demonstrate their solution. Points out the importance of checking work. Aids in bringing to light any possible generalizations.	Offers to discuss her or his solution. Listens or discusses other's solutions. Attempts to identify generalizations. Attempts to analyze what worked and why.

telling students what to do. For example, these teachers were uncomfortable at first in responding to questions like "Does this mean to add?" or "Is this right?" with "I can't tell you, try to decide for yourself" or "Compare your solution with Billy's." The final phase, Problem and Solution Discussion, was the least successful phase of the three for teachers. Often they were unsure of how directive they should be in modeling good strategies or in accepting student efforts, no matter how naive. Frequently they were unable to identify any possible generalizations.

Despite the lack of success with Phase 3 of the instructional scheme, student problem-solving behavior improved dramatically. In order to measure behavior changes, a posttest consisting of two process problems was administered to samples of experimental and control students. Members of the MPSP staff who were not involved in the study were asked to analyze the students' work in terms of: (*a*) use of an appropriate strategy; (*b*) use of relevant information in the problem; and (*c*) evidence of planning and evaluating. These evaluators were asked to sort the papers into three piles: papers of experimental students, papers of control students, and indeterminate papers. Sixty-four percent of the experimental students' papers were placed in the correct pile, whereas only 27% of the papers of control students were placed in the experimental group pile. Though this sort of evaluation was informal and fraught with shortcomings, it did suggest that solving problems does improve problem-solving ability.

After making some modifications in the instructional procedures, a second teaching experiment was conducted using two fourth-grade, three fifth-grade, and one sixth-grade classes (all different from classes involved previously). Class sizes were about the same as earlier, with the exception of the sixth-grade class, which had 16 students. Teachers had the same general characteristics as those in the earlier study. During a 4-week period, the six classes worked on 15 problems as time permitted. Fourteen of the problems were chosen as pairs, equivalent with respect to one or more factors. These factors were: context, mathematical structure, type of information provided, and size of number in the problem statements. Pairing of problems was done to facilitate an unobtrusive examination of student growth.

As in the first study, the Discussion phase remained problematic for teachers. This is perhaps to be expected because it is difficult to teach students to analyze and generalize, and teachers are rarely required by the mathematics (or other) textbooks they use to engage students in these types of activities. Consequently, it was unreasonable to expect much success without providing teachers with extensive in-service training.

Notwithstanding the difficulties inherent in the instructional scheme, the results of the second study, like the first, were very encouraging. Students were enthusiastic and evidence was found of changes in their manner of attacking problems. Also, from an analysis of students' work, there was a discernible change from the first to the second problem of the paired problems. Interviews with both teachers and students supported this observation. It was common to hear remarks from students like, "We've done this one before except the days are different." In some cases, as much as 4 weeks had elapsed between the first and the second problems, yet memory for both contextual details and appropriate solution-strategies was high. In most cases, the second problem had become a routine exercise.

In general each of the six goals was achieved to some extent. At the same time, evidence was found that problem-solving ability is facilitated significantly by solving problems. The fact that student growth was widespread for both studies in spite of the failure of the Problem and Solution Discussion phase provides additional weight to this claim. Formal instruction was never consciously employed, yet virtually every child improved in either or both of the cognitive and affective domains.

A wealth of ''soft'' but valuable information was gained as a result of this year-long effort. If the two studies just described are considered pilot studies, a framework has been established for conducting a carefully designed, longitudinal study. Because funding for the MPSP was terminated at the end of the second year, a more formal test of the conjectures had to be postponed. Such a test was completed in the Parkersburg, West Virginia schools under the direction of a former member of the MPSP staff. This study was a multiyear project involving several hundred elementary- and middle-school students. Experimental classes were compared with control classes using both quantitive and qualitative measures. Also, the development of students' cognitive processes over time was monitored (Charles, 1982; Charles and Lester, Note 11).

Other Studies

Five research studies growing out of the MPSP have been or are being conducted. Of these, the West Virginia study was mentioned in the last section and is not discussed further.

A main feature of a study by Moses (1977) was an investigation of a possible explanation of Conjecture 5 (i.e., children's initial problem representations are syntactically formed). She suspected that a reason for the lack of success by many students in solving problems stems from the fact that school mathematics instruction is based primarily on analytic thought processes. Furthermore, several years of ''analytic'' instruction tends to repress any natural tendency to use visual solution processes. Moses developed an instructional unit designed to improve students' abilities to perceive relationships in a story and to manipulate objects mentally. The subjects for her study were students in two fifth-grade classes. Over a period of 9 weeks, she provided instruction in the study of three-dimensional objects via concrete representations, the study in two dimensions of drawings of two- and three-dimensional objects, and the study of the transformation of word problems into two- or three-dimensional representations. Results indicated that instruction in visual processes improved spatial ability (as measured by a variety of spatial ability tests) and that this instruction had a greater effect on solving spatial problems than on solving analytical problems. In fact, students performed slightly less well on the analytic problems as a result of the instruction. However, instruction did not affect overall problem-solving performance. Moses pointed out that the instructional period probably needed to be much longer than 9 weeks and that visually oriented instruction should begin earlier than fifth grade if there is to be much hope of lasting changes in the ways children represent and organize problem information.

Studies by Putt (1978) and Proudfit (1979) investigated Conjectures 3 and 4. Among several other things, both studies compared the effects of specific problem-solving instruction with the effects of no special instruction (in which students were simply given problems to solve). Subjects in both investigations were fifth graders. Both Putt and Proudfit focused primarily on qualitative information obtained from students' work on problems. There were two very interesting results of Putt's research. First, his instruction was very effective in getting students to ask themselves questions that might help them better understand a problem. Second, students who were not taught any specific strategies used a wider variety of strategies for some problems than students who were taught specific strategies. The first result suggested that systematic instruction can help students learn how to evaluate their level of understanding of a problem. The second result points to a potential danger in teaching students a few "tools of the trade." That is, they tend to rely primarily on the acquired skills and their behavior becomes mechanical and inflexible. In general, Putt's research supported Conjecture 3 but offered little new insight about Conjecture 4.

Proudfit (1979) found that the only significant difference between her instruction group and a "practice-only" group was in the ability of the students to evaluate their answers and solution processes. Students who were taught to examine their performance during the four stages of problem solving (understanding, planning, implementing, and evaluating) were more capable of evaluating their activities than students in the practice-only group. Also, she found that for both groups the majority of errors made were understanding errors that did not seem to be attributable to reading difficulties.

To summarize, it appears that children can be taught certain skills and strategies associated with successful problem-solving (Conjecture 3). However, whether or not such instruction is superior to simply having students try to solve problems without formal instruction is still open to debate.

Harik (1979, 1981) was not interested in problem-solving instruction, but rather in the influence of certain task variables on task difficulty and on the problem-solving processes used. Her research was motivated to some extent by (*a*) the observation that trial-and-error (particularly *guessing*) is a dominant strategy among novice problem-solvers (Conjecture 1) and (*b*) appreciation of the power of guessing as a legitimate problem-solving process. Harik developed a set of word problems embodying algebraic systems of simultaneous equations for which guessing was a nearly unavoidable process. The goal was to examine closely the nature of students' guessing behavior. Subjects in her study were 60 average-to-above-average seventh-graders who had no previous instruction in algebra. Consequently, they were not expected to be able to translate the problems into simultaneous equations. The problems varied with respect to a structural variable, the size of the search space, and a content variable, the number of unknowns and conditions. Each student was asked to think aloud while solving six word-problems. A graphical coding scheme was developed to portray patterns and to elucidate different types of guessing behaviors.

Among the most important results of Harik's research were the following:

1. Three types of guessing moves were identified—probabilistic moves, certainty moves, and manipulative moves. A *probabilistic move* involves a guess on the unknown, calculations to determine if the guess satisfies the conditions, and a conclusion drawn from the guess. A *certainty move* is one in which a final or preliminary result is reached by means of deduction and/or induction. A *manipulative move* is a series of manipulations of numbers and operations in such a way that they produce, in the problem solver's mind, a reasonable answer (Harik, 1981).
2. Problems with small search-spaces were significantly easier than were problems with large search-spaces.
3. Problems with three unknowns and three conditions were much harder than were problems with two unknowns and two conditions.
4. Subjects were goal oriented and did not persist in pursuing blind alleys.
5. Although it was neither taught nor encouraged, 97% of the subjects used guessing.
6. Only 1% of the subjects displayed an essentially algebraic approach to the solution of the problems.
7. Although most subjects were able to coordinate simultaneous conditions, about one-third of the problem solutions indicated a lack of coordination (Conjecture 2).

Thus, Harik's research provides valuable insight into the nature of the problem-solving behavior of generally inexperienced problem-solvers. By designing problems that would elicit trial-and-error (guessing) behavior, she was able to determine that in most instances guesses were clearly not random (76% of all moves were probabilistic moves). In addition, she found that many students as old as 12 to 13 years of age still failed to coordinate multiple conditions.

Harik's study involved far more than an investigation of guessing behavior and the ability to coordinate multiple conditions in a problem, and the study represents a thoughtful adaptation of techniques of state–space analysis to the study of an important class of mathematical problems, namely, word problems involving multiple unknowns and variables. Like the research of Charles, Moses, Proudfit, and Putt, it illustrates an attempt at one institution to initiate a series of systematic, coordinated investigations of some of the fundamental questions in mathematical problem-solving research.

Issues for Future Research

In the preceding sections I have provided a bird's-eye view of current trends in mathematical problem-solving research. In this final section, I raise several issues associated with these trends. I hope discussing these issues will facilitate an exchange of ideas regarding the directions mathematical problem-solving research will take in the near future.

Need for Theory

Past mathematical problem-solving research has suffered from the absence or neglect of theory. The shift toward a theory orientation raises two questions:

1. Are any existing psychological theories adequate to explain mathematical problem-solving, or must special theories be developed? For example, do the principles of information processing, based largely on "puzzle problems" (e.g., Tower of Hanol, missionaries and cannibals), generalize to mathematical problem-solving?
2. Are there certain fundamental problem-solving skills and processes that are the exclusive domain of mathematics?

With respect to Question 1, it is of course too soon to tell. A number of studies by mathematics educators have recently been completed or are currently underway that attempt to determine the generalizability of psychological research. An answer to the second question will come only after considerably more research is conducted, based on general theories of problem solving. One especially promising trend is the renewed interest in general models of cognitive abilities. Luria's (1966, 1973) model of the functional organization of the brain and the related factor-analytical model proposed by Das, Kirby, and Jarman (1975, 1979) are especially noteworthy inasmuch as they contain descriptions of the processes that underlie cognitive abilities. Also, Sternberg's componential theory of intelligence is an information-processing theory that may prove to be particularly useful in research related to problem-solving instruction (Sternberg, 1981; Sternberg & Weil, 1980).

Within the mathematics education community, Garofalo (1982) at Indiana University plans to extend his work of investigating the structure of arithmetical performance to other types of mathematical activity. In addition, at the University of Georgia, Kilpatrick recently has begun a large-scale study of mathematical abilities and one of his students has undertaken a comprehensive review of all factor-analytic research on mathematical abilities (Deguire, 1980). Research of this type will go a long way toward linking mathematical thinking to, and distinguishing it from, other types of thinking.

Diversity of Research Tasks

Problem-solving research has been a confused area of inquiry largely because widely diverse types of tasks are used, clearly affecting the generalizability of results. With this in mind, the following questions arise:

1. Is there a need for a core of research tasks (i.e., a collection of research instruments) to be used by all problem solving researchers?
2. What research questions are addressed best by using problems from the standard mathematics curriculum and what questions are best addressed by puzzle problems and the like?
3. How much emphasis should be placed on research involving real-world, applied problem-solving?
4. Is there a need for a systematic classification of problem types that could be used by researchers to aid communication of the most appropriate tasks for a particular purpose?

It is impossible to draw any confident conclusions about problem-solving research results if the tasks used are not clearly described. In addition, the same problems must be used in several studies. I suggest that research tasks be chosen on the basis of:

1. a consideration of the task variables involved (viz., syntax, content, context, structure and process variables; Goldin & McClintock, 1979);
2. the extent to which the tasks are well suited to the research questions;
3. the relevance of the tasks for the mathematics curriculum; and,
4. the suitability of the tasks for the subjects in question (or vice versa).

The National Collection of Research Instruments for Mathematical Problem Solving (Goldin, McClintock, & Webb, 1977) represents a first step toward establishing a core of research tasks. A primary objective of this collection is to provide a resource of mathematical tasks that have been used in at least one research study. Problems are documented with respect to data on subject, treatment, and task variables.

Regarding the second question, it is not at all obvious that results based on puzzle-problem research can be extended to problem-solving involving standard mathematics problems. Indeed, there is some evidence to suggest that very different things occur when mathematical problems rather than puzzle problems are employed (e.g., Silver, 1981). Problems like the Tower of Hanoi and missionaires and cannibals are rich as research tasks because they lend themselves to state–space and other forms of systematic analysis, which in turn make it possible to study the relationships between various task-structure variables. Problems of this sort should be used to assist in the generation of hypotheses that can then be tested using more standard mathematical tasks.

How much emphasis to put on research involving real-world applied problem-solving is difficult to determine. Lesh *et al.* (1979) considers applied mathematical problem-solving to occur when ordinary people attempt to solve real problems involving substantive mathematical content in real (or at least realistic) situations, but there has been an appalling lack of research related to applied problem-solving. Bell (1979) echoed this observation with the claim that he has found in the mathematical problem-solving research literature: ''very little that is applicable to improving instruction even for conventional 'word problems' and practically nothing related to 'applied problem solving' '' (p. 5). If Bell's statement is accurate, and it seems to be, the plea made by Lesh for more research must be heeded not only by researchers but also by curriculum developers and teacher educators.

Competency versus Performance Models

Considerable attention has been paid to identifying traits of successful (expert, good) problem-solvers for the purpose of developing competency models of problem-solving behavior (i.e., how competent problem-solvers behave). Should more emphasis be placed on developing performance models of behavior?

This issue is closely related to the need for theory. The ''structure of cognitive processes and abilities'' studies that are becoming more common in mathematics education indicate more concern for performance models of problem-solving behavior. Far too little is known about the nature of the problem representations that ''typical'' individuals form, the types of misconceptions they have, the errors they make, and the heuristic strategies they naturally employ. Also, there is a danger in devoting exclusive attention to creating competency models. An underlying assumption of at least some of the competency-model research is that less capable problem-solvers will become more proficient if they are taught the skills and processes used by the experts. Although it is important to identify those behaviors that distinguish good from poor problem-solvers, it is far from axiomatic to think that such behaviors can be taught (e.g., can individuals be taught to curtail thinking, or to notice problem structure?).

Regarding research toward the development of performance models, the work of Matz (Note 8) in high-school algebra and Brown and Burton (1978) in ''basic math'' (essentially elementary-school arithmetic) is particularly promising and worth consideration by mathematics educators. More of this sort of diagnostic model-building research needs to be undertaken by mathematics educators. Psychologists and computer scientists cannot be entrusted with the sole responsibility for the development of models of the sort created by Matz, and Brown and Burton; they simply are not sensitive to the same issues, nor are they interested in pursuing answers to the same questions that motivate the research of mathematics educators.

The foregoing comments should not be interpreted as implying that no compentency-model research should be undertaken by mathematics educators. The excellent work of Larkin and her associates with university physics students and professors has led to the creation of models of physics problem-solving that serve to describe what experts do differently from novices (Larkin, 1980; Larkin *et al.*, 1980).

Among other things, Larkin's work suggests that a main difficulty in solving physics problems lies not with applying principles, but rather with selecting which principles to apply. This is a major source of the difference between expert and novice physics problem-solvers. Research of this sort is valuable insofar as it serves to pinpoint specific deficiencies in the problem-solving behavior of novice or poor problem-solvers, while at the same time indicating what good solvers do.

Problem-Solving Instruction

There is little agreement among mathematics educators regarding how best to improve students' problem-solving performance beyond the obvious fact that attempting to solve problems is a necessary ingredient.

Mathematics educators are eager for research to provide answers to instruction-related questions. Schoenfeld's (1979, 1980) "managerial strategy" approach and the "metacognition" approaches of Silver *et al.* (1980), Stengel, *et al.* (Note 10), and Charles (1982) represent promising developments in recent problem-solving instruction research. Schoenfeld's approach with college students is to stress specific mathematical problem-solving heuristics and managerial strategies that are useful in mathematical problem-solving. The heuristic strategies are (*a*) drawing a diagram, (*b*) looking for an inductive argument if there is an integer parameter, (*c*) arguing by contradiction or contrapositive, (*d*) trying a similar approach with fewer variables, and (*e*) establishing subgoals. A *managerial strategy* is an efficient means for choosing approaches to problems, for avoiding blind alleys, and for allocating problem-solving resources (Schoenfeld, 1980). Schoenfeld recognized the need to do more than equip students with a collection of skills and strategies. They must also be taught how to marshal their resources to make reasonable decisions about which strategies to try and when to try them (see also Schoenfeld, Chapter 10, present volume.)

Metacognition instruction typically emphasizes having students discuss and think about the processes they use to solve problems in an effort to make them view problem solving as a multifaceted complex of processes and to make them aware that many problems allow multiple solution methods. This approach has particular promise for elementary- and middle-school students because it is in the early years of mathematics learning that students need to develop an understanding of the nature of problem solving and an awareness that mathematics is much more than a collection of facts, algorithms, and formulas.

Role of the Teacher

The teacher's role in problem-solving instruction has been almost totally ignored as a research variable by mathematics educators. The teacher is an important "environment factor" and must be taken into consideration in any instruction-related research.

This issue is, of course, related to the preceding issue but it is listed separately for emphasis. No approach to problem-solving instruction is likely to succeed if the teacher does not make students sense the importance of problem solving or does not reward students' problem-solving efforts. Unfortunately, beyond these rather general observations relatively little is known about the way in which the teacher affects problem-solving behavior. An important reason why so little knowledge has accrued from instruction research is that teacher variables have been ignored or factored out. Perhaps an alliance is needed between mathematics educators involved in teaching-strategies research and those interested in problem-solving instruction.

The Nature of Problem-Solving Improvement

Until quite recently, researchers have relied almost exclusively on quantitative measures of problem-solving performance. Current trends indicate wholesale desertion of traditional practices for new, promising, but untried, qualitative masures. Three questions associated with this issue come readily to mind.

1. Is it enough to use the number of correct solutions or time to solution as the measure of performance? Alternatively, is it enough to look for the use of heuristic strategies and skills and other processes?
2. What sort of transfer of learning should be expected? For example, should broad transfer be expected as a result of instruction or should transfer be anticipated only to analogous problems? Also, what kinds of processes and skills should be expected to transfer after instruction?
3. What is the potential of paper-and-pencil tests for providing valuable information about processes used during problem solving?

Quantitative and statistical procedures (e.g., factor analysis, regression analysis, interaction analysis) are proven valuable tools for educational research. Before the use of these techniques is abandoned, current methods of qualitative analysis must be refined to achieve the level of conceptual clarity and rigor that is characteristic of quantitative analysis. Also, quasi-qualitative analyses that are restricted to counting the number of times particular processes are exhibited do little to provide insight into the true nature of problem-solving behavior. Much work is needed on the development of reliable and valid methods for collecting truly qualitative data.

Regarding the first question, it seems obvious that multiple criteria for performance, both quantitative and qualitative, should be employed. The second question is more difficult to answer because it is largely a matter of the researcher's theoretical position. This once again points to the importance of theory-based research. The final question is partly a matter of theoretical position and partly a methodology issue. Although this question is not of interest to most psychologists, it is a real and important issue for mathematics educators who are interested in classroom practices. There is nearly unanimous agreement that standard problem-solving tests (e.g., Stanford, Metropolitan) provide insufficient data about problem-solving behavior and, for the most part, include only routine translation problems. Substantial effort has been devoted to developing better paper-and-pencil instruments for group administration (e.g., Proudfit, Note 2; Romberg & Wearne, Note 3; and Schoen & Oehmke, Note 4). Yet, although paper-and-pencil test-development may be a noble cause, such tests have little value as research tools at the present time. The evolution of sound theories of mathematical problem-solving will result only from the systematic and prolonged observation of students.

Looking Back

Anyone who has taken a serious interest in mathematical problem-solving knows that the "Looking Back" stage (as described by Polya, 1957) may be the most neglected stage in the process of solving problems. An obvious aspect of this stage is to look over all the steps in the solution to make sure all conditions have been satisfied and all work has proceeded in a rational fashion. In looking back over the preceding pages it may be that not all the conditions of the task set at the beginning of this chapter have been satisfied. If this is the case, perhaps the failure can be attributed to the complexity of the task. Another aspect of Polya's fourth stage is to ask if the problem could be attacked in another way. Other existing characterizations of the mathematical problem-solving literature may provide additional ideas.

Perhaps the most important part of the Looking Back stage is to attempt to identify the key features of the solution effort that may prove to be useful in future problem-solving. The key ingredients of this chapter are: need for theory, interest in covert behavior, high quality of current research efforts, and issues for the future.

With respect to the first ingredient, it would be a mistake to devote undue attention to the development of special theories of mathematical problem-solving. There are at least two alternatives: (*a*) consider mathematical problem-solving in terms of general theories of problem solving, or (*b*) develop general theories of mathematics learning and consider problem solving as a type of learning within these theories. Regarding the first alternative, studying mathematical problem-solving in a broader context may make it possible to identify what distinguishes *mathematical* problem-solving from other types. As a result, such questions as "What causes students to have difficulty in solving problems?'' may become easier to answer. The second alternative is attractive because it allows for the consideration of problem solving as a natural part of all mathematics learning. This alternative makes it possible to look at, for example, concept learning, algorithm learning, and formula application from a problem-solving perspective. Greeno (1980) makes the point that the

> reason that the distinction between knowledge-based performance and problem solving has eroded is that we now can characterize the performance of individuals who solve problems; and when we carefully consider the performance that occurs in more routine situations, we find that the essential characteristics of real problem solving are there also (p. 10).

Greeno continues: ''It is seriously misleading to label performance in some situations as problem solving and in other situations in which the same kind of cognitive processes occur as not involving problem solving'' (p. 12). Both these alternatives will require greater cooperation among researchers of widely different backgrounds, interests, and perspectives.

The second major point of this chapter is that, despite the growing interest in processes, there remains too much emphasis on the results of problem solving and on overt behavior. Instead, there should be more direct attention given to determining the cognitive demands being made upon the problem solver. In order to illustrate this point, consider the following example. A popular elementary school problem is the following:

> There were 8 people at a party. If each person shook every other person's hand once, how many handshakes would there be?

The little research that has been done using this problem typically has focused on the types of strategies students employ to solve it (e.g., trial-and-error, simplifying the problem, acting out the situation), on syntactical factors that influence comprehension, or on whether or not students can generalize the solution. Nowhere has there been an analysis of what cognitive processes problem solvers are being asked to use in order to solve this problem. If such an analysis occurred, it might be possible to identify potential sources of difficulty, to match problems to the developmental level of students, and otherwise to design better problem-solving instruction. Also, the results of the most recent National Assessment of Educational Progress indicate that even when students possess adequate reading and computational skills, many of them are still unable to solve problems involving the use of these skills (Carpenter, Corbitt, Kepner, Lindquist, & Reys, 1981). Apparently it is not enough to equip students with a collection of skills.

Bundy (Note 12), an artificial-intelligence scientist, has observed that knowledge of concepts and acquisition of skills is not enough for successful problem-solving. Bundy's approach is to analyze tasks, not just to develop a hierarchy of prerequisite knowledge, but also to identify the mental processes required of the problem solver. More research is needed that looks closely at the thought processes involved in searching for and selecting those skills and ideas, previously mastered in isolation, that are needed to solve a problem.

The research described in this chapter emphasizes only one important aspect of current mathematical problem-solving research, namely, that it is being thoughtfully and systematically conducted. The range of types of problem-solving investigations in recent years has been as wide as the quantity has been large. The diversity in research interests among mathematics educators was illustrated clearly at a recent conference on issues in mathematical problem-solving research held at Indiana University (Lester & Garofalo, 1982). Five working groups were formed around five broad areas: knowledge structures, instruction, learning, measurement and assessment, and metacognition. Research is currently underway or is being planned in each of these areas.

Finally, it appears that several of the issues raised in the preceding section also are applicable to other disciplines within education (e.g., science education, reading) and to psychology. Although it is necessary and inevitable for educational and psychological researchers to pursue different questions and perhaps to employ different methodologies, mutual benefit would accrue from more open dialogue about research issues and from cooperative efforts among researchers in different fields. It is my hope that this chapter is a step in that direction.

Reference Notes

1. *Report of the Snowmass Conference on the K–12 Mathematics Curriculum.* Bloomington, Indiana: Indiana University, Mathematics Education Development Center, June, 1973.
2. Proudfit, L. *The development of a process evaluation instrument* (Tech. Rep. 5, Mathematical Problem Solving Project). Bloomington, Indiana: Indiana University, Mathematics Education Development Center, 1977.
3. Romberg, T. A., & Wearne, D. *Romberg-Wearne mathematics problem solving test.* Madison, Wisconsin: University of Wisconsin, Research and Development Center for Cognitive Learning, 1975.
4. Schoen, H. L., & Oehmke, T. *Iowa problem solving test.* Iowa City: University of Iowa, 1980.
5. Chi, M., Feltovich, P. J., & Glaser, R. *Representation of physics knowledge by experts and novices* (Tech. Rep. 2). Pittsburgh: University of Pittsburgh Learning Research and Development Center, 1980.
6. Schoenfeld, A. H., & Herrmann, D. J. *Problem perception and knowledge structure in expert and novice mathematical problem solvers.* Manuscript submitted for publication, 1981.
7. Webb, N. L. *A review of the literature related to problem solving strategies used by students in grades 4, 5, and 6.* (Tech. Rep. 1, A). Mathematical Problem Solving Project. Bloomington, Indiana: Mathematics Education Development Center, Indiana University, 1977.
8. Matz, M. *Towards a process model for high school algebra errors.* (Working Paper 181). Cambridge, Massachusetts: MIT, Artificial Intelligence Laboratory, 1979.
9. LeBlanc, J. F., & Kerr, D. R. *The mathematical problem solving project: Problem solving strategies and applications of mathematics in the elementary school* (Final Rep.). Bloomington, Indiana: Mathematics Education Development Center, Indiana University, 1977.

10. Stengel, A., LeBlanc, J. F., Jacobson, M., & Lester, F. K. *Learning to solve problems by solving problems* (Tech. Report 2, D). Mathematical Problem Solving Project. Bloomington, Indiana: Mathematics Education Development Center, Indiana University, 1977.
11. Charles, R. I. and Lester, F. K. *An evaluation of a process-oriented mathematical problem-solving instructional program in grades five and seven.* Manuscript submitted for publication, 1983.
12. Bundy, A. *Analysing mathematical proofs* (Research Rep. 2). Department of Artificial Intelligence, University of Edinburgh, 1975.

References

Asher, J. J. Towards a neo-field of problem solving. *Journal of General Psychology,* 1963, *68,* 3–8.

Barnett, J. The study of syntax variables. In G. Goldin & C. E. McClintock (Eds.), *Task variables in mathematical problem solving.* Columbus, Ohio: ERIC/SMEAC, 1979.

Bell, M. S. Applied problem solving as a school emphasis: An assessment and some recommendations. In R. Lesh, D. Mierkiewicz, & M. Kantowski, (Eds.), *Applied mathematical problem solving.* Columbus, Ohio: ERIC/SMEAC, 1979.

Bronfenbrenner, U. The experimental ecology of education. *Educational Researcher,* 1976, *5*(9), 5–15.

Brown, A. Knowing when, where, and how to remember: A problem of metacognition. In R. Glaser (Ed.), *Advances in instructional psychology.* Hillsdale, New Jersey: Erlbaum, 1978.

Brown, J. S., & Burton, R. R. Diagnostic models for procedural bugs in basic mathematical skills. *Cognitive Science,* 1978, *2,* 155–192.

Brownell, W. A. Problem solving. In *Psychology of Learning,* 41st Yearbook of the National Society for the Study of Education (Part 2). Chicago: The Society, 1942.

Caldwell, J. *The effects of abstract and hypothetical factors on word problem difficulty in school mathematics.* Unpublished doctoral dissertation, University of Pennsylvania, 1977.

Caldwell, J., & Golding, G. A. Variables affecting word problem difficulty in school mathematics. *Journal for Research in Mathematics Education,* 1979, *10*(5), 323–336.

Carpenter, T. P., Corbitt, M. K., Kepner, H., Lindquist, M. M., & Reys, R. E. *Results and implications of the second mathematics assessment of the National Assessment of Educational Progress.* Reston, Virginia: National Council of Teachers of Mathematics, 1981.

Charles, R. I. An instructional system for mathematical problem solving. In S. L. Rachlin (Ed.), *MCATA monograph on problem solving.* Calgary, Canada: University of Calgary, 1982.

Chase, W. G., & Simon, H. A. Perception in chess. *Cognitive Psychology,* 1973, *4,* 55–81.

Das, J. P., Kirby, J., & Jarman, R. F. Simultaneous and successive syntheses: An alternative model for cognitive abilities. *Psychological Bulletin,* 1975, *82,* 87–103.

Das, J. P. Kirby, J., & Jarman, R. F. *Simultaneous and successive cognitive processes.* New York: Academic Press, 1979.

Davis, G. A. Current status of research and theory in human problem solving. *Psychological Bulletin,* 1966, *66*(1), 3–54.

DeGuire,L. *A review of the factor-analytic literature on mathematical abilities.* Paper presented at the 4th International Congress on Mathematics Education, Berkeley, California, August, 1980.

Dodson, J. W. *Characteristics of successful insightful problem solvers.* Report 31 of The National Longitudinal Study of Mathematical Abilities. Stanford, California: School Mathematics Study Group, 1972.

Duncan, C. P. Recent research on human problem solving. *Psychological Bulletin,* 1959, *56,* 397–429.

Ericsson, K. A., & Simon, H. A. Verbal reports as data. *Psychological Review,* 1980, *87*(3), 215–251.
Flavell, J. H. Metacognitive aspects of problem solving. In L. B. Resnick (Ed.), *The nature of intelligence*. Hillsdale, New Jersey: Erlbaum, 1976.
Garofalo, J. *Behavior regulation, simultaneous synthesis and the factor structure of arithmetic*. Unpublished doctoral dissertation, Indiana University, 1982.
Goldberg, D. J. *The effects of training in heuristic methods in the ability to write proofs in number theory*. Unpublished doctoral dissertation, Teacher's College, Columbia University, 1974.
Goldin, G. A. Structure variables in problem solving. In G. A. Goldin & C. E. McClintock (Eds.), *Task variables in mathematical problem solving*. Columbus, Ohio: ERIC/SMEAC, 1979.
Goldin, G. A. The measure of problem-solving outcomes. In F. K. Lester & J. Garofalo (Eds.), *Mathematical problem solving: Issues in research*. Philadelphia, Pennsylvania: The Franklin Institute Press, 1982.
Goldin, G. A., & Luger, G. F. *State-space representations of problem structure and problem solving behavior* (Tech. Report). University of Pennsylvania, 1974.
Goldin, G. A., & Luger, G. F. *Problem structure and problem solving behavior*. Proceedings of International Joint Conference on Artificial Intelligence, Tbilisi, USSR. Cambridge, Massachusetts: MIT Press, 1975.
Goldin, G. A., & McClintock, C. E. (Eds.). *Task variables in mathematical problem solving*. Columbus, Ohio: ERIC/SMEAC, 1979.
Goldin, G., McClintock, C. E., & Webb, N. *Task variables in problem solving research*. Panel presentation at the Annual Meeting of the Annual Meeting of the National Council of Teachers of Mathematics, Cincinnati, Ohio, April, 1977.
Gorman, C. J. A critical analysis of research on written problems in elementary school mathematics. *Dissertation Abstracts,* 1968, *28*(11), 4818–4819A.
Greeno, J. G. Nature of problem solving abilities. In W. K. Estes (Ed.), *Handbook of learning and cognitive processes: Human information processing* (Vol. 5). Hillsdale: New Jersey Erlbaum, 1978.
Greeno, J. G. A theory of knowledge for problem solving. In D. T. Tuma & F. Reif (Eds.), *Problem solving and education: Issues in teaching and research*. Hillsdale, New Jersey: Erlbaum, 1980.
Harik, F. Heuristic behavior associated with problem tasks. In G. Goldin & C. E. McClintock (Eds.), *Task variables in mathematical problem solving*. Columbus, Ohio: ERIC/SMEAC, 1979.
Harik, F. *The influence of problem structure on problem difficulty and problem solving processes*. Unpublished doctoral dissertation, Indiana University, 1981.
Hatfield, L. F., & Bradbard, D. A. (Eds.), *Mathematical problem solving: Papers from a research workshop*. Columbus, Ohio: ERIC/SMEAC, 1978.
Jerman, M. E. Individualized instruction in problem solving in elementary school mathematics. *Journal for Research in Mathematics,* 1973, *4*(1), 6–19.
Jerman, M. E. Problem length as a structural variable in verbal arithmetic problems. *Educational Studies in Mathematics,* 1974, *5,* 109–123.
Jerman, M., & Rees, R. Predicting the relative difficulty of verbal arithmetic problems. *Educational Studies in Mathematics,* 1972, *4,* 306–323.
Kantowski, M. G. Processes involved in mathematical problem solving. *Journal for Research in Mathematics Education,* 1977, *8*(3), 163–180.
Kantowski, M. G. The teaching experiment and Soviet studies of problem solving. In L. L. Hatfield & D. A. Bradbard (Eds.), *Mathematical problem solving: Papers from a research workshop*. Columbus, Ohio: ERIC/SMEAC, 1978.
Kilpatrick, J. Analyzing the solution of word problems in mathematics: An exploratory study (Doctoral dissertation, Stanford University, 1967). *Dissertation Abstracts,* 1968, *28*(11), 4380–A.
Kilpatrick, J. Problem solving and creative behavior in mathematics. In J. W. Wilson & L. R. Carry

(Eds.), *Reviews of recent research in mathematics ecucation* (Vol. 19). Stanford, California: School Mathematics Study Group, 1969.

Kilpatrick, J., Wirszup, I. (Eds.). *Soviet studies in the psychology of learning and teaching mathematics* (14 vols.). Stanford, California: School Mathematics Study Group, 1969–1975.

Krutetskil, V. A. *The psychology of mathematical abilities in school children.* Chicago: University of Chicago Press, 1976.

Kulm, G. *The classification of problem solving.* Columbus, Ohio: ERIC/SMEAC, 1979.

Kulm, G., Campbell, P. F., Frank, M., & Talsma, G. *Analysis and synthesis of mathematical problem solving processes.* Paper presented at the Annual Meeting of the National Council of Teachers of Mathematics, St. Louis, April, 1981.

Larkin, J. H. Skilled problem solving in physics: A hierarchical planning model. *Journal of Structural Learning,* 1980, *6,* 271–297.

Larkin, J., McDermott, Simon, D. P., & Simon, H. A. Expert and novice performance in solving physics problems. *Science,* 1980, *208,* 1335–1542.

Lesh, R., Mierkiewicz, D., & Kantowski, M. (Eds.), *Applied mathematical problem solving.* Columbus, Ohio: ERIC/SMEAC, 1979.

Lester, F. K. Mathematical problem solving in the elementary school: Some educational and psychological considerations. In L. L. Hatfield & D. A. Bradbard (Eds.), *Mathematical problem solving: Papers from a research workshop.* Columbus, Ohio: ERIC/SMEAC, 1978.

Lester, F. K. A procedure for studying the cognitive processes used during problem solving. *Journal of Experimental Education,* 1980, *48*(4), 323–327. (a)

Lester, F. K. Research in mathematical problem solving. In R. Shumway (Ed.), *Research in mathematics education.* Reston, Virginia: National Council of Teachers of Mathematics, 1980. (b)

Lester, F. , & Garofalo, J. (Eds.). *Mathematical problem solving: Issues in research.* Philadelphia, Pennsylvania: The Franklin Institute Press, 1982.

Lucas, J. The teaching of heuristic problem solving strategies in elementary calculus. *Journal for Research in Mathematics Education,* 1974, *5*(1), 36–46.

Lucas, J. F., Branca, N., Goldberg, D., Kantowski, M. G., Kellogg, H., & Smith, J. P. A process-sequence coding system for behavioral analysis of mathematical problem solving. In G. Goldin & C. E. McClintock (Eds.), *Task variables in mathematical problem solving.* Columbus, Ohio: ERIC/SMEAC, 1979.

Luger, G. F. State-space representation of problem-solving behavior. In G. Goldin & C. E. McClintock (Eds.), *Task variables in mathematical problem solving.* Columbus, Ohio: ERIC/SMEAC, 1979.

Luria, A. R. *Higher cortical functions in man.* New York: Basic Books, 1966.

Luria, A. R. *The working brain: An introduction to neuropsychology.* New York: Basic Books, 1973.

McClintock, C. E. Heuristic processes as task variables. In G. A. Goldin & C. E. McClintock (Eds.), *Task variables in mathematical problem solving.* Columbus, OH: ERIC/SMEAC, 1979.

Meyer, R. A. Mathematical problem solving performance and intellectual abilities of fourth-grade children. *Journal for Research in Mathematics Education,* 1978, *9*(5), 334–338.

Moses, B. E. *The nature of spatial ability and its relationship to mathematical problem solving.* Unpublished doctoral dissertation, Indiana University, 1977.

National Council of Supervisors of Mathematics. Position paper on basic mathematical skills. *Arithmetic Teacher,* 1977, *25*(1), 19–22.

Polya, G. *How to solve it* (2nd ed.). New York: Doubleday, 1957.

Polya, G. *Mathematical discovery: On understanding, learning and teaching problem solving* (Vol. 1). New York: Wiley, 1962.

Polya, G. *Mathematical discovery: On understanding, learning and teaching problem solving* (Vol. 2). New York: Wiley, 1965.

Post, T. R., & Brennan, M. L. An experimental study of the effectiveness of a formal versus an informal presentation of a general heuristic process on problem solving in tenth-grade geometry. *Journal for Research in Mathematics Education,* 1976, *7*(1), 59–64.

Proudfit, L. *The examination of problem solving processes by fifth-grade children and its effect on problem solving performance.* Unpublished doctoral dissertation, Indiana University, 1979.

Putt, I. J. *An exploratory investigation of two methods of instruction in mathematical problem solving at the fifth-grade level.* Unpublished doctoral dissertation, Indiana University, 1978.

Reed, S. K., Ernst, G. W., & Banerji, R. The role of analogy in transfer between similar problem states. *Cognitive Psychology,* 1974, *6,* 436–450.

Schoenfeld, A. H. Teaching problem solving in college mathematics: The elements of a theory and a report on the teaching of general mathematical problem-solving skills. In R. Lesh, D. Mierkiewicz, & M. Kantowski (Eds.), *Applied mathematical problem solving.* Columbus, Ohio: ERIC/SMEAC, 1979.

Schoenfeld, A. H. Teaching problem-solving skills. *The American Mathematical Monthly,* 1980, *87*(10), 794–805.

Silver, E. A. Student perceptions of relatedness among mathematical verbal problems. *Journal for Research in Mathematics Education,* 1979, *10,* 195–210.

Silver, E. A. Recall of mathematical problem information: Solving related problems. *Journal for Research in Mathematics Education,* 1981, *12*(1), 54–64.

Silver, E. A., Branca, N. A., & Adams, V. M. Metacognition: The missing link in problem solving? In *Proceedings of the 4th International Conference for the Psychology of Mathematics Education,* Berkeley, California, August, 1980.

Simon, D. P., & Simon, H. A. Individual differences in solving physics problems. In R. Siegler (Ed.), *Children's thinking: What develops?* Hillsdale, New Jersey: Erlbaum, 1978.

Sternberg, R. J. Factor theories of intelligence are all right almost. *Educational Researcher,* 1981, *9*(8), 6–13, 18.

Sternberg, R. J., & Weil, E. M. An aptitude-strategy interaction in linear syllogistic reasoning. *Journal of Educational Psychology,* 1980, *72,* 226–234.

Suydam, M. N. Research related to the mathematics learning process. In *Forschung zum Prozess des Mathematiklernens: Materialen und Studien.* Bielefeld, W. Germany: Institut fur Diedaktik der Mathematik, Universitat Bielefeld, 1976.

Vos, K. The effects of three instructional strategies on problem solving behaviors in secondary school mathematics. *Journal for Research in Mathematics Education,* 1976, *7*(5), 264–275.

Waters, W. Concept acquisition tasks. In G. A. Goldin & C. E. McClintock (Eds.), *Task variables in mathematical problem solving.* Columbus, Ohio: ERIC/SMEAC, 1979.

Webb, N. L. *An exploration of mathematical problem solving processes.* Unpublished doctoral dissertation, Stanford University, 1975.

Webb, N. L. Content and context variables in problem tasks. In G. A. Goldin & C. E. McClintock (Eds.), *Task variables in mathematical problem solving.* Columbus, Ohio: ERIC/SMEAC, 1979.

CHAPTER 9

Conceptual Models and Applied Mathematical Problem-Solving Research*

Richard Lesh, Marsha Landau,
and Eric Hamilton

This chapter defines and illustrates a theoretical construct, the *conceptual model,* as an adaptive structure central to research at the interface of two NSF (National Science Foundation)-funded projects. The Rational Number (RN) project (Behr, Lesh, & Post, Note 1) investigates the nature of children's rational-number ideas in Grades 2–8. The Applied Problem Solving (APS) project (Lesh, Note 2) investigates successful and unsuccessful problem-solving behaviors of *average-ability students,* working on problems that involve easy to identify *substantive mathematical content,* and on *realistic problem-solving situations,* in which a variety of outside resources (including calculators, resource books, other students, and teacher-consultants) are available. Lesh (1981) gives a rationale for these emphases. The kinds of problems we focus on in this chapter are much "smaller" than those that have received our greatest attention in the APS project, more closely resembling the "word problems" that appear in elementary and middle school textbooks.

The chapter consists of three major sections. The first section particularizes the discussion of conceptual models by referring specifically to rational-number concepts.

The second section discusses results from interviews in which 80 fourth through eighth graders solved problems presented in a number of formats. The

*This research was supported in part by the National Science Foundation under grants SED 79–20591 and SED 80–17771. Any opinions, findings, and conclusions expressed in this chapter are those of the authors and do not necessarily reflect the views of the National Science Foundation.

ISBN No. 0-12-444220-X

interviews revealed that: (*a*) during the solution process, subjects frequently change the problem representation from one form to another (e.g., written symbols to spoken words, spoken words to pictures or concrete models, etc); and, (*b*) at any given stage, two or more representational systems may be used simultaneously, each illuminating some aspects of the situation while deemphasizing or distorting others. A major conclusion drawn from the data is that purportedly realistic word-problems often differ significantly from their real-world counterparts in their difficulty, the processes most often used in solutions, and the types of errors that occur.

The third section of the chapter presents results from the written testing program of the RN project (see Chapter 4). Item difficulty depended on the structural complexity of the underlying conceptual models, and on the types of representational translations that the item required. Baseline information was derived comparing fourth through eighth graders' abilities to translate within and between representational modes (written language, written symbols, pictures) on sets of structurally related tasks.

The chapter concludes with remarks about how the reported research advances our understanding of the growth and development of children's conceptual models and contributes to the theoretical framework for current work on the RN and APS projects.

Conceptual Models

A *conceptual model* is defined as an adaptive structure consisting of (*a*) within-concept networks of relations and operations that the student must coordinate in order to make judgments concerning the concept; (*b*) between-concept systems that link and/or combine within-concept networks; (*c*) systems of representations (e.g., written symbols, pictures, and concrete materials), together with coordinated systems of translations among and transformations within modes; and (*d*) systems of modeling processes; that is, dynamic mechanisms that enable the first three components to be used, or to be modified or adapted to fit real situations.

For a given mathematical concept, the first two components of a student's conceptual model make up what might be called the student's *understanding* of the idea; within- and between-concept systems define the underlying structure of the concept. The third component includes a variety of qualitatively different systems for representing these understandings—using written symbols, spoken language, static figural models (e.g., pictures, graphs, diagrams), manipulative models (e.g., concrete materials), or real-world "scripts." The fourth component contains processes for (*a*) changing the real situation to fit existing understandings, (*b*) changing existing understandings to fit the situation, and (*c*) changing the model to fill gaps, eliminate internal inconsistencies, and resolve conflicts within the model itself.

Representational systems differ from one another because they emphasize or deemphasize different aspects of the underlying structure of the concept. They also differ in generative power and in their ability to manipulate relevant ideas and data simply and economically in various situations. For example, sometimes a picture is worth a thousand words; sometimes language is clearer and more efficient.

The distinction between *understandings* and *representations of understandings* is quite important in mathematics. Some major advances in mathematics have resulted from the creation of clever or powerful representations (e.g., Cartesian coordinates and decimal notation) that initially functioned primarily as externalized models of ideas (i.e., structures) that were already known. Later these representations provided new tools for generating new ideas.

The first three components of a conceptual model contain most of the "actions" commonly associated with condition–action pairs in computer-simulated information-processing models of cognition. These actions transform information *within* the model, but their application does not lead to the development of new, more refined, or higher order conceptual models. Conceptual models are *closed* (in a mathematical sense) under the operations in the first three components. The fourth component consists of dynamic mechanisms that enable the first three components to develop and adapt to everyday applications.

Although between-concept systems (part *b* of the definition) and modeling processes (part *d*) are the components of conceptual models that are most salient for the types of problems that have been the major focus of the APS project, an adequate description of those components depends on a firm understanding of within-concept networks (part *a*) and systems of representations (part *c*), which are priority foci of the RN project. Therefore, this chapter focuses on parts *a* and *c*. Detailed treatments of parts *b* and *d* will appear in future publications from the APS project.

Background for Current Research on Applied Problem-Solving

The APS project is unusual among problem-solving projects in mathematics education because, rather than emanating from instructional development or research on problem solving itself, it grew out of research on *concept formation*. The goals of that research included tracing the development of selected mathematical ideas and identifying task characteristics that influenced students' abilities to use the ideas in particular situations. Many of our theoretical perspectives evolved during investigations of mathematical abilities that are deficient in "learning disabilities" subjects (Lesh, 1979b), social and affective factors that influence problem-solving behavior (Cardone, 1977; Lesh, 1979a), the development of spatial and geometric concepts in children and adults (Lesh & Mierkiewicz, 1978), and the role of representational systems in the acquisition and use of rational-number concepts (see Chapter 4 of this volume).

A central question that the APS project is designed to address is, "What is it, beyond *having* an idea, that enables an average ability student to *use* it in realistic everyday situations?" We believe that many of the most important applied problem-solving processes contribute significantly to both the *meaningfulness* and the *usability* of mathematical ideas. It is not necessarily the case that students first learn an idea, then add some general problem-solving processes, and finally (if ever) use the idea and processes in real situations, that is, those in which some knowledge about the situation is needed to supplement the underlying mathematical ideas and processes. Rather, there is a dynamic interaction between the content of mathematical ideas and the processes used to solve problems based on those ideas. This assumption has both practical and theoretical implications. Applications and problem solving are unlikely to be fully accepted in the school mathematics curriculum unless teachers and other practitioners are convinced that they play an important role in the *acquisition* of basic mathemetical ideas. We believe that applications and problem solving should not be reserved for consideration only *after* learning has occurred; they can and should be used as a context within which the learning of mathematical ideas takes place.

The task of selecting or designing problems to use in research on applied mathematical problem-solving can be approached from at least three directions. First, one can start with important elementary mathematical ideas, sort out the different interpretations that these ideas can have in realistic situations, and identify problem situations in which these interpretations occur. This is the perspective of a recent project headed by Usiskin (Note 4) and Bell as well as parts of our RN project. One conclusion from these projects is that textbook word-problems typically represent only a narrow sample of idea interpretations and problem types that should be addressed.

A second perspective, and the one that has characterized the largest portion of our efforts on the APS project, starts with realistic everyday situations in which mathematics is used and attempts to identify the processes, skills, and understandings that are most important in their solution. Again, one conclusion is that the problem types, ideas, and skills that appear to be most critical are quite different from those that have been emphasized by mathematics education spokespersons for "basic skills" or "problem solving," and by research on textbook word-problems (Lesh, 1983).

A third approach, which is the one taken in the second section of this chapter, is to start with typical textbook word-problems and create concrete or real-world situations characterized by the same mathematical structures. From this perspective, the goal is not to investigate responses to word problems as an end in itself; rather, the goal is to understand problem solving in realistic everyday situations.

Although an abundance of research has been conducted related to problem solving (see, for example, the literature review in Lester, Chapter 8 of this volume), there has been very little applied problem-solving research that investigates the development of conceptual models or that incorporates any of the three approaches to task selection described above.

Before describing the concrete or realistic word-problem isomorphs used in the interviews reported in the second section of this chapter, we first characterize 18 realistic problems, *not* based on word problems, which have provided the research sites for most of the investigations in the APS project.

Most of our problems have been designed to require 10–45 minutes for solution by average-ability seventh-graders, and to resemble problem-solving situations that might reasonably occur in the everyday lives of the students or their families. Realistic ''outside'' resources were available, including other students, calculators, resource books and materials, and teacher consultants. Solution attempts were therefore not blocked by deficient technical skills (e.g., computation) or memory capabilities (e.g., recall of measurement facts).

All our problems were based on straightforward uses of easy-to-identify concepts from arithmetic, measurement, or intuitive geometry. No tricks were needed; the most direct solution path was a correct path, although it was not necessarily the most efficient or elegant.

In contrast with simple, one-step word-problems, a variety of solutions and solution paths were possible, varying in complexity and sophistication. The relevant ideas seldom fit into neat disciplinary categories; most of our problems required the retrieval and integration of ideas and procedures associated with a number of distinct topics, including several arithmetic operations, various number systems (e.g., rational numbers, negative numbers, decimals), several qualitatively distinct measurement systems (e.g., length, time), or intuitive ideas from geometry, physics, or other subject-matter areas.

Many of our problems were designed so that the critical solution stages would be ''nonanswer giving'' stages. For example, in many realistic problem situations, problem formulation or trial answer refinement are crucial in the solution process. Thus, for many of our problems, the goal was not to produce a numerical ''answer;'' instead it was to make nonmathematical decisions, comparisons, or evaluations using mathematics as a tool.

In typical textbook word-problems, only two or three numbers usually appear, and the most common errors result when youngsters use one of their ''number crunching routines'' to produce an answer, sensible or not. When ''too much'' information is given, (for example, three numbers may appear instead of two) the student is expected to ignore the irrevelant information. When "not enough" is given, the student is expected to conclude that the problem cannot be done. In more realistic situations, and in many of the problems that were the foci of our APS research, there was *simultaneously* "too much" and "not enough" information. Often, there was an overwhelming amount of information, all relevant, and the main difficulty was to select and organize the information that was "most useful" in order to find an answer that was "good enough." Furthermore, the given information may have consisted of both qualitative and quantitative information that had to be combined in some sensible way. In other problems, not enough information may have been provided, but a usable answer still had to be found. It may have been necessary to identify or generate additional information or data as part of a solution attempt when not all the information was given initially. Thus, in most of our problems, conceptual models serve as "filters" to select information from real situations, and as "interpreters" to organize or transform data, or to fill in (or compensate for) missing information. In such situations, a model always distorts or deemphasizes some aspects of the real situation in order to clarify or emphasize others and a major goal of our APS research is to model students' modeling behaviors (Lesh, 1982).

In much the same way that our attempts to understand students' problem behaviors are embodied in the creation of models of their cognitive processing, we assume that our subjects' attempts to understand the problems we pose are embodied in the creation and adaptation of conceptual models. Below, each of the four components of a conceptual model (within-concept networks, between-concept systems, representational systems, and modeling mechanisms) is reconsidered in greater detail, with brief references to past research.

Within-Concept Networks

To most mathematicians, mathematics is the study of *structure,* the content of mathematics consists of *structures,* and to *do* mathematics is to *create* and *manipulate* structures. It is our hypothesis that these structures (whether they are embedded in pictures, manipulative materials, spoken language, or written symbols), and the processes used in manipulating and creating these structures, comprise the "conceptual models" that mathematicians and mathematics students use to solve problems.

In past research on the development of number and measurement concepts (Lesh, 1976), spatial and geometric concepts (Lesh & Mierkiewicz, 1978), and rational-number concepts (Lesh, Landau, & Hamilton, 1980), the similarities and differences between formal axiomatic structures and children's cognitive structures have been investigated. Formal axiomatic systems were used to generate tasks and clinical interview questions to help identify the nature of children's primitive conceptualizations of mathematical ideas. For example, the relationships that a child uses to think about and compare ratios are similar to the relations that define a ''complete ordered field'' in which the elements are equivalence classes of ordered pairs of whole numbers. On the other hand, when a rational number is interpreted as a position on a number line, many of the relationships that children notice and use are simplified and restricted versions of the structural properties that define the metric topology of the rational-number line—such properties as betweenness, density, distance, and (non)completeness. According to the ''number line'' interpretation the set of rational numbers is regarded (intuitively at first) as a subset of the set of real numbers, whereas, using the ''ratio'' interpretation, rational numbers are thought of as extensions of the whole numbers. A youngster's rational-number conceptual model has a within-concept network associated with each rational-number subconstruct, for example, ratio, number line, part–whole, operator, rate, and indicated quotient.

Between-Concept Systems

One of the most important properties of mathematical ideas, whether they occur as conceptual models used by children or as formal systems used by mathematicians, is that they are embedded in well-organized *systems* of ideas, so that part of the meaning of individual ideas derives from relationships with other ideas in the system or from properties of the system as a whole (Lesh, 1979b). In rational-number conceptual models, these between-concept systems have three components: (*a*) within-concept networks associated with different rational-number types (e.g., part–whole fractions, ratios, rates, decimals, and operator transformations); (*b*) links between those networks (including understandings of the ''sameness'' and/or ''distinctness'' of the rational-number types); and (*c*) operations that, among other things, enable the transformation of a given rational number into different forms. Further, between-concept systems link rational-number ideas with other concepts such as measurement, whole-number division, and intuitive geometry concepts related to areas and number lines.

For most youngsters, between-concept systems associated with rational numbers are poorly organized and unevenly formalized. The between-concept systems derive some of their meaning from within-concept networks, and, in turn, the within-concept networks derive some of their meaning from the between-concept systems in which they are embedded. For example, the transformation of a simple fraction to a percent gives the fraction a meaning that includes proportion ideas.

That a whole structure and its parts each derive some meaning from the other has profound implications for human learning, development, and problem solving. The resulting chicken-and-egg dilemma concerning the ''whole structure'' versus ''parts within the whole'' dictates that cognitive growth must involve more than quantitative additions to knowledge or processing capabilities—qualitative reorganizations must also occur. That is, as various ideas, relationships, and operations evolve into a whole system with properties of its own, elements within the system achieve a new status by being treated as parts of the whole.

Representational Systems

An introduction to the role of representational systems in applied problem-solving is given in Lesh (1981), which includes a discussion of translation processes that contribute to the meaningfulness of mathematical ideas. Figure 9.1 shows some of the translations among different representational modes and some of the contributions to understanding ideas that those translations can make.

Though this depiction of modes and translations is neither exclusive nor exhaustive, it has suggested useful areas of investigation for the RN and the APS projects. Just as mathematical ideas are embedded in larger between-concept systems that contribute to the meaningfulness of those ideas, so are those ideas and between-concept systems embedded in representational systems such as those shown.

When we say a student *understands* a mathematical concept, part of what we mean is that he or she can use the kinds of translation processes depicted in Figure 9.1. For example, when we say a student *understands* fractions, we mean, in part, that he or she can express fraction ideas presented with circular regions using rectangular regions, or using written symbols.

Modeling Mechanisms

In applied problem-solving, important translation and/or modeling processes include (*a*) simplifying the original problem situation by ignoring "irrelevant" characteristics in a real situation in order to focus on other characteristics; (*b*) establishing a mapping between the problem situation and the conceptual model(s) used to solve the problem; (*c*) investigating the properties of the model in order to generate information about the original situation; and (*d*) translating (or mapping) the predictions from the model back into the original situation and checking whether the results "fit."

As the results we present in this chapter show, Figure 9.2 represents a useful, though oversimplified, conceptualization of the problem-solving process. For example, different aspects of the problem may be represented using different representational systems, and the solution process may involve mapping back

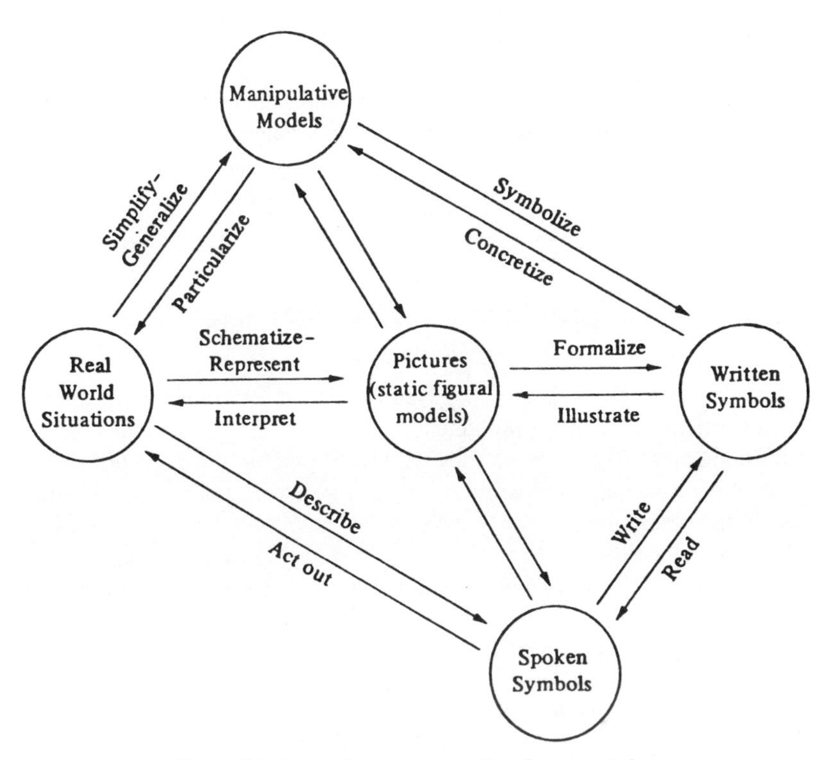

Figure 9.1 Translations among modes of representation.

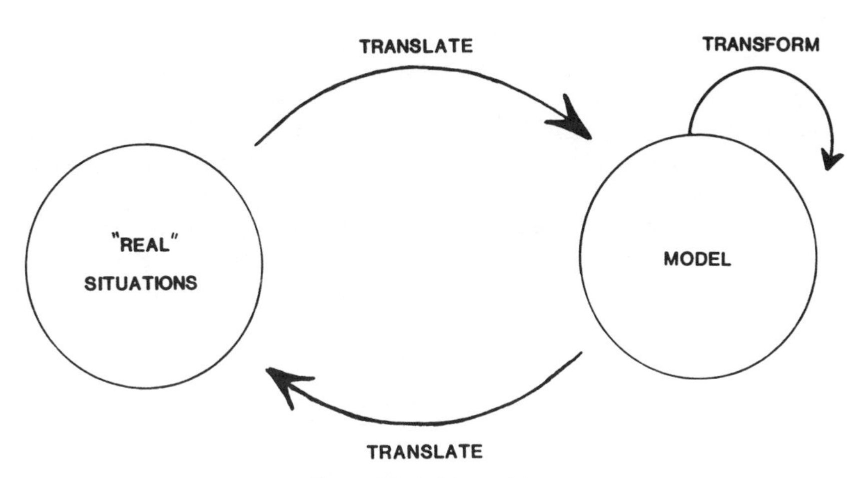

Figure 9.2 Problem solving.

and forth among several systems—perhaps using pictures as an intermediary between the real situation and written symbols.

In the next two sections of this chapter, we report data relating to two of the four aspects of conceptual models, namely within-concept networks and representational systems. Our goal is to lay a groundwork for future data presentations and elaborations relating to between-concept systems and modeling mechanisms.

Rational-Number Task Interviews

This section reports the results of interviews designed to explore the differences in student responses to sets of problems based on the same arithmetic structures but varying in presentation format. Standard word-problems were included in the interview to permit the direct comparison of students' responses to word problems with their responses to corresponding real or more realistic problems.

Some generalizations emerged from the interviews taken as a whole; other important conclusions were found by examining in detail the results of individual problems. The chief findings were that (*a*) word problems differ from their real-world counterparts with respect to difficulty, the predominant representational mode selected for solution, and most frequent error types; (*b*) varying the tasks on any of a number of dimensions (e.g., number size, context, type of manipulative material present) is accompanied by variations in the subjects' performance, suggesting that most of our subjects had unstable conceptual models relating to these tasks; and (*c*) subjects use a number of modes of representation, either simultaneously or sequentially, as they proceed through problem solutions.

The Subjects

Individual interviews were conducted with 80 subjects, 16 from each grade level, fourth through eighth. There were 46 girls and 34 boys, with approximately equal numbers of boys and girls at each grade level. At each grade, the students were chosen to represent a range of levels of rational-number understanding, as measured by the battery of written tests (described in the third section of this chapter) that were developed by the RN project. Among the 16 youngsters interviewed at each grade, 4 were selected from each quartile.

The Interviews

Each 45–60-minute interview involved the 11 basic problems described below. Some of the problems were presented as "word problems," typed on index cards, and handed to the student. Materials such as papers, pencils, clay pies, Cuisenaire rods, and counters were available within easy reach on the interview table; their use was neither encouraged nor discouraged. The "concrete problems" or "real problems" were presented orally using the materials described in each problem.

The problems were always presented in the order shown below. The entire interview is reproduced here so the reader may refer to the wording and order of the problems.

The Problems

1. The Chocolate-Eggs Addition Problems:
   - *E:* Present 6 eggs in a 12-pack carton, and say, "This is one-half carton of eggs."
   - *E:* Remove the preceding carton, present 4 eggs in a 12-pack carton, and ask, "How much is this?"
   - *E:* Repeat the preceding question, using cartons containing: 3 eggs, 8 eggs, 5 eggs, 9 eggs, and 10 eggs.
   - *E:* Present 2 eggs in one carton, and 3 eggs in a second carton, and ask, "How much is this altogether?"
   - *E:* Repeat the preceding question, using cartons containing 4 eggs and 5 eggs, respectively.
2. The Chocolate-Eggs Multiplication Problem (Presented Orally):
   - *E:* Present one dozen chocolate eggs in a 12-pack carton. Then, pose (orally) the following question, "Mike took $\frac{3}{4}$ of a carton of eggs to a picnic. He and his friends ate $\frac{2}{3}$ of the eggs. How many eggs did they eat altogether?"
   - *E:* Immediately, say, "I'll repeat the question slowly." (Repeat orally—more than once, if requested.)
3. The Addition Word-Problem:

   Jim's family ordered two pizzas for supper, one with sausage and one with mushrooms. Jim ate $\frac{1}{4}$ of the mushroom pizza, and $\frac{1}{5}$ of the sausage pizza. How much pizza did he eat altogether?
4. The Multiplication Word-Problem:

   Yesterday, Karen ate $\frac{1}{4}$ of a chocolate cake. Today, she ate $\frac{1}{3}$ of the remaining cake. How much did she eat altogether?
5. The Cuisenaire Rod Addition-Problem:

   Present a set of rods, glued onto a posterboard, where a "unit" rod is created using a $\frac{3}{8} \times \frac{3}{8}$ inch wooden stick, cut 12 cm long.

Then, present a pile of loose rods, in the following lengths:

2 12-cm rods to represent the *unit,* or 1
2 10-cm rods to represent the fraction $\frac{5}{6}$
2 9-cm rods to represent the fraction $\frac{3}{4}$
2 8-cm rods to represent the fraction $\frac{2}{3}$
2 7-cm rods to represent the fraction $\frac{7}{12}$
3 6-cm rods to represent the fraction $\frac{1}{2}$
2 5-cm rods to represent the fraction $\frac{5}{12}$
6 4-cm rods to represent the fraction $\frac{1}{3}$
6 3-cm rods to represent the fraction $\frac{1}{4}$
10 2-cm rods to represent the fraction $\frac{1}{6}$
16 1-cm rods to represent the fraction $\frac{1}{12}$

E: Point to the 12-cm rod that is glued onto the posterboard and say, "This is one unit long."

E: Then, point in turn to each of the other rods on the posterboard and (in each case) say, "How long is this?"

E: Give *S* one of the loose 9-cm rods and say, "How long is this?" Then, repeat the question using the 7-cm rod, and the 6-cm rod.

E: Next, remove the posterboard from the table, give *S* the 5-cm rod, and say, "How long is this?" Then, repeat the question using the 8-cm rod.

E: Remove the 8-cm rod, put a 6-cm rod and a 4-cm rod end to end on the table in front of *S*, and say, "How long is this (pointing to the length of the two rods together) altogether?"

E: Remove the preceding rods, put a 9-cm rod and an 8-cm rod end to end on the table in front of *S*, and say, "How long is this (pointing to the length of the two rods together) altogether?"

6. The Concrete (Pizza) Addition-Problem:

E: Present precut parts of 6-inch clay "pizzas" on cardboard "plates," like those in Figure 9.3a, and say "Pretend that these are pieces of pizza."

E: Present the $\frac{1}{2}$ piece and say, "How much is this?"

E: Present the $\frac{1}{3}$ piece and say, "And, how much is this?"

E: Present the $\frac{1}{2}$ piece and the $\frac{1}{3}$ piece simultaneously, as in Figure 9.3b, and say, "Now, how much is this altogether?"

E: Remove the preceding pieces, present the $\frac{3}{4}$ piece, and say, "How much is this?"

E: Remove the $\frac{3}{4}$ piece, present the $\frac{5}{6}$ piece, and again ask, "How much is this?"

7. The "Realistic" (Cake) Addition-Problem:

E: Present precut parts of 6 × 8-inch clay "cakes" on (scored)

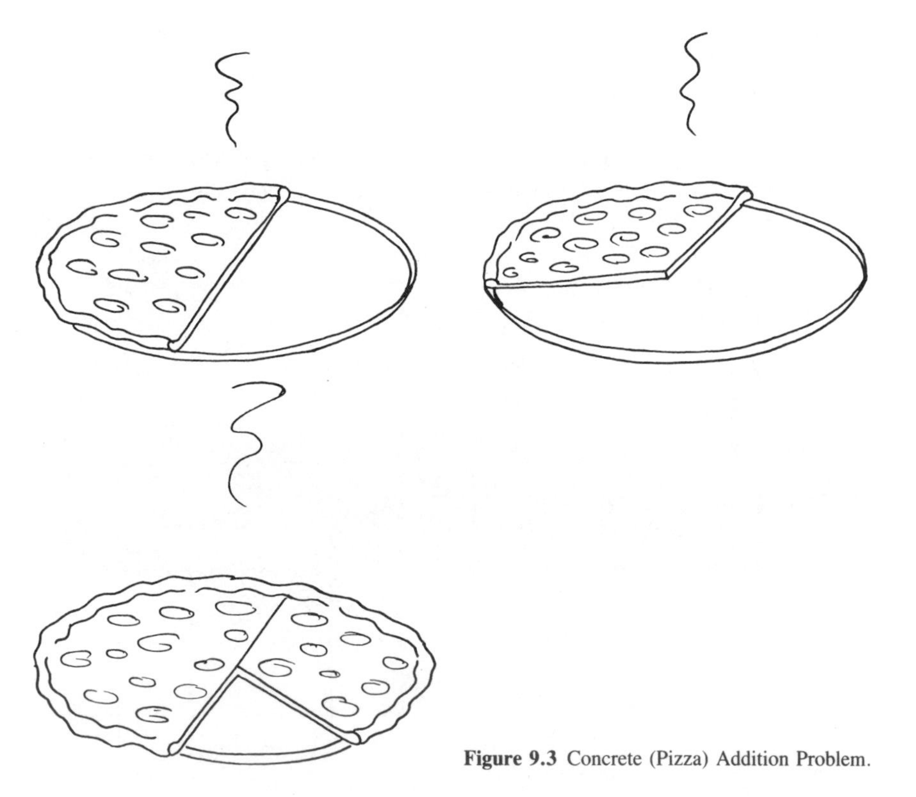

Figure 9.3 Concrete (Pizza) Addition Problem.

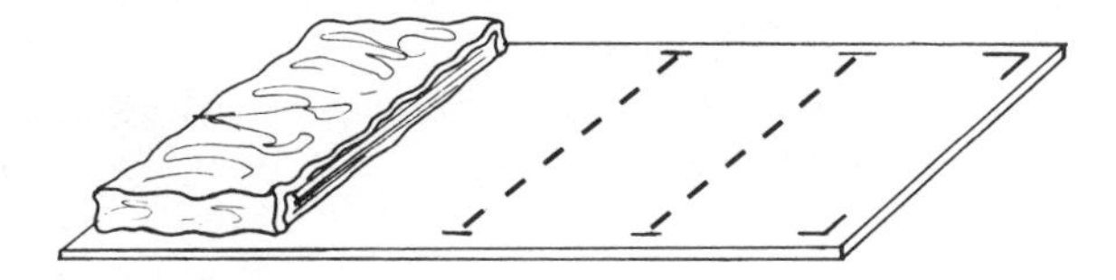

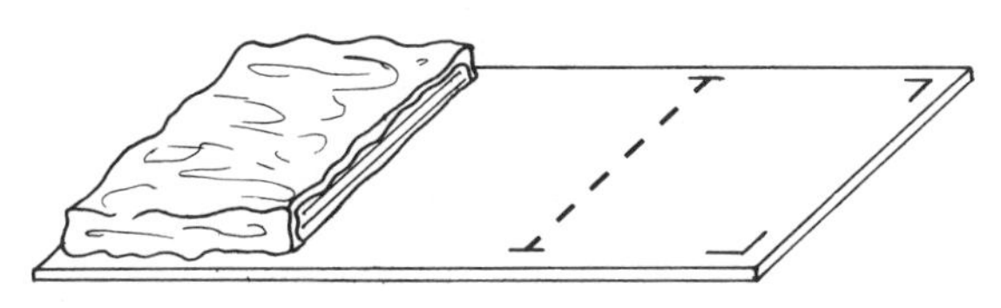

Figure 9.4 "Realistic" (Cake) Addition Problem.

cardboard "Plates," like those in Figure 9.4, and say, "Pretend that these are pieces of cake."

E: Present the $\frac{1}{4}$ piece and say, "Yesterday, I ate this (pointing) much cake. How much did I eat?"

E: Next, put the $\frac{1}{4}$ piece into a covered box (because it was eaten) on the table, slightly to the left side in front of *S*. Then, slightly to the right side in front of *S*, present the $\frac{1}{3}$ piece, and say, "Today, I will eat this (pointing) much. How much will I have eaten altogether (pointing to both the "hidden" $\frac{1}{4}$ piece and the $\frac{1}{3}$ piece)?"

8. The Area Problems:

E: Show *S* two red shapes, like the ones shown in Figure 9.5, drawn on a "checkerboard" grid.

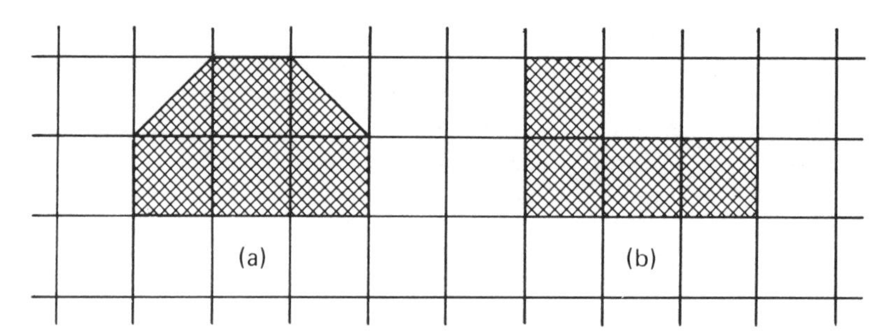

Figure 9.5 Area Problems - Part I.

Point to the first shape (Figure 9.5a) and ask, "How many squares are in this shape?" Then, point to the second shape (Figure 9.5b) and repeat the question.

E: Show *S* a red triangle on a sheet of white paper (Figure 9.6a); then cover the shape with a clear acetate grid (Figure 9.6b), point to the covered shape (Figure 9.6c) and ask, "How many squares are in this shape?"

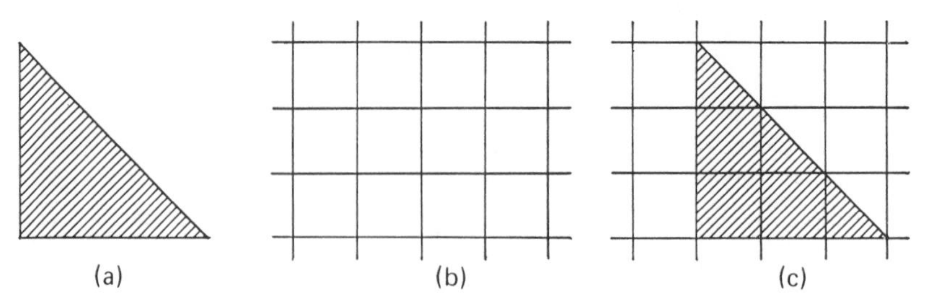

Figure 9.6 Area Problems - Part II.

E: Repeat the preceding question, only hand the sheet of acetate to *S* and use a red trapezoid shape, like the one shown in Figure 9.7.

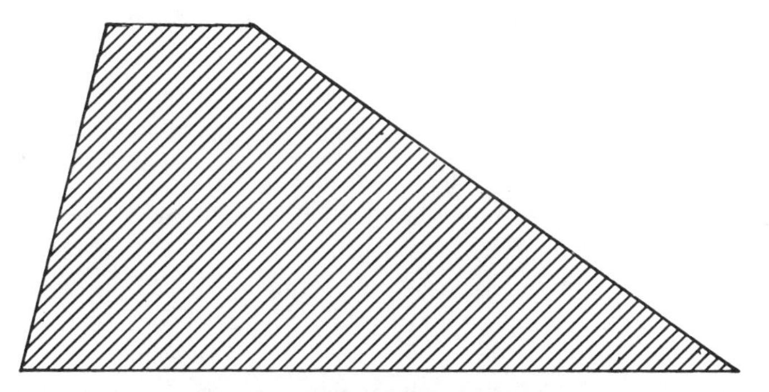

Figure 9.7 Area Problems - Part III.

9. The Plain Hershey Candy-Bar Multiplication-Problem:

E: Give *S* a whole plain Hershey candy bar, saying, "The next problem we are going to do has to do with this candy bar. But, we will use this piece of brown paper to stand for the candy bar. Then, as a reward for answering all of these questions for me, you can keep the candy bar. Is that OK?"

Take away the candy bar, and give *S* a piece of brown paper, like the appropriate one shown in Figure 9.8. Use the piece of paper that is marked into 10 squares to represent the plain candy bar, or for Problem 10 below, use the unmarked piece to represent the almond candy bar.

Figure 9.8 Candy-Bar Multiplication Problems.

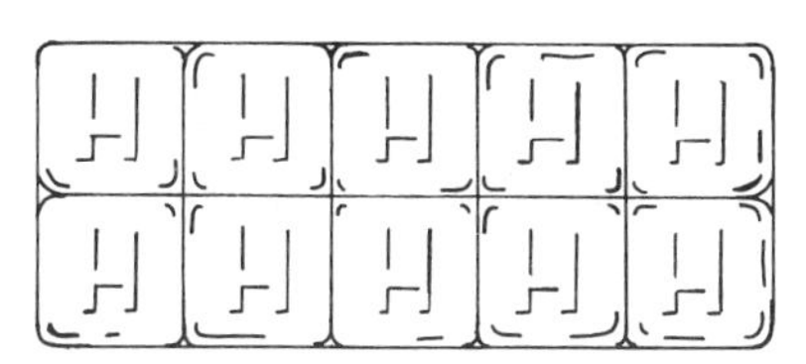

E: Say, "Yesterday, Bob's mother gave him a plain hershey candy bar in his lunch. He ate part of it and saved $\frac{1}{2}$ for today. Today, he ate $\frac{2}{3}$ of the remaining candy. How much of the whole candy bar did he eat altogether?"

E: Immediately, say, "I'll repeat the question slowly." (Repeat orally—more than once if requested.)

10. The Almond Hershey Candy-Bar Multiplication-Problem

The almond candy-bar problem was identical to the plain candy-bar problem, except that an unmarked piece of paper was used, and the orally presented statement of the question was:

"Yesterday, Ann's mother gave her a plain Hershey candy bar in her lunch. She ate part of it and saved $\frac{3}{4}$ for today. Today, she ate $\frac{1}{3}$ of the remaining candy. How much of the whole candy bar did she eat today?" (Note: The order of presentation of Questions 9 and 10 was reversed for 50 of the children in each quartile at each grade level.)

11. The Pencil-and-Paper Multiplication Computation-Problems:

One at a time, present the following two written computation problems. Each problem is written at the top of an otherwise blank sheet of paper.

$$\frac{1}{2} \times \frac{1}{3} =$$
$$\frac{2}{5} \times \frac{3}{4} =$$

Results and Discussion

Results of the interviews are presented below. First, the differences in the difficulty of the problems are displayed in a table of success rates for each problem in each grade. Then several of the problems are discussed with respect to (*a*) solution processes used, in particular, the predominant mode of representation in which a solution was reached; (*b*) the types of errors that typically occurred; and (*c*) the difficulty of the item compared with that of related items. The Addition Word-Problem (Problem 3) is discussed in greatest detail; discussion of other problems focuses on results we view as most important, interesting, or unexpected. Many of the more obvious and predictable results are summarized in table form.

The Addition Word-Problem

PROCESSES

All the solution attempts began by translating Problem 3 into written symbols, for example, $\frac{1}{4} + \frac{1}{5} =$ ___. Seventy percent of the students wrote the correct

TABLE 9.1

Success Rates on Rational Number Interview Problems

Problem	Percentage correct at each grade level					
	Gr 4	Gr 5	Gr 6	Gr 7	Gr 8	Total
Addition word problem						
$\frac{1}{4} + \frac{1}{5}$	.19	.25	.50	.44	.50	.38
Concrete addition-problems						
Name $\frac{1}{2}$	1.00	1.00	1.00	1.00	1.00	1.00
Name $\frac{1}{3}$	.94	.94	1.00	1.00	1.00	.97
Name $\frac{1}{2} + \frac{1}{3}$	.13	.06	.25	.19	.25	.17
Name $\frac{3}{4}$	.63	.75	.88	.88	.88	.80
Name $\frac{5}{6}$	.13	.19	.38	.19	.38	.25
Chocolate-eggs problems						
Name $\frac{1}{3}$	.63	.63	.75	.81	.88	.74
Name $\frac{1}{4}$	.69	.63	.75	.81	.88	.75
Name $\frac{2}{3}$	.50	.56	.63	.63	.69	.60
Name $\frac{5}{12}$	.69	.81	.81	.88	.94	.83
Name $\frac{3}{4}$	.56	.56	.56	.63	.63	.59
Name $\frac{5}{6}$	.38	.44	.56	.56	.63	.51
Name $\frac{1}{6} + \frac{1}{4}$	.31	.31	.50	.56	.56	.45
Name $\frac{1}{3} + \frac{1}{5}$	.31	.31	.44	.56	.56	.44
Cuisenaire rods problems						
Name $\frac{5}{6}$	.38	.38	.50	.69	.69	.53
Name $\frac{2}{3}$	.31	.38	.50	.69	.69	.53
Name $\frac{1}{2} + \frac{1}{3}$	.13	.19	.44	.56	.56	.38
Name $\frac{3}{4} + \frac{2}{3}$	.06	.19	.38	.44	.50	.31
Area problems						
Triangle	.63	.63	.75	.81	.81	.73
Trapezoid	.13	.19	.25	.38	.38	.26
Word problem						
$(1 - \frac{1}{4}) \times (\frac{1}{3})$	.06	.25	.38	.31	.44	.29
Plain Hershey bar						
$\frac{1}{2} \times \frac{2}{3}$	.00	.06	.19	.25	.25	.15
Hershey bar with almonds						
$\frac{3}{4} \times \frac{1}{3}$	.25	.31	.38	.44	.44	.36
Chocolate-eggs problem						
$\frac{3}{4} \times \frac{2}{3}$	.31	.25	.44	.63	.69	.46
Compute						
$\frac{2}{5} \times \frac{3}{4}$	.69	.75	.75	.94	.88	.80
Compute						
$\frac{1}{2} \times \frac{1}{3}$	.75	.81	.88	1.00	.94	.88

symbolic expression horizontally; 16% wrote it vertically. The remaining 14% of the subjects made errors in recording, usually omitting one or more of the symbols +, /, or =. It is surprising that, although this problem immediately followed two problems involving concrete materials, none of the students initially used materials or drew a picture to represent the problem.

ERRORS

The number of students at each grade level who gave each response type is shown in Table 9.2. Note that all the students who responded correctly (37% of the total) used an algorithmic least-common-denominator (LCD) method.

DIFFICULTY

Some of the most interesting observations about the Addition Word-Problem stem from a comparison of responses to this item with those to a question posed later (following the structured interview) that asked the subjects to ''act out'' the problem using clay circular regions to represent the pizzas. Students who had obtained the answer $\frac{2}{9}$ for their written calculation of the sum were able to look at the clay pizzas and recognize that $\frac{2}{9}$ was incorrect. However, when confronted with their written work on the problem, about half of these subjects maintained that $\frac{2}{9}$ was still correct. The discrepancy between the results obtained from the two different representations apparently did not trouble these children; several explicitly stated something like, ''That's okay! These are pizzas and those are numbers—they aren't the same.'' Such comments seem to indicate either a belief that mathematical computations (in symbolic form) need not agree with real-world observations (of clay pizzas) or that mathematics is simply unpredictable, so sometimes one obtains one answer and sometimes another for the same problem.

The presence of concrete materials in the follow-up problem did not enhance performance. Whereas 30 children had obtained the correct answer to the word problem (all of them using written symbols), a total of only 20 children arrived at the correct answer when the same problem was posed using concrete materials. Three students attempted to solve the problem using the concrete materials; they all obtained incorrect results. All 20 children who were correct were among the 27 children who persisted in using written symbols to obtain an answer. This means that 7 subjects who had previously obtained the correct answer using written symbols were no longer able to solve the problem using written symbols after it was presented in concrete form. They became confused and reverted to adding both numerators and denominators rather than using the LCD approach as before. This outcome runs counter to the widespread belief that materials make a problem easier to solve because it is more meaningful and real.

TABLE 9.2

Addition Word-Problem Response Types

Response	Number at each grade level					Total	%
	Gr 4	Gr 5	Gr 6	Gr 7	Gr 8		
Correct answer using LCD	3	4	8	7	8	30	38
Incorrect answer using LCD	1	2	1	3	2	9	11
Tried to find LCD, then tried other methods (e.g. materials)	0	2	2	2	2	8	10
Added both numerators and denominators	10	6	4	4	4	28	35
No response or gave up	2	2	1	0	0	5	6
	16	16	16	16	16	80	100

The interviewer probed to find whether this difference in performance was the result of rote execution of a meaningless algorithm in the first case. Some of the students reported that, when they wrote $\frac{1}{4} + \frac{1}{5} =$ after reading the word problem, they were thinking about parts of circles, and that when they wrote their solution they again thought about circular regions. For these subjects, the symbols were meaningfully related to stored images of concrete objects—but the images were abandoned in favor of a more powerful symbolic procedure for actually carrying out the computation. The follow-up request that the subjects act out the problem using the concrete materials required a demonstration of the symbolic algorithm relating it to the pieces of clay pizzas. Most of the students recognized their inability to respond to the question using the materials and worked out answers using pencil and paper. We believe it is likely that the LCD procedure, although *present,* was still an unstable element of the rational-number conceptual models of the 7 students who were no longer able to reach the correct result using written symbols. Attending to the concrete materials somehow contributed to the breakdown of an effective symbolic method for solution.

Although concrete materials often provide a useful representation for some stages in solving a problem, it may be extremely difficult to carry out the entire solution in terms of them. Good students eventually learn to select an appropriate representational system to fit each particular part of a problem situation or each specific stage in the overall solution; this understanding builds slowly and requires coordinating complex within-concept networks and between-concept systems.

The Concrete (Pizza) Addition-Problem

In the Problem 6 sequence, subjects were asked to identify circular clay pieces cut in the sizes $\frac{1}{2}$, $\frac{1}{3}$, $\frac{3}{4}$, and $\frac{5}{6}$. The addition question was to find the sum of $\frac{1}{2} + \frac{1}{3}$

TABLE 9.3

Concrete (Pizza) Problem

	Number correct at each grade					
Size of piece	Gr 4	Gr 5	Gr 6	Gr 7	Gr 8	Total[a]
$\frac{1}{2}$	16	16	16	16	16	80 (100)
$\frac{1}{3}$	15	15	16	16	16	78 (98)
$\frac{1}{3} + \frac{1}{2}$	2	1	4	3	4	14 (18)
$\frac{3}{4}$	10	12	14	14	14	64 (80)
$\frac{5}{6}$	2	3	6	3	6	20 (25)

[a]Numbers in parentheses are percentages of correct responses.

immediately after identifying the two (differently colored) component pieces, which were then put together on a single plate. Table 9.3 shows the number of students at each grade level who responded correctly to each question.

PROCESSES

Identifying $\frac{1}{2}$ and $\frac{1}{3}$ caused no difficulties for the subjects. Of the other fractions, $\frac{3}{4}$ was obviously much easier to recognize than was $\frac{5}{6}$. Either the missing piece was identified as $\frac{1}{4}$, so the piece itself had to be $\frac{3}{4}$, or the missing piece was used as a measure for cutting the remainder of the pizza. (Note that this procedure would be useful only when the missing piece represents a unit fraction.)

An immediate verbal response (apparently based on perceptual cues, and accompanied by no overt actions) was given by 80% of the children for the $\frac{3}{4}$ piece and by 63% of the children for the $\frac{5}{6}$ piece. When the interviewer probed (''How did you figure that out?''), nearly all these students *did* draw lines on the clay or on the plate, but the action typically seemed to be a justification rather than the source of the original response. This was especially apparent for the $\frac{5}{6}$ piece.

Table 9.4 shows the number of students at each grade level using each of the following five types of procedures to name the $\frac{5}{6}$ piece: (*a*) relatively passive perceptual cues were used with no lines drawn overtly; (*b*) lines were drawn in a seemingly trial-and-error way, apparently to fit some previous perceptual cues; (*c*) the missing piece was used as a standard for cutting the remaining pizza, with the hope that cuts of this size would divide the remaining pizza into a whole number of equal parts; (*d*) the remaining pizza was cut into equal-size pieces, presumably with the hope that cuts of this size would also fit the missing piece; and (*e*) the plate was divided into equal-sized pieces (as though the missing piece

TABLE 9.4

Procedures for Identifying $\frac{5}{6}$

Procedure	Gr 4	Gr 5	Gr 6	Gr 7	Gr 8	Total[a]
	Number at each grade					
Perceptual cues only	3	2	1	1	1	8 (10)
Trial-and-error drawing	5	4	2	4	3	18 (23)
Focus on missing piece	4	5	5	3	5	22 (28)
Focus on pizza	4	5	7	8	6	30 (38)
Focus on plate	0	0	1	0	1	2 (03)
	16	16	16	16	16	80 (100)

[a]Numbers in parentheses are percentages of total subjects.

had not yet been removed), with the hope that cuts of this size would simultaneously fit both the remaining pizza and the missing piece.

Overall, only 25% of the students correctly identified the $\frac{5}{6}$ piece. Among these 20 students, 15 used a procedure that appeared to be based on the missing piece, 3 seemed to focus on the pizza (i.e., their first cuts were not at all the same size as the missing piece), and only 2 successfully used trial and error guided by some sort of perceptual cues.

Table 9.5 shows the number of students at each grade level using various solution procedures for the concrete addition-problem ($\frac{1}{2} + \frac{1}{3}$). Twelve of the 14 successful subjects used a written symbolic procedure; only 2 students obtained a correct answer using a concrete procedure.

Because the concrete addition-problem is more complex than the identification of single pieces, it was more difficult to sort solution procedures into distinct

TABLE 9.5

Predominant Procedures for Finding $\frac{1}{2} + \frac{1}{3}$

Procedure	Gr 4	Gr 5	Gr 6	Gr 7	Gr 8	Total[a]
	Number at each grade					
Perceptual only	2	1	1	2	1	7 (09)
Trial and error	2	1	2	0	0	5 (06)
Missing piece	1	1	2	0	1	5 (06)
Pizza	5	6	2	4	5	22 (28)
Plate	0	0	0	0	0	0 (00)
Written/symbolic	6	7	9	10	9	41 (51)
	16	16	16	16	16	80 (100)

[a]Numbers in parentheses are percentages of total subjects.

categories. Many students went back and forth among (or combined parts of) several of the six basic procedures shown in Table 9.5. For example, some of the students who were classified as using a written symbolic procedure actually began by trying to find an answer using a concrete procedure. Interestingly, however, no student who began by using a written symbolic procedure later switched to one of the concrete procedures.

Some interesting differences appear in a comparison of Tables 9.4 and 9.5. Even though the addition problem was presented in concrete form, 51% of the subjects used paper-and-pencil procedures to solve it. The presence of the two differently colored pieces for the addition problem seemed to draw subjects' attention to the pizza that was *present,* so more solution attempts were based on the focus-on-pizza procedure rather than the missing-piece procedure, which produced such good results for the single $\frac{5}{6}$ piece.

ERRORS

Among the errors that occurred in relation to concrete procedures for the identification of the $\frac{5}{6}$ piece, 63% involved too many cuts, 37% too few cuts. In nearly all cases, errors were related to the fact that pieces were not all cut the same size (nor the same size as the missing piece). The fraction name given as an answer often did not correspond to the visual representation. Some students counted n pieces present and gave $1/n$ as an answer. Others counted m spaces in the missing piece and n pieces present and gave the answer m/n (instead of $n/(m + n)$), indicating some confusion between part–whole (fraction) and part–part (ratio) relationships.

For the concrete addition-problem $\frac{1}{2} + \frac{1}{3}$, the most frequent error committed by subjects using a concrete procedure was to divide each of the two pieces in either halves or thirds and conclude that the sum was $\frac{4}{5}$ or $\frac{6}{7}$.

DIFFICULTY

For the concrete addition-problem $\frac{1}{2} + \frac{1}{3}$, the overall success rate was 17.5% (14 of the 80 students), compared with 25% for the identification of the $\frac{5}{6}$ concrete piece and 37.5% for the addition word-problem ($\frac{1}{4} + \frac{1}{5}$, discussed above). Because the concrete problem was presented after the word problem in the interview, and $\frac{1}{2}$ and $\frac{1}{3}$ are generally better understood than $\frac{1}{4}$ and $\frac{1}{5}$, respectively, the greater difficulty with the concrete problem is somewhat surprising. It is therefore interesting to trace the procedures and the accuracy of the 80 students on the word problem and then on the concrete problem (see Figure 9.9).

Out of the 33 students who used concrete materials to solve the concrete addition problem only 2 were successful. On the other hand, 12 of the 40 who

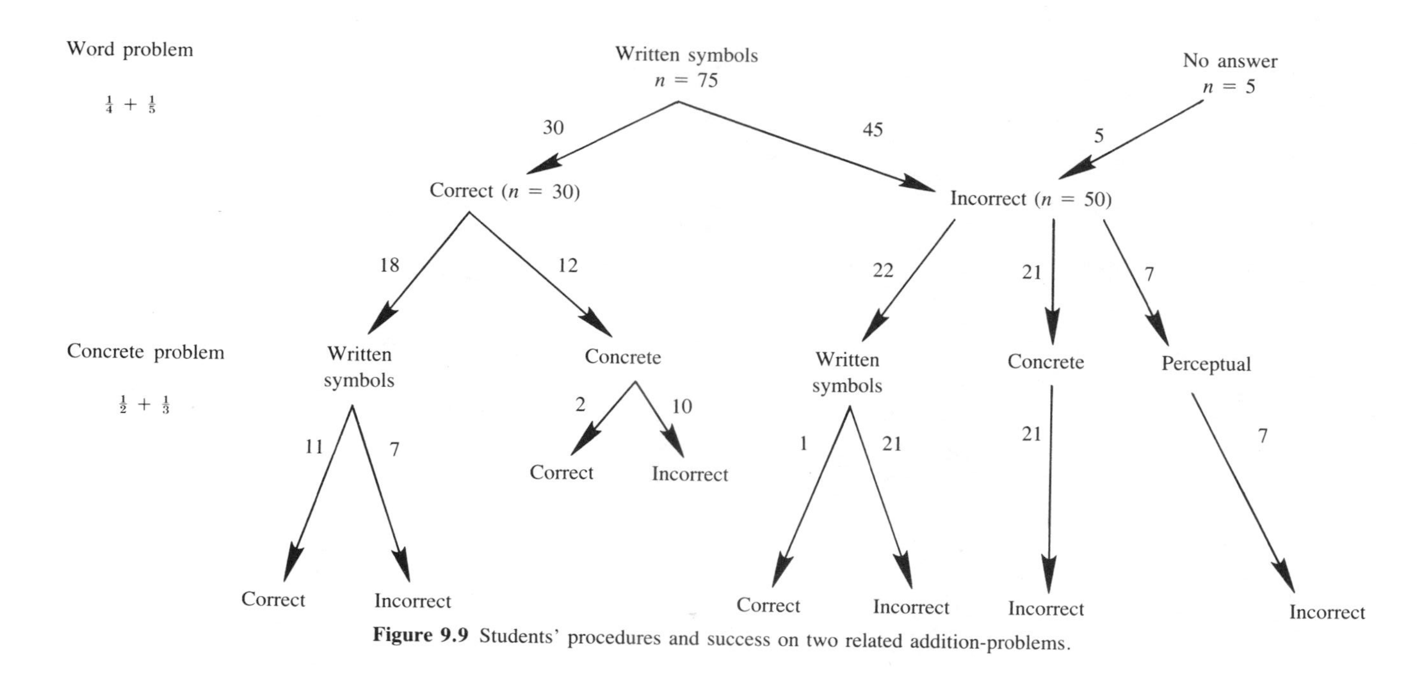

Figure 9.9 Students' procedures and success on two related addition-problems.

used written symbols on the concrete problem were correct. Eleven of these 12 had previously obtained the correct answer using written symbols on the addition word-problem. Notice that switching from a written symbolic representation to a concrete representation that matched the mode in which the problem was presented proved to be a poor strategy: only 2 out of the 12 who switched were correct, compared to 11 correct out of the 18 who persisted in using written symbols.

Still, pencil-and-paper solution procedures appear to have been somewhat more difficult when they were applied to the "concrete" situation than to information given in the "word" problem. This is consistent with the results of the follow-up questions to the "word" problem in which giving the students additional concrete aids actually made the "word" problem more difficult for some students. Perhaps, for some students, even when a problem is immediately converted to a written expression such as $\frac{1}{2} + \frac{1}{3} =$ __, what is going on in the student's mind may be slightly different for a "word" problem than for a "concrete" problem.

The "Realistic" (Cake) Addition-Problem

A problem that involves concrete materials, like the one in the preceding section, is not necessarily *real* in the sense that it would be likely to occur in an everyday situation. It is unlikely that someone would actually put two pieces of pizza together and ask, "How much altogether?" Even if the question were asked, reasonable answers would vary from saying, "That much," pointing to the pizza, to "Almost a whole pizza"; it would be surprising to see someone take out pencil and paper to calculate a response. In real situations, quantitative information usually is given for some purpose; information is stated or recorded at a level of precision that is reasonable in terms of both the source of the information and the use to which it will be put. Estimation, approximation, and rounding off reflect important properties of the models we use to describe real situations. Unfortunately, successful performance on textbook word-problems often requires the suspension of everyday criteria for evaluating the sensibility of realistic questions and responses, with students learning, instead, to give answers that teachers and textbooks expect, sensible or not.

Another characteristic that distinguishes many real problems from their word problem counterparts is that, in real problem-situations, the relevant information is not necessarily all given in the same representational mode. For example, in real addition-situations that involve fractions, the two (or more) items to be added may not occur as two written symbols, two spoken words, or two cakes; the addends may be one piece of cake and one written symbol, one fraction word and one written symbol, or (in our "realistic" addition-problem) one fraction word and one piece of cake. This is because, when symbols (written, spoken, or concrete–pictorial) are used to represent something, it is often because the thing

TABLE 9.6

Procedures for the "Real" Addition Problem

Procedure	Number at each grade					Total[a]
	Gr 4	Gr 5	Gr 6	Gr 7	Gr 8	
Drew a picture	5	4	2	2	3	16 (20)
Used a gesture	2	1	1	2	3	9 (11)
Used written symbols	7	8	10	9	9	43 (54)
Miscellaneous	2	3	3	3	1	12 (15)
	16	16	16	16	16	80 (100)

[a]Numbers in parentheses are percentages of total subjects.

being represented is not present spatially or temporally. Thus, part of the difficulty for students in these multiple-mode situations is to translate both addends (as well as the answer) into a single representational system. Our "realistic" problem exemplified this type of problem, virtually ignored in textbooks, which inherently involves more than one representational system. In the problem the subject first identifies $\frac{1}{4}$ of a cake, which is presumably eaten, thus hidden, then the subject is shown another $\frac{1}{3}$ of the cake to be eaten, and is finally asked how much cake will have been eaten altogether.

PROCESSES

Table 9.6 shows the number of students at each grade who used each of four basic response types for the "realistic" addition problem. None of the concrete solution attempts was successful. Forty-three students (54%) used pencil-and-paper procedures, which was slightly higher than for the concrete problem. Drawing pictures and using gestures (accounting for 31% of the responses), as well as many of the responses in the *miscellaneous* category, were never used on either the concrete addition-problem or the addition word-problem discussed above.

In this problem, because the $\frac{1}{4}$ piece was hidden, some representation of the hidden piece was required. The type of representation chosen (i.e., picture, gesture, symbol) directly influenced the solution procedures and errors that were made.

ERRORS

Students who drew a picture of the hidden piece often (in 8 out of 16 cases) drew a picture of the piece that was visible. Apparently, they were uncomfort-

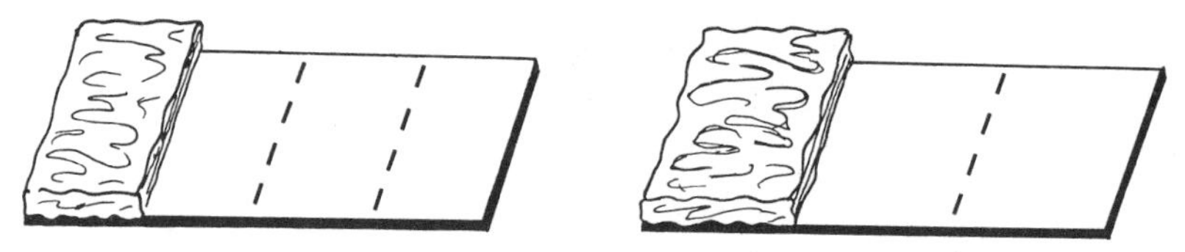

Figure 9.10 Cake Addition Problem.

able having one of the addends as a picture while the other was a "real" piece of cake (i.e., having addends in two different representational systems). The most common (incorrect) responses were: (*a*) $\frac{2}{7}$: the shaded pieces (see Figure 9.10) were counted, and then the pieces (again, see Figure 9.10) were counted; (*b*) $\frac{2}{5}$: the shaded pieces were counted, and then the unshaded pieces were counted. (Note: This latter response points again to the confusion, quite common in children's primitive rational number thinking, between the part–whole relationships that are relevant for *fraction* situations and the part–part relationships between distinct quantities that are relevant for *ratio* situations.)

Students who used a gesture (usually accompanied by verbal descriptions) to represent the hidden piece, often used their hand to indicate how much more of the plate would be covered by the hidden piece. Figure 9.11 illustrates roughly the procedure that was used. The most common (incorrect) response was $\frac{1}{2}$; that is, the size of the hidden piece was distorted significantly in order to fit an answer that the students considered "nice." (Note: The predominance of $\frac{1}{2}$ in the rational-number thinking of children is well known. For example, see Kieren & Southwell, 1979.)

Students who used a written symbol to represent the hidden piece made errors and gave answers similar to those discussed in connection with the concrete problem. There was again evidence that some students shifted from one representation to another, and from internal to external representations, during the solution process.

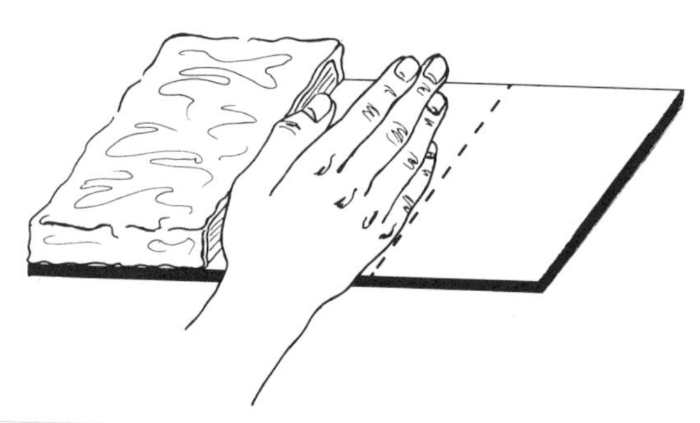

Figure 9.11 Indicating the hidden piece.

DIFFICULTY

None of the subjects who used concrete procedures obtained a correct result. Of the 43 children who used written symbols, 19 were correct, which represents 24% of the total. There was a higher success rate among subjects using written symbols for this problem than for either the concrete addition or the addition word-problems discussed above. Perhaps adding $\frac{1}{4} + \frac{1}{3}$ was easier because it is one of the most popular demonstration problems appearing in textbooks to illustrate the LCD method for adding fractions, and so was familiar to the students.

The Chocolate-Eggs Addition-Problem

Given a carton of 12 eggs as a unit, subjects were presented various numbers of eggs and asked to identify the fraction represented by each ($\frac{1}{3}$, $\frac{1}{4}$, $\frac{2}{3}$, $\frac{5}{12}$, $\frac{3}{4}$, $\frac{5}{6}$). They were then asked to find two sums ($\frac{1}{6} + \frac{1}{4}$ and $\frac{1}{3} + \frac{5}{12}$) for which the two addends were displayed as sets of eggs in two separate cartons.

PROCESSES

Virtually none of the students used paper and pencil for the egg addition-problems. Characteristics of the materials themselves apparently facilitated the higher rate of correct responses (see Table 9.1). First, the carton itself was always present as a frame to remind subjects of the whole unit. Focusing on the carton as the whole and the eggs as the parts, with distinct vocabulary to refer to them, made it easier for subjects to keep both in mind while making judgments about part–whole relationships. The fact that the whole consists of discrete objects rather than a continuous quantity (such as an area or length) made it easier to subdivide into unit fractions consisting of equal-sized sets of eggs. These characteristics of the materials also made part–part relationships easier to notice, which caused difficulties for some subjects.

ERRORS

Compared with errors occurring on the other concrete addition problems, part–part errors were more common on the egg problems. For example, for naming $\frac{5}{12}$ (the easiest identification problem), the most frequent incorrect answer was $\frac{5}{7}$. When the subjects' attention was focused on the *part,* that is, the five eggs in the carton, they seemed to lose track of the *whole,* responding with the ratio of eggs to empty spaces. This was, therefore, another situation in which part–part ratio ideas were sometimes confused with part–whole fraction ideas.

Another source of confusion was the relationship between the number of objects and the sizes of unit fractions. For example, $\frac{1}{3}$ was more difficult to identify than $\frac{5}{12}$ because *four* eggs had to be recognized as *one*-third. Similarly, $\frac{2}{3}$ was still more difficult, because *eight* eggs had to be recognized as *two*-thirds. In each case the whole carton had to be partitioned into unit fractions consisting of equal sets of objects, then some number of these sets had to fit the part in the part–whole relationship.

For the egg addition problems, the presentation of the two addends in separate cartons was a source of confusion that was not present in other questions. Thus, $\frac{9}{24}$ was one popular incorrect answer for $\frac{1}{3} + \frac{5}{12}$. Another common incorrect response was $\frac{9}{15}$, the ratio of eggs to empty spaces in the two cartons. In both error types the students lost track of the unit.

DIFFICULTY

Table 9.1 shows that the chocolate-egg problems were considerably easier than problems involving the same fractions but different materials or modes of representation. In spite of the increase in the salience of part–part notions, the characteristics of these materials, described above, are apparently responsible for the subjects' greater success.

The Cuisenaire Rods Addition-Problems

For the Cuisenaire rods addition-problems, like the egg problems, the parts and wholes were separate and easy to identify, and appropriate unit fractions were fairly easy to recognize; these characteristics, in addition to the subjects' familiarity with the rods, contributed to relatively good performance (see Table 9.1). On the other hand, rods are continuous rather than composed of discrete pieces, and there is no ever-present frame to maintain the size of the unit as there is for the egg problems.

The typical solution procedure for the Cuisenaire rod problems consisted of three steps. The first step was to find an appropriate unit fraction piece that fit a whole number of times into both the 12-cm unit rod (call it n) and the number of pieces that fit the length being measured (call it m) were counted. Finally, if these steps were completed correctly, the answer was m/n.

Thus, unlike the case with the egg problems, for which the answers could almost be read from the materials, the Cuisenaire rod problems required the chaining of several (relatively simple) steps. The part–whole judgments needed to find an acceptable unit fraction piece depend on within-concept networks in the subject's rational-number conceptual model; the measuring and labeling of lengths of the part and the whole depend on between-concept systems (relating measurement and fraction ideas) and representational systems (to accommodate parallel processing of the concrete objects as visual stimuli while attaching spoken symbolic labels during counting).

The Area Problems

For the area problems, the goal was to find how many square units there were in a nonrectangular figure; specific fractions could not be abstracted and plugged into a written algorithmic procedure. These problems were very much embedded in the graphical pictorial representation that was given to the subjects.

Nearly all the students used the following procedure on the area problems: First, they counted all the whole squares. Then, they looked for parts of other squares that would fit together to make wholes.

Finding pieces to make wholes seemed to be a relatively simple task for most students. When errors were made on the area problems, they usually resulted from memory overload. That is, the student would forget how many whole pieces had been counted when he or she went on to find pieces to make up more wholes, or would lose track of which pieces had already been used in making previous wholes. Throughout this counting process, it was striking that virtually none of the students used any recording system either to count or to keep track of the counted pieces. They relied entirely on internal memory. Still, their errors tended to be within $+1$ or -1 of the correct answer.

The Multiplication Problems

Six different multiplication problems were given. One problem was a word problem whose solution can be characterized by the expression $\frac{1}{4} + \frac{1}{3} \times (1 - \frac{1}{4})$. Two of the problems involved Hershey candy bars, one using a plain candy bar whereas the other used a candy bar with nuts; this was an important distinction because the plain bar is scored into 10 squares. The fourth problem was also a concrete problem, involving chocolate eggs in a 12-pack carton. The last two problems were straightforward computation problems, $\frac{1}{2} \times \frac{1}{3}$ and $\frac{2}{5} \times \frac{3}{4}$.

The success rates for the six problems were included in Table 9.1. The computation problems were the easiest, then the problem using the eggs, the candy bar with nuts, the word problem, and the plain candy bar.

The Multiplication Computation Problems

The multiplication computation was quite easy, although probably for the wrong reasons. Recall that the most common (incorrect) computation procedure for addition involved adding the numerators and denominators. For multiplication, a comparable procedure yields a correct answer.

The Multiplication Word-Problem

PROCESSES

For the multiplication word-problem, which was slightly more difficult than the earlier addition word-problem, the solution procedures were entirely different. For the addition problem, no one drew pictures, and the only successful procedures involved pencil-and-paper computations. For the multiplication problem, the attempted solution procedures are shown in Table 9.7. Sixty-five percent of the students used spoken language and/or drew pictures; almost half of these were successful. None of the students using exclusively pencil-and-paper procedures was successful.

ERRORS

On the word problem, if a picture was used, errors occurred most often when subjects drew two pictures, one showing $\frac{1}{4}$ and another showing $\frac{1}{3}$. The students did not know what to do with the two separate pictures, and often resorted to doing "some number thing" with some of the numbers 1, 3, 4, $\frac{1}{3}$, or $\frac{1}{4}$. Similar errors were committed by students who initially translated the word problem into written symbols. The fractions most frequently written were $\frac{1}{4}$ and $\frac{1}{3}$, not (for example) $\frac{3}{4}$, the amount remaining after the first piece of cake was removed.

Concrete Multiplication Problems

Looking at success rates in Table 9.1, one of the most interesting results is the radical difference in difficulty among the three concrete problems, two of which were closely related. The explanation for this fact has been discussed in connection with several of the addition problems. Slight differences in materials often greatly hinder or facilitate a student's ability to use a system of rational number relations and operations. *Having* a conceptual model and *being able to use it in a given situation* are quite different. A student's ability to use a given conceptual model depends considerably on the *stability* (i.e., degree of coordination) of the constituent structures. For many of our fourth- through eighth-grade subjects, rational-number concepts were apparently relatively unstable, particularly with respect to the task of translating from one representational system to another.

TABLE 9.7

Solution Procedures Used on the Multiplication Word Problem

Procedure	Number at each grade level					
	Gr 4	Gr 5	Gr 6	Gr 7	Gr 8	Total[a]
Used written symbols	4	5	6	6	5	26 (34)
Used concrete materials	0	0	0	0	1	1 (01)
Used pictures (with little language)	3	2	1	0	1	7 (09)
Used spoken language						
only	2	1	2	4	4	13 (16)
and pictures	5	6	4	3	4	22 (28)
and written symbols	2	2	3	3	12	22 (28)

[a]Numbers in parentheses are percentages of total subjects.

PROCESSES AND DIFFICULTY

Tables 9.8, 9.9, and 9.10 display the number of subjects who arrived at correct and incorrect answers using each solution procedure. The use of the concrete materials that were part of the problem presentation was least helpful for the plain candy-bar problem in which the division of the rectangle into 10 squares probably interfered with attempts to divide the bar into thirds. The concrete materials were most facilitative for the chocolate-eggs problem in which the materials lend themselves to a ready representation of the fractions in the problem; that is, $\frac{1}{4}$ of 12 eggs is 4 eggs, $\frac{3}{4}$ is 9 eggs, $\frac{2}{3}$ of 9 eggs is 6 eggs, or $\frac{1}{2}$ of the whole carton. For the candy-bar problem with nuts, the unmarked rectangle neither contributed to nor interfered with the subject's ability to impose fourths.

TABLE 9.8

Plain Hershey Bar Solution Procedures[a]

Procedure	Correct	Incorrect	Overall	Success (%)
Manipulated materials (using little language)	8	44	52	15
Overtly used spoken language	3	8	11	27
Did it "in their heads" (probably using internal language)	1	7	8	14
Used written symbols	0	9	9	00
	12	68	80	15

[a]For all grades.

TABLE 9.9

Almond Hershey Bar Solution Procedures[a]

Procedure	Correct	Incorrect	Overall	Success (%)
Manipulated materials (using little language)	19	34	53	36
Overtly used spoken language	7	7	14	50
Did it "in their heads" (probably using internal language)	3	5	8	38
Used written symbols	0	5	5	00
	29	51	80	36

[a]For all grades.

Further, the plain candy-bar problem was the one for which a language representation would be least helpful. Finding $\frac{2}{3}$ of $\frac{1}{2}$ can be quite a different problem from $\frac{1}{2}$ of $\frac{2}{3}$, the commutative property notwithstanding. For $\frac{1}{2}$ of $\frac{2}{3}$, subjects can find an answer thinking of the problem as analogous to $\frac{1}{2}$ of two objects, for which the result is one object, or, in this case, $\frac{1}{3}$. This type of language representation, implemented either overtly or covertly, would facilitate reaching a correct result in the candy-bar (with nuts) problem ($\frac{1}{3}$ of $\frac{3}{4}$) and the chocolate-egg multiplication problem ($\frac{2}{3}$ of $\frac{3}{4}$). It was not helpful for the plain candy-bar problem ($\frac{2}{3}$ of $\frac{1}{2}$).

Apart from comparisons among the concrete multiplication problems, it is of interest to compare solution procedures on these concrete problems with those on the concrete addition-problems discussed above. The most striking difference is the drastically reduced rate of written symbolic procedures on the multiplication problems, accompanied by a much greater reliance on concrete procedures.

TABLE 9.10

Chocolate Eggs Multiplication[a]

Procedure	Correct	Incorrect	Overall	Success (%)
Manipulated materials (using little language)	25	21	46	54
Overtly used language (with no overt use of pictures)	8	12	20	40
Did it "in their heads" (probably using internal language)	4	6	10	40
Used written symbols	0	4	4	00
	37	43	80	46

[a]For all grades.

Language representations, useful for some of the multiplication problems, did not occur on the addition problems.

Summary of Major Conclusions from the Interviews

The data collected in the Rational Number Task Interviews support the following generalizations, some of which are based on the data as a whole, and some of which stem from results on one or more particular items. Reference is made to the problem or problems that were especially relevant.

1. Realistic word-problems often differ significantly from isomorphic concrete or real-world problems in difficulty, preferred solution processes, and the types of errors most commonly made. In fact, virtually any small change in the task had a noticeable impact on performance for most of our fourth- through eighth-grade subjects. This variability is regarded as an indication of the instability of subjects' rational-number conceptual models.
   a. Problems presented in terms of concrete materials are not necessarily easier than those presented orally or using written language and symbols. For a number of children, performance *declined* when they were encouraged to use concrete aids to help in solving problems. Compare, for example, the results on the addition word-problem with its concrete follow-up question and with the concrete addition-problem. Some subjects were able to execute the written LCD procedure correctly on the word problem but were unable to do so (even when they again were working in written symbols) after concrete materials had been brought into the problem situation.
   b. The representational system favored by subjects was distinctly problem-specific. For example, subjects were much more inclined to translate the addition word-problem into written symbols than the multiplication word-problem. Even within the class of concrete problems, we found, for example, that the chocolate eggs were *used* for finding answers whereas the circular regions were not. The fact that the specific fractions involved in the problems may have influenced the results only strengthens the point.
   c. Some successful solution procedures are closely tied to specific problem-representations and are not consistently "called up" for solving an isomorphic problem presented in a different representational mode. For example, the missing piece strategy used by subjects to identify the $\frac{5}{6}$ circular region was apparently unavailable for finding the sum of $\frac{1}{2} + \frac{1}{3}$ even when circular pieces were placed next to each other on the same plate.

d. There was a great deal of evidence of confusion between part–whole fraction ideas and part–part ratio ideas; to whatever extent this was present in the subjects irrespective of the stimulus, it appeared that certain materials (e.g., the chocolate eggs) elicited more inappropriate part–part responses than other materials.

e. The mode of representation was important not only in the initial problem presentation, but also when selected by the subject. For example, on the "realistic" addition-problem, the mode in which the subject chose to represent the hidden $\frac{1}{4}$ piece seemed to affect the course of the solution dramatically. On particular multiplication problems, for which the denominator of the first factor was a divisor of the numerator of the second factor, a language-based representation apparently facilitated correct responses.

2. Representation of a problem situation is a very active process.

a. During the course of a solution attempt, subjects often changed their external representation of the problem from one mode to another. For example, very few subjects solved the concrete addition-problem in the concrete mode; most changed to written symbols.

b. Subjects often used two or more systems of representation simultaneously. The most obvious occurrences were in response to the multiplication word-problem for which a large number of subjects used spoken language and pictures, and the Cuisenaire rod addition-problems, for which many subjects monitored their solution steps by narrating their actions.

c. Internal and external representations of the problem situation interacted with each other and influenced the course of the solution. On several problems (for example, in identifying concrete, pizza circular regions) subjects responded based on perceptual cues, then constructed post hoc explanations of their answers for the interviewer. For the "realistic" addition-problem, the strength of the understanding of the fraction $\frac{1}{2}$ led many children to adjust their external representation of $\frac{1}{4} + \frac{1}{3}$ so that $\frac{1}{2}$ would look like an appropriate response.

3. Some beliefs children have about mathematics and about how to respond to mathematics problems affect performance.

a. Most of our subjects seemed to be working toward precise answers as the goal of solving problems; there were few instances of estimates or approximations. This tendency was most apparent in response to the more typical textbook word-problems and least apparent on the area problems.

b. Several children were unconcerned about obtaining different results for the addition word-problem when they used concrete materials versus paper-and-pencil procedures.

Rational-Number Written-Test Results

Written language, written symbols, and several types of pictorial representations were used in assessing children's rational-number understandings in the set of written tests developed in the RN project. These tests, the subject of this section, provided the baseline data about paper-and-pencil modes of representations and fraction and ratio within-concept networks that the interviews described above were designed to extend and explore.

Three paper-and-pencil tests were developed for the RN project: the Assessment of Rational Number Concepts (CA), the Assessment of Rational Number Relationships (RA), and the Assessment of Rational Number Operations (OA). The first tests basic fraction and ratio concepts. The second tests understanding of relationships between rational numbers, involving ordering, equivalent rational forms, and simple proportions. The third tests abilities to perform addition and multiplication operations with fractions. This section of the chapter identifies the rational-number characteristics tested and summarizes some general results from two of the tests (the OA and the CA). Further information on the Rational Number testing program appears in Lesh and Hamilton (1981).

The written tests were administered to about 1000 students in Grades 2 through 8 in Evanston, Minneapolis, DeKalb, and Pittsburgh between early November and late January of the 1980–1981 school year. The tests were prepared in two parallel versions, with most of the students taking Version I. The tests were administered in a modularized form so that a core of items was given to all grades, younger students did fewer items than older students, and the most difficult items were done only by older students.

Two of the tests, the CA and the OA, are referred to extensively in this section. Each item from those tests is reproduced in Appendix 9.B, with data on the fourth through eighth graders who took the texts. In all, 650 fourth through eighth graders took the CA and 608 took the OA. The items for each test are arranged in order of percentage correct for the students who were given the item. The purpose of this arrangement is to enable a broad overview of order of difficulty for the items on each instrument. A sample item from the Concepts test is given in Figure 9.12.

Beneath the item are two lines of information, which can be interpreted as follows: ''Item C–8'' refers to the overall rank (by proportion correct) of this item on this test. Thus, out of the 60 items on the Concepts test, this one had the eighth highest score. Next, ''(5)'' refers to the original number of the item on the test (and corresponds to the number next to the item's question stem). ''Gr. 4–8, $n = 650$'' means that Grades 4–8 were given the item, involving 650 students. Next is a set of letters and numbers corresponding to the answer choices: ''a)2 b)596 . . . '' This means that 2 students out of the 650 selected choice *a*, 596 chose *b*, and so forth. ''*na*/1'' means that one student gave no answer.

5. Which picture shows two-thirds shaded?

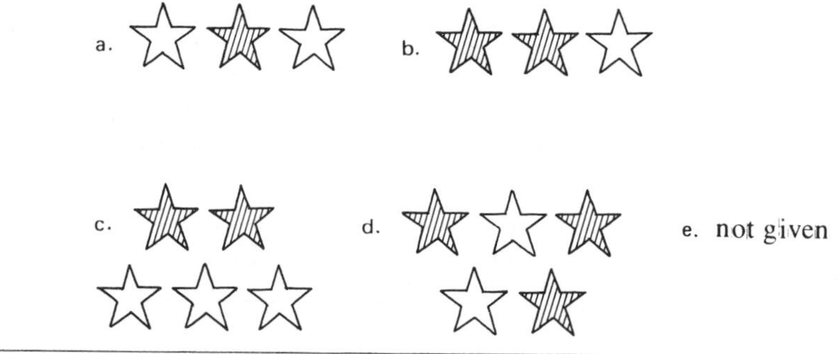

Item C–8 (5): .916, Grades 4–8, n = 650.
a) 2 b) 596 c) 32 d) 12 na) 1
G4–.797 (7/43) G5–.938 (7/43) G6–.949 (6/60) G7–.969 (6/60) G8–.981 (4/60)

Figure 9.12 Item C-8.

On the second line, two results for each grade that took the item are given: the proportion correct for that grade, and the item rank for that grade. Thus, "G4–.797 ($\frac{7}{43}$)" means that 79.7% of the fourth graders did this item correctly, and this item was the seventh *easiest* out of 43 items for the fourth graders. Item ranks varied across grades on the items. Notice, for example, that this item was the sixth easiest (out of 60 items) for the sixth and seventh graders, but the fourth easiest (out of the same 60 items) for the eighth graders.

Two caveats are in order. First, because the ordering of these items collapses *across* all grade levels, the order of difficulty of an item is somewhat different from its order for all grades that did that item. Second, all items are included in the list, even though not all grades did all items. Thus, for example, the rank order of 7 on item 8 for the fifth graders is out of 43 items, whereas the rank of 6 for the sixth graders is out of 60 items (which include all 43 items done by the fifth graders).

Items on each test are identified along several dimensions, including type of representational translation, rational-number size, and rational subconstruct (fraction or ratio). A list of characteristics for each test is included in Appendix 9.B, along with a generating scheme for the test items.

Technical Information

Cronbach-alpha reliabilities across the tests averaged 0.881 for all the tests, excluding the short 15-item OA given to fourth graders, which had an alpha reliability of 0.489. Within-subject reliabilities averaged .850, excluding the fourth grade OA, which had a within-subject reliability of 0.318. One external validity measure involved the correlation between eighth-grade performance on six OA items and six corresponding National Assessment items administered to 13-year-olds in 1979. Although item stems in each pair were similar or identical, answer sets were different in that the OA limited responses to five given choices or *omit,* whereas the National Assessment items required students to write in their answers. Students scored higher on all OA items, given answer choices, than their National Assessment counterparts on comparable items with no answer choices given. The high correlation between the six score pairs (0.918), and the fact that one would expect an incrementally higher score for items that give answer choices compared with those that do not, are evidence of the comparability of the eighth-grade sample with the National Assessment sample.

Discussion of CA and OA Results

A plethora of observations from these results are readily made. This section will identify and discuss a few of these, including:

1. A difficulty ordering of translations between representational modes.
2. The easiest and most difficult problems from each test.
3. Notable jumps in performance from one grade to a later grade, including examples of apparent declines from one grade to another.
4. Item pairs that suggest the "nonadditivity" of some rational-number understandings.

Ordering of Translations between Representational Modes

Representational translations on the CA were

1. Symbols to written language (e.g., C–3).
2. Written language to symbols.
3. Picture to pictures (e.g., C–12).
4. Written language to pictures (e.g., C–17).
5. Picture to written language (e.g., C–20).
6. Symbols to pictures (e.g., C–32).
7. Picture to symbols (e.g., C–11).

Because each item on the test can be characterized by one of these seven translations, one can think of the CA as consisting of seven "subtests." Considering the 43 items that were done by all five grades, the order in which the translations are listed above proved to be the order of their increasing difficulty

TABLE 9.11

Translation Success Rate (%) for 43 Items on CA Instrument

Translation type	Grade(s)						
	4	5	6	7	8	4–8	6–8
Symbol to Written	93.2	95.0	98.3	98.9	95.4	95.8	97.5
Written to Symbol	64.2	74.6	85.2	85.1	88.8	77.9	86.3
Picture to Picture	69.9	80.2	80.5	85.3	86.5	79.4	83.9
Written to Picture	67.3	78.3	78.3	80.4	84.5	76.8	81.0
Picture to Written	56.9	66.8	72.6	81.3	83.2	70.3	78.7
Symbol to Picture	52.0	65.8	71.3	77.5	81.4	67.7	76.5
Picture to Symbol	50.0	60.1	67.3	72.2	74.5	63.1	71.1

for each grade, with only minor exceptions. The fourth and fifth graders found the written-to-symbol subtest more difficult than the picture-to-picture translations and more difficult than the written-language-to-picture translations. Also, the seventh graders did slightly better on the picture-to-written translations than on written-to-picture translations. Table 9.11 shows the representation translation success rates for each grade.

Such an ordering is plausible. The easiest translations are those that involve simply reading a rational number in two different modes, requiring little or no conceptual processing of the meaning of the rational number (e.g., C–4). The relative lack of familiarity with the symbol notation provides a reasonable explanation for the apparent difficulty the early fourth and fifth graders experienced with the items involving four symbolically expressed rationals in the answer set (e.g., the written-language-to-symbol subtest). Next are those that require mapping one picture to another that is isomorphic with respect to the fraction shaded (e.g., C–24). Following that are translations between pictures and written language, followed by translations between pictures and symbols. The written language representations of rationals appear to be easier to process than the symbolic coding of rationals. As will be discussed later, this may be attributed to the fact that symbolic representations do not encode rational components (i.e., numerator and denominator) differently, though they have different meanings, whereas written language representations do express each component in a different form. For both translation types, those involving written language and those involving symbols, the translations *to* pictures were more difficult than *from* pictures. The former translations include four pictures in the item answer set, whereas for the latter, only one picture appears in the item stem. Translations *to* pictures thus demanded more visual processing than did translations *from* a single picture in the item stem. Thus, the data support the implication that a written or symbolic expression is easier to process than is a pictorial representation.

Easiest and Most Difficult Items

The single exception to the assertion that "a written or symbolic expression (involving no conceptual processing) is easier to express than a pictorial representation" is in item C-1. This item asked students to identify $\frac{1}{2}$ as the shaded fraction of a circle. Earlier research has affirmed the primacy of "halfness" in children's earliest rational-number understandings (e.g., Kieren, 1976). This was also the case on the OA. Item 01, involving "giving away half of six puppies," was the easiest item for all grades except sixth, for which it was second easiest. Otherwise, the easiest items on OA involved the interpretation of concrete situations with simple part–whole ideas, expressed in written language.

The most difficult items on both tests required what could be called *second order* processing of the part–whole idea. For all grades, C-60 was the most difficult CA item. It required interpreting each third of the configuration of nine circles as a *whole,* as per the given information. Understandably, the most popular distractor was $\frac{7}{9}$, selected by 36% of the students. Furthermore, this item elicited many more part–part responses ($\frac{7}{2}$, selected by 16% of the students) than other items involving discrete objects (such as C-5, for which less than 7% of students selected either of two possible part–part distractors in the answer set).

A similar item was the fourth most difficult item on the CA, given only to sixth through eighth graders. C-57 required students to interpret each large rectangle as a third, with three large rectangles thus comprising a whole. For this item, nearly 40% of the students selected the response interpreting each large rectangle as a whole, rather than as a third partitioned into fourths (or twelfths) of the whole.

"Fractions involving fractions" also proved to be among the most difficult on the OA. Less than 1 in 6 seventh graders and barely 1 in 5 eighth graders interpreted O-32 as involving a fraction of a fraction requiring multiplication. Even fewer students correctly interpreted a similar item, O-34. Barely 1 in 10 seventh graders and 1 in 6 eighth graders could tell how many thirds equal $\frac{1}{5}$; more than half of the seventh and eighth graders simply answered "cannot be done" (O-33). The prevalence of this response is ironic because rational numbers are the first number system children encounter that is closed with respect to all four basic operations. The most difficult item on the OA was O-35, requiring the student to process a fourth of a half and a half of a half and then find their sum. Fewer than 1 in 12 seventh and eighth graders selected the correct response for this item. For this particular item, a visual estimate of the fraction shaded in each half of the picture was required, and this surely contributed to its difficulty.

Interestingly, however, immediately preceding this item on the original test was O-14, for which the student also had to compute a fourth of a half, and a half of a half, and then their sum. On this item, however, we showed the whole rectangle partitioned, and explained each of the possible responses. Nearly one-half selected the correct response on that item, though hardly any could do the next problem, O-35, which was very similar and which had the same answer.

Jumps from One Grade to Another

The three largest performance jumps from one grade to another occur between the same two grades, fifth and sixth, for items O-5, O-6, and O-12. The average increase from Grade 5 to Grade 6 on these items was 43.2%, compared with an average increase of only 11.2%. All three of the largest jumps involved addition or subtraction of fractions with like denominators. One can infer a significant instructional effect between early fifth grade and early sixth grade with respect to this skill. These items provide an interesting contrast with O-7, another fraction addition-problem with the same denominator. Fifth graders performed significantly better on this item, for which the denominator was expressed in written language rather than in symbols. The most popular distractor for $\frac{1}{3} + \frac{1}{3}$ was $\frac{2}{6}$. Understanding the numerator of each addend as a cardinal value is familiar to the students. The denominators, however, look like standard cardinal numbers, but their meaning is quite different. Problems such as O-5, with representations that preclude attaching the same meaning to numerator and denominator, do not require counterintuitive processing or algorithms for standard fraction symbols. Attaching different meanings or algorithms to numbers with the same form, but different position in the fraction, is the principal instructional achievement discerned by the OA, and it occurs between early fifth and early sixth grade.

Jumps "backward" from one grade to the next also merit comment. On the CA, sixth graders outperformed seventh graders on four out of five consecutive items on the original test (52–55, and C-7, C-56, C-34, and C-37 on the reordered version in Appendix 9.A). Three items required converting representations from mixed to improper form, and vice versa.

The greatest proportion of "downward jumps" occurred between sixth and seventh grade. On the CA, seventh graders as a whole averaged 71.3%, versus 67.8% for the sixth grade. On the OA, sixth graders actually scored slightly better as a group than did seventh graders, by a margin of 50.8 to 49.9% (considering only the 28 items done by both grades). This suggests the possibility that fraction operation skills from Grade 6 to Grade 7 are unstable and that instruction barely maintains or only slightly improves their level.

Task Variable Additivity

A natural question would be the feasibility of devising a hierarchy of rational-number task variables that would enable prediction of performance on rational-number tests. Such a hierarchy would involve variables $v1 \ldots vn$, with additive properties such as

$$(v1 + v3) - (v1 + v4) = d \rightarrow (v2 + v3) - (v2 + v4) = d$$

or

$$(v1 + v2) > (v1 + v3) \rightarrow (v4 + v2) > (v4 + v3),$$

where the sum $(vi + vj)$ represents the expected performance level for a rational-number task comprising two task variables, vi and vj, where the difference, d, between two tasks is the expected performance difference for a student population, and where order, $>$, is defined by order of difficulty among tasks.

Several examples from the CA and the OA suggest that such a hierarchy is not plausible. They indicate that the impact of changing a single variable in a pair of items may either be much greater than or possibly even opposite to a similar or identical variable change for another pair of items.

EXAMPLE 1

The task variable change is from multiplying 2 fractions of the form $\frac{p}{q}$ ($v1$), to multiplying a whole number by a fraction ($v2$). Sample item pairs are (O–18, O–8) and (O–19,O–34).

Students found whole-number–fraction multiplication significantly more difficult than fraction–fraction multiplication with a symbol-only representation, such as Items O–18 and O–8. However, in the context of a brief word-problem, the same variable change produced the opposite result, with students finding the fraction–fraction more difficult than the whole-number–fraction multiplication. In this case, if $v3$ is symbol mode computation, and $v4$ is word-problem context, we would have:

$$\begin{array}{c} \text{O–18} \quad \text{O–8} \\ (v3 + v2) > (v3 + v1) \text{ but} \\ (v4 + v2) < (v4 + v1) \\ \text{O–19} \quad \text{O–34} \end{array}$$

whereas, under the assumption of additivity, if $v2 > v1$, and if $(v3 + v2) > (v3 + v1)$, it would follow that $(v4 + v2)$ would be greater than $(v4 + v1)$.

EXAMPLE 2

The task variable change is from continuous pictorial representation ($v1$), to discrete object representation ($v2$). Sample item pairs are (C–39, C–60) and C–17, C–8). Students found item C–17, with a continuous representation, more difficult than the discrete item representation in C–8. C–39 contains a continous representation of the whole, and proved to be one of the easier items on the CA. In contrast, C–17, representing a whole with three discrete objects, was the most difficult item on the CA. The intrinsic nature of a discrete object representation of the whole in C–60 elicited a special distractor, the one that interpreted the entire configuration as a whole. For these two item-pairs, if $v3$ is written-to-picture identification of a simple fraction, and $v4$ is, given a whole, identify a fraction greater than 1, we have

$$\underset{\text{C–17}}{(v1 + v3)} < \underset{\text{C–8}}{(v2 + v3)}$$

but

$$\underset{\text{C–39}}{(v1 + v4)} > \underset{\text{C–60}}{(v2 + v4)}$$

Conceptual-Model Processes in the Written Tests

Although applied problems evoke richer transformations in conceptual models than those observed on the CA and OA, it is helpful to consider such processes on those instruments. Such a discussion, pertaining to any particular test item, would involve three components and the interactions among them: the conceptual model brought to bear on the problem, the item stem, and the item answer-choice set.

The "interactions" depicted in Figure 9.13 involve such processes as imposing meaning in the direction of the arrow. For example, a conceptual model imposes a meaning on the representation "$\frac{1}{2}$."

The item stem or answer set may distract a subject who has an unstable conceptual model. For example, on Item C-51, a rational-number conceptual model that has not fully differentiated cardinality from part–wholeness, and is, therefore, unstable with respect to that difference, would be easily distracted by choices such as answer (*d*). In general, the more stable the conceptual model, the less either the problem stem or the answer set will influence the model by refining or distracting it.

The conceptual model directs two different process types when the student is doing these kinds of paper-and-pencil items. The first is (within-stem) or (within-answer-set) processing, for which rational-number meanings are imposed or

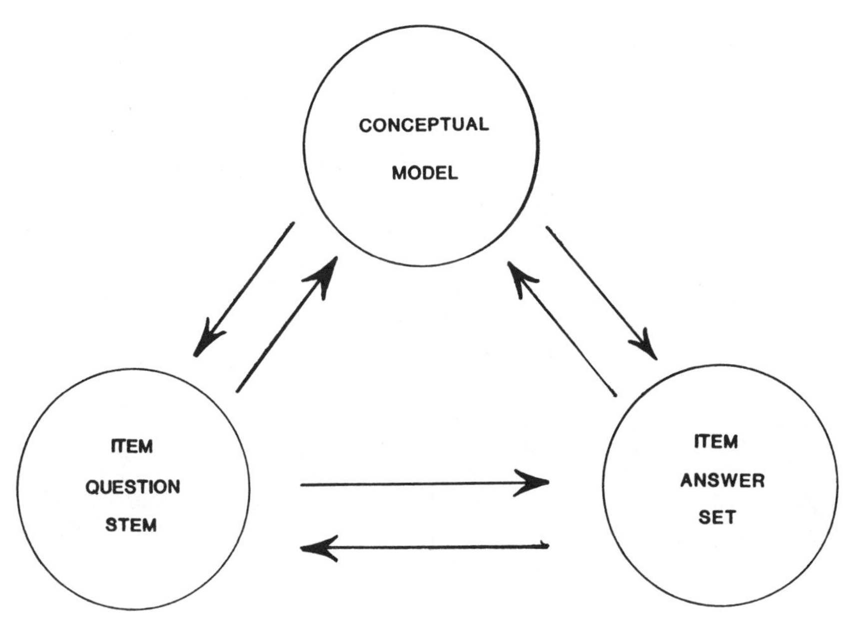

Figure 9.13 Components affecting written-test-item response.

attached to each representation in the stem or answer set. The second processing type involves the translations between the stem and the answer set (that contribute to the meaningfulness of within-stem or within-answer-set processing), and which compares meanings imposed on the stem with meanings imposed on the answer set until a "fit" is achieved. The conceptual model effectively encodes and processes both the stem and the answer set and imposes a structure on each. Translations back and forth between the stem and the answer set continue until an isomorphism is established between the stem and one choice in the answer set. Every time the model translates the stem structure to the answer choice set and fails to make an isomorphism, it reprocesses the representations in each and perhaps modifies the meaning or interpretation attached to each representation. In this sense, the translation contributes to the within-stem and within-answer-set processing.

The role of translations, vis-à-vis the amount of processing required to read symbol or written-language representations, merits comment. First, the easiest translations were those that required no meaningful rational number conceptual understandings (e.g., C-3). The data suggest that youngsters can do this without part–whole understandings. It may be that written or symbolic expressions, such as those in C-3, are intuitive and unstable for some children, as are some of their pictorial understandings of part–whole relationships. It may be, however, that when a symbol, for which a student has an unstable or intuitive model, appears with four pictures that illustrate part–whole understandings that the student can identify only intuitively, the understandings in the two modes intersect and the student selects the correct response. The translation stabilizes the understandings associated with the two representations by implicitly saying to the student "whatever is in Mode A means the same thing that is in Mode B, so the overlap of understandings you have between the modes is the correct understanding within each mode." In this way, the multiple-choice items serve to refine or stabilize a model.

An example in which this may occur is Item C–5, which was intended to distract students with part–part understandings. It appears to have had the opposite effect. Students who did not come to the item with models differentiating part–part from part–whole had the opportunity for the item answer-set to facilitate such a differentiation, by giving two examples of part–part (*c* and *d*) and only one example of part–whole (*b*), with the explicit assumption that only one answer is correct. Students could therefore deduce the correct response and learn something in the process.

Thus, it is possible to view each response as a function of six variables:

1. Within-item-stem processing, directed by the conceptual model.
2. Within-answer-set processing, directed by the conceptual model.
3. Translations and matching, from item stem to answer set.
4. Interaction between conceptual model stability and available distractors (i.e., in the answer set).
5. Modification of the conceptual model by previous items or the current problem.
6. Error term.

Concluding Remarks

Research on the acquisition and use of mathematical concepts and processes is influenced by the researchers' theoretical perspectives, whether or not they are explicitly stated. The investigators' beliefs about how cognitive structures are organized and interrelated in the minds of their subjects shape the questions that are chosen to be addressed, the methods and tasks that are selected, and what is regarded as important among the data that result. The chief goal of this chapter has been to define and illustrate our present understanding of a theoretical construct, the conceptual model, that underlies our program of research on applied mathematical problem-solving. The rational-number data presented here, collected from written tests and structured interviews, are highly relevant to our applied problem-solving research because the rational-number system is one of the most sophisticated systems familiar to middle-school youngsters and because the variety of subconstructs provides a rich context for applications appropriate to our subjects.

We have focused on within-concept networks and systems of representations, the two components of conceptual models that are most elementary and crucial to an understanding of the construct. These components have been described in relation to rational-number conceptual models both because they are more discernible in that context than they would be if we tried to describe them in more complex applied problem-solving situations, and because it is apparent how our interest in the growth and use of these components provided a framework and direction for our research. The other two components of conceptual models, between-concept systems and dynamic (modeling) mechanisms, are addressed in our current applied problem-solving research.

We have described three approaches to the construction of tasks for research focusing on the growth and development of conceptual models used by middle-school students to solve realistic problems involving mathematical ideas. Tasks may originate in the mathematical ideas themselves, or they may be created to be isomorphic to typical textbook word-problems, or they may be derived from real problem-situations in which the use of mathematics arises naturally. The research included in this chapter has taken the second of these approaches. Furthermore, it has emphasized the rational-number ideas themselves more than the processes for *using* the ideas. The bulk of our earlier research takes the first approach, whereas the APS project also utilizes the third.

Both the interviews and the written tests focused on fraction and ratio ideas, two of the within-concept networks subsumed by general rational-number understanding, and attended particularly to various representations of these ideas and translations among these representations. The written tests assessed children's abilities to translate within and among paper-and-pencil modes of representation: static figures, written symbols, and written language. The interviews made it possible to assess translations involving concrete materials and more realistic representations, in addition to those mentioned above.

We have investigated other translations in modified testing situations in which the stimulus was not written (either spoken or displayed using concrete materials) but answers were written, and in interviews, in which the stimulus and response could involve spoken language, written symbols and language, and concrete materials (Landau, Hamilton, & Hoy, 1981). The second phase of the RN project is examining the role of spoken language as an intermediary between problem situations and written symbolism (Behr *et al.*, Note 3); the use of pictorial representations as an aid in problem solving is also being investigated (Landau, Note 5). Thus, we have a more general interest in modeling processes, of which paper-and-pencil translations emphasized on the written tests are only one type.

The two components of the conceptual model and the method for building tasks that have been investigated and used in the research reported here provide needed underpinnings for our current work in the APS project, which addresses the remaining two components of the conceptual model within the context of tasks originating in real situations. The between-concept systems of the rational-number conceptual model must be addressed in the APS research because, in so many of the larger, applied problems appropriate for our seventh-grade subjects, the ideas—fraction, ratio, proportion, percent—are inextricably connected to what the subjects know about measurement, area, and number lines, as well as to real-world understandings.

When the goal of research is to investigate the processes, skills, and understandings that enable youngsters to use their current understandings of a mathematical idea, it is appropriate to use the kind of small problems discussed in this chapter. However, when the goal is to study the mechanisms by which learners modify and adapt their understandings in the course of solving problems, it becomes more important to focus on the larger kinds of problems that characterize our current applied problem-solving research. The modeling mechanisms students use for creating and refining various interpretations of problem situations will be emphasized in future reports on findings from the APS project.

Appendix 9.A

Reordered Versions of the CA and OA Instruments

Information lines key: The format for the two lines of information following each item is explained below using a reduced picture of Item C–24 and its information lines as an illustration.

"C–24" means this item had the 24th highest composite score (for all the students who took the item) out of the 60 items on the Concepts Assessment (CA). An "O" preceding the number signifies that the item came from the 35-item Operations Assessment (OA).

"(9)" refers to the original item number, and is the number appearing next to the item question stem.

".750, Grades 4–8, n = 650" gives the composite score for the students who took this item (650 students in Grades 4–8).

The number of respondents for answer choice.

The number of students who did not answer the item.

9. Which picture shows the same fraction as the shaded part of this line segment?

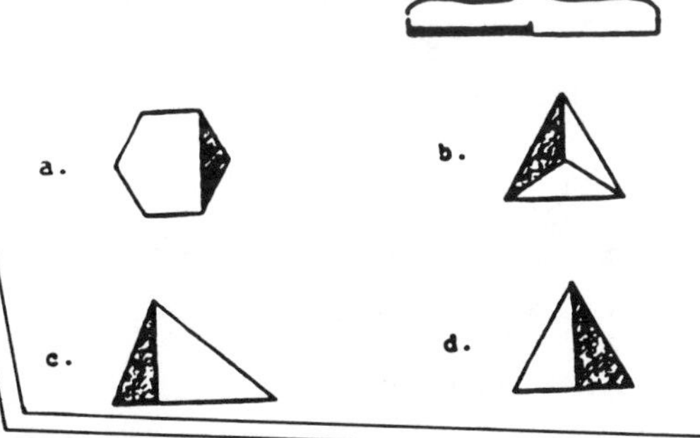

Item C–24 (9): .750, Grades 4–8, n = 650. a) 40 b) 23 c) 18 d) 488 e) 69 na) 11
G4–.644 (22/43) G5–.753 (23/43) G6–.762 (25/60) G7–.836 (21/60) G8–.816 (30/60)

This item was the 22nd easiest item out of 43 for the fourth graders. It was the 30th easiest out of 60 items for the eighth graders. Note: The 60 items done by the sixth through eighth graders include the 43 items done by the fourth and fifth graders.

64.4% of the fourth graders did this item correctly.

➢ These directions preceded the original CA:

ASSESSMENT OF RATIONAL-NUMBER CONCEPTS

Directions: Read each question and set of answers carefully. Select the choice that you think answers the question. Mark the appropriate space ρn your answer sheet. If a question does not have the correct answer given, mark space "e" on your answer sheet. If you do not know how to do a problem, leave the answer space blank.

1. What fraction of this circle is shaded?

a. 2. b. $\frac{1}{2}$ c. 1 d. $\frac{1}{4}$ e. not given

Item C–1 (1): .975, Grades 4–8, n = 650.
a) 2 b) 634 c) 5 d) 5 e) 3 na) 1
G4–.938 (2/43) G5–.975 (1/43) G6–1.00 (1/60) G7–1.00 (1/60) G8–.981 (4/60)

17. Which fraction says "two-fourths"?
a. 6 b. 24% c. 8 d. $\frac{2}{4}$ e. not given

Item C–2 (17): .964, Grades 4–8, n = 650.
a) 0 b) 5 c) 8 d) 627 e) 8 na) 2
G4–.957 (1/43) G5–.950 (4/43) G6–.991 (2/60) G7–.959 (8/60) G8–.972 (6/60)

18. Which fraction says $\frac{2}{3}$?
a. five b. six c. two-thirds
d. twenty-three percent e. not given

Item C–3 (18): .958, Grades 4–8, n = 650.
a) 0 b) 4 c) 623 d) 14 e) 8 na) 1
G4–.932 (3/43) G5–.950 (4/43) G6–.983 (3/60) G7–.989 (2/60) G8–.954 (9/60)

10. Which fraction says "three-fourths"?
a. 34 b. $\frac{3}{4}$ c. $3\frac{1}{4}$ d. $\frac{4}{3}$ e. not given

Item C–4 (10): .955, Grades 4–8, n = 650.
a) 5 b) 621 c) 7 d) 10 e) 3 na) 4
G4–.901 (4/43) G5–.950 (4/43) G6–.974 (5/60) G7–.989 (2/60) G8–.990 (1/60)

8. How long is the snake?

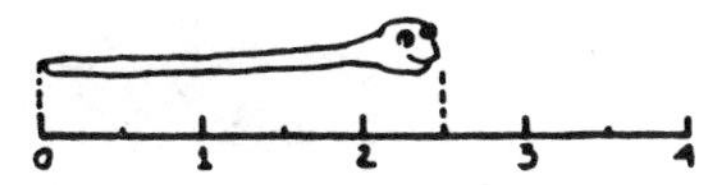

a. two b. two and one-half c. three
d. three and a half e. not given

Item C–5 (8): .947, Grades 4–8, n = 650.
a) 1 b 616 c) 6 d) 11 e) 13 na) 1
G4–.852 (6/43) G5–.969 (2/43) G6–.983 (3/60) G7–.979 (4/60) G8–.990 (1/60)

The following preceded the first ratio item on the original CA, and applies to Items C–6, C–19, C–30, C–35, C–40, C–43, C–52, and C–54 on this reordered version.

Many questions that follow use the word *ratio*. In the picture below, the ratio of circles to squares is 4 to 3.

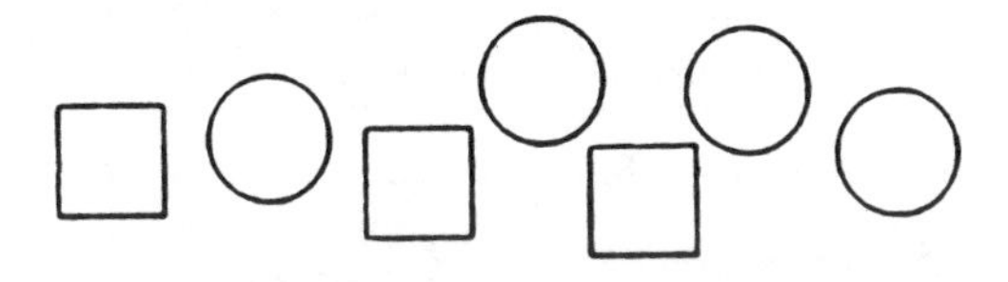

24. Which picture shows the ratio of two circles to three triangles?

a. 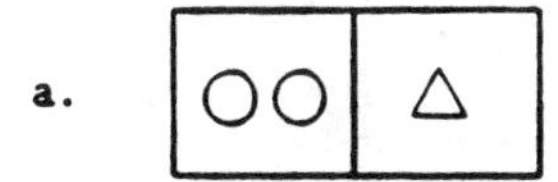

b. 

c. 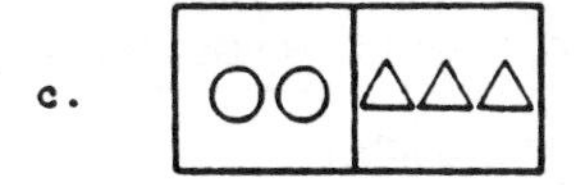

d. 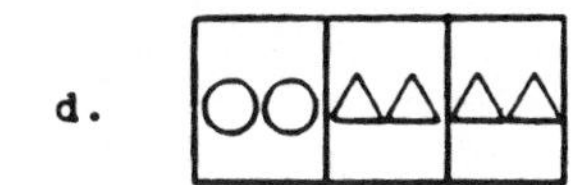

e. not given

Item C–6 (24): .941, Grades 4–8, n = 650.
a) 9 b) 11 c) 612 d) 4 e) 8 na) 6
G4–.883 (5/43) G5–.956 (3/43) G6–.932 (8/60) G7–.969 (6/60) G8–.990 (1/60)

52. Which picture shows the same fraction shaded as this picture?

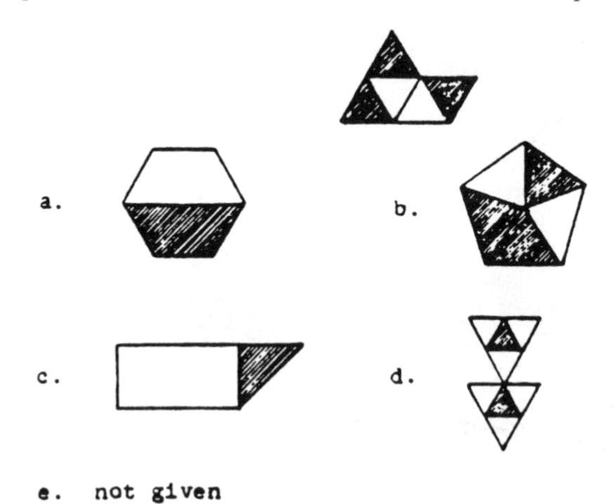

Item C–7 (52): .916, Grades 6–8, n = 325.
a) 5 b) 298 c) 3 d) 5 e) 8 na) 6
— — G6–.923 (9/60) G7–.857 (19/60) G8–.963 (7/60)

5. Which picture shows two-thirds shaded?

Item C–8 (5): .916, Grades 4–8, n = 650.
a) 2 b) 596 c) 32 d) 12 e) 7 na) 1
G4–.797 (7/43) G5–.938 (7/43) G6–.949 (6/60) G7/.969 (6/60) G8–.981 (4/60)

16. What fraction of the set of triangles is shaded?

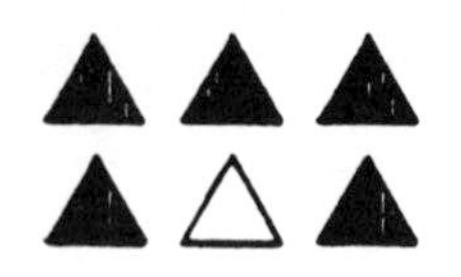

a. $\frac{3}{2}$ b. $\frac{5}{2}$ d. $\frac{3}{5}$ e. not given

Item C–9 (16): .899, Grades 4–8, n = 650.
a) 8 b) 20 c) 9 d) 22 e) 585 na) 6
G4–.766 (8/43) G5/.913 (8/43) G6–.949 (6/60) G7–.979 (4/60) G8–.954 (9/60)

6. Which picture shows $\frac{1}{2}$ shaded?

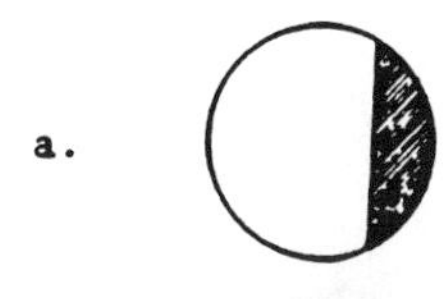

b. 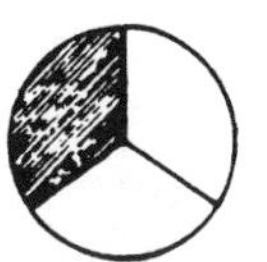

c. 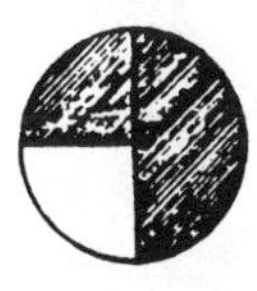

d. 

e. not given

Item C–10 (6): .876, Grades 4–8, n = 650.
a) 23 b) 16 c) 11 d) 25 e) 570 na) 3
G4–.760 (10/43) G5–.895 (10/43) G6/.898 (11/60) G7–.948 (11/60) G8–.935 (13/60)

3. What fraction of this picture is shaded?

a. $\frac{3}{2}$ b. $\frac{5}{2}$ c. $\frac{2}{5}$ d. $\frac{3}{5}$ e. not given

Item C–11 (3): .861, Grades 4–8, n = 650.
a) 40 b) 8 c) 30 d) 560 e) 7 na) 5
G4–.742 (12/43) G5–.913 (8/43) G6–.838 (16/60) G7–.908 (15/60) G8–.944 (11/60)

20. Which picture below shows the same fraction shaded as this set of circles?

a. 

b. 

c. 

d. 

e. not given

Item C–12 (20): .855, Grades 4–8, n = 650.
a) 19 b) 4 c) 23 d) 556 e) 45 na) 3
G4–.766 (8/43) G5–.888 (11/43) G6/.864 (14/60) G7–.887 (17/60) G8–.899 (19/60)

32. What fraction of this line segment is shaded?

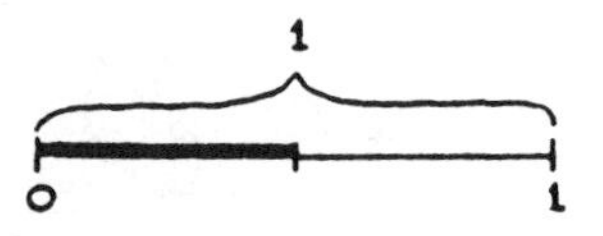

a. one-half b. one
c. two d. one-third e. not given

Item C–13 (32): .836, Grades 4–8, n = 650.
a) 544 b) 68 c) 8 d) 12 e) 10 na) 8
G4–.748 (11/43) G5–.820 (15/43) G6–.813 (20/60) G7–.959 (8/60) G8–.908 (17/60)

19. Which number goes with the point?

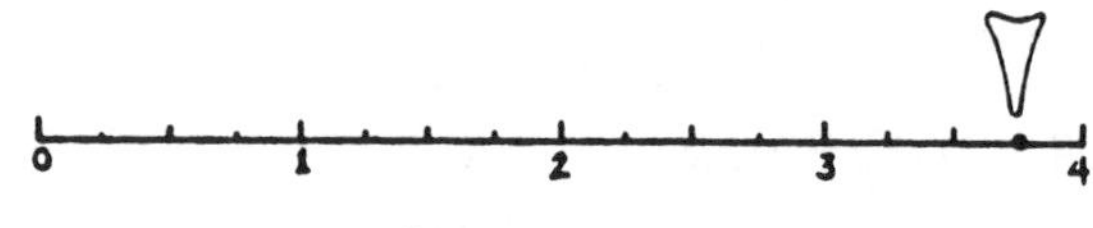

a. $\frac{14}{4}$ b. $3\frac{4}{5}$ c. $\frac{15}{5}$ d. $3\frac{3}{4}$ e. not given

Item C–14 (19): .835, Grades 4–8, n = 650.
a) 10 b) 23 c) 8 d) 543 e) 52 na) 13
G4–.705 (13/43) G5–.777 (13/43) G6–.872 (12/60) G7–.959 (8/60) G8–.963 (7/60)

33. What fraction of the eggs are circled?

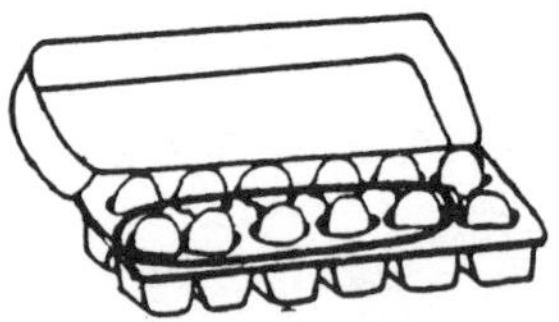

a. $\frac{5}{12}$ b. $\frac{5}{7}$ c. $\frac{5}{8}$ d. $\frac{7}{12}$ e. not given

Item C–15 (33): .830, Grades 4–8, n = 650.
a) 540 b) 65 c) 7 d) 8 e) 18 na) 12
G4–.668 (20/43) G5–.833 (14/43) G6–.872 (12/60) G7–.948 (11/60) G8–.917 (16/60)

31. What fraction of this rectangle is shaded?

a. $\frac{1}{5}$ b. 4 c. $\frac{1}{4}$ d. $\frac{3}{2}$ e. not given

Item C–16 (31): .829, Grades 4–8, n = 650.
a) 539 b) 9 c) 70 d) 5 e) 21 na) 6
G4–.674 (19/43) G5–.839 (13/43) G6–.847 (15/60) G7–.938 (13/60) G8–.926 (14/60)

4. Which picture shows three-fourths shaded?

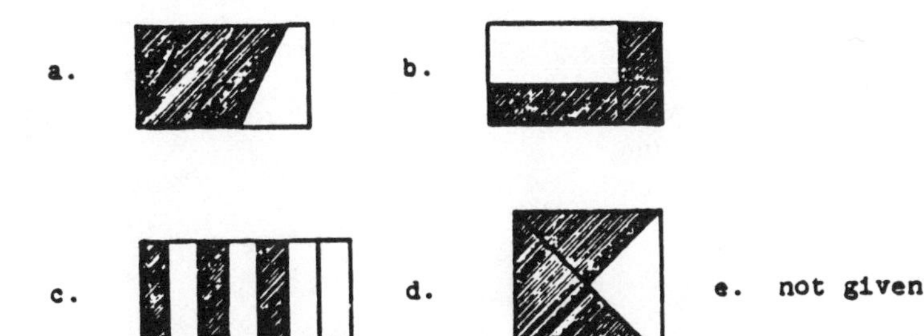

Item C–17 (4): .824, Grades 4–8, n = 650.					
a) 44	b) 12	c) 34	d) 536	e) 21	na) 2

G4–.680 (17/43) G5–.851 (12/43) G6–.830 (17/60) G7–.897 (16/60) G8–.926 (14/60)

➢ The following explanation and picture preceded a set of five items on the original CA, and applies to Items C–18, C–25, C–31, C–45, and C–47 on this reordered version.

In 44–48, each picture is a fraction of a whole. Tell which answer says how many more parts like the one shown below must be added to make the whole. A sample problem is given below.

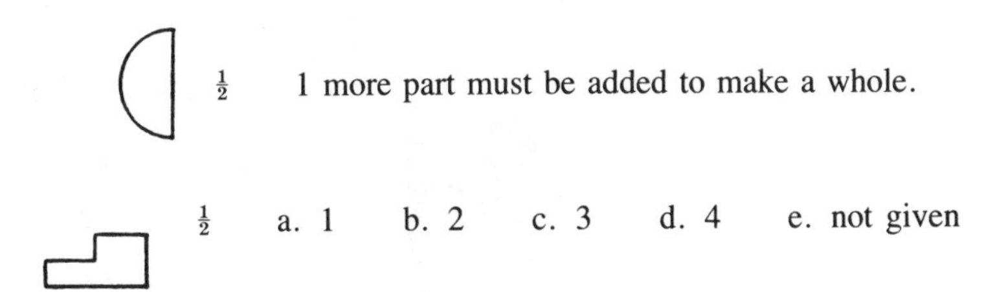

Item C–18 (47): .806, Grades 6–8, n = 325.

a) 262 b) 23 c) 12 d) 6 e) 13 na) 9

— — G6–.737 (27/60) G7–.785 (28/60) G8–.899 (19/60)

➢ For the following item, see the comments and illustration preceding Item C–6.

28. What picture shows the same ratio as the ratio of shaded to unshaded triangles in this picture?

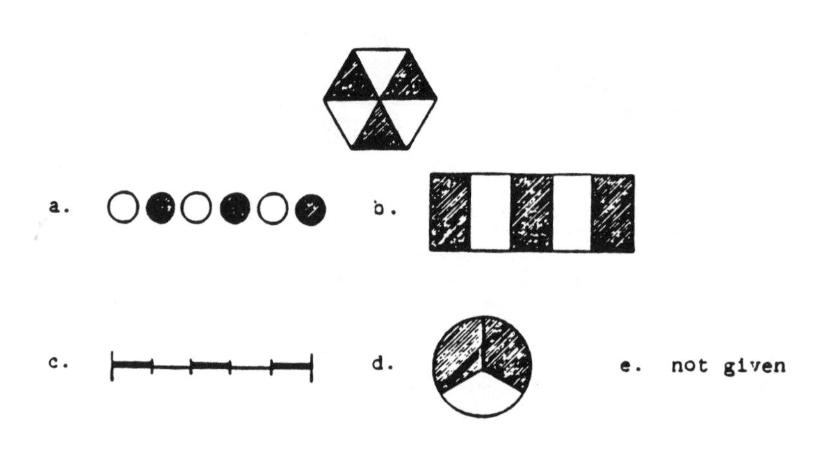

Item C–19 (28): .780, Grades 4–8, n = 650.
a) 507 b) 55 c) 23 d) 17 e) 31 na) 17
G4–.687 (16/43) G5–.765 (21/43) G6–.788 (22/60) G7–.836 (21/60) G8–.880 (21/60)

2. What fraction of this picture is shaded?

a. four-thirds b. three-fourths c. one-third
d. one-half e. not given

Item C–20 (2): .778, Grades 4–8, n = 650.
a) 26 b) 506 c) 64 d) 3 e) 46 na) 5
G4–.650 (21/43) G5–.783 (17/43) G6–.830 (17/60) G7–.806 (26–60) G8–.880 (21/60)

13. Which picture is divided into equal size parts?

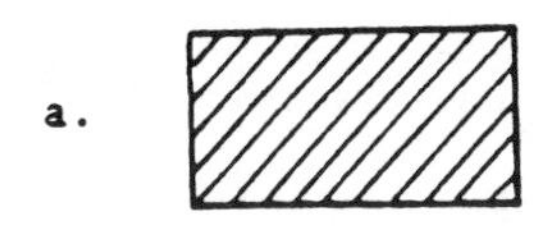

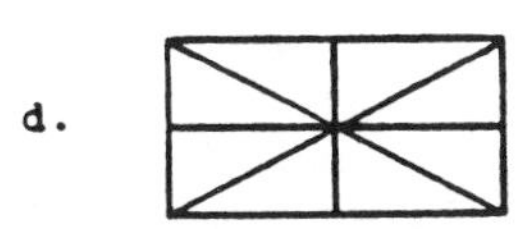

e. not given

Item C–21 (13): .776, Grades 4–8, n = 650.
a) 21 b) 38 c) 19 d) 505 e) 63 na) 3
G4–.680 (17/43) G5–.783 (17/43) G6–.788 (22/60) G7–.816 (25/60) G8–.862 (24/60)

29. What fraction of this circle is shaded?

a. $\frac{1}{2}$ b. $\frac{1}{5}$ c. $\frac{1}{3}$ d. $\frac{3}{1}$ e. not given

Item C–22 (29): .761, Grades 4–8, n = 650.
a) 36 b) 28 c) 495 d) 21 e) 57 na) 13
G4–.582 (23/43) G5–.771 (20/43) G6–.830 (17/60) G7–.887 (17/60) G8–.825 (28/60)

27. Which picture shows $\frac{3}{4}$ shaded?

e. not given

Item C–23 (27): .756, Grades 4–8, n = 650.
a) 37 b) 5 c) 28 d) 78 e) 492 na) 10
G4–.582 (23/43) G5–.765 (21/43) G6–.813 (20/60) G7–.836 (21/60) G8–.871 (23/60)

9. Which picture shows the same fraction as the shaded part of this line segment?

a. 

b. 

c. 

d. 

e. not given

Item C–24 (9): .750, grades 4–8, n = 650.
a) 40 b) 23 c) 18 d) 488 e) 69 na) 11
G4–.644 (22/43) G5–.753 (23/43) G6–.762 (25/60) G7–.836 (21/60) G8–.816 (30/60)

➢ For the following item, see the directions and illustration preceding Item C–18.

44. ◇ $\frac{1}{3}$ a. 1 b. 2 c. 3 d. 4 e. not given

Item C–25 (44): .744, grades 6–8, n = 325.
a) 18 b) 242 c) 42 d) 8 e) 22 na) 3
— — G6–.728 (28/60) G7–.775 (30/60) G8–.733 (38/60)

11. Which container measures cups in thirds?

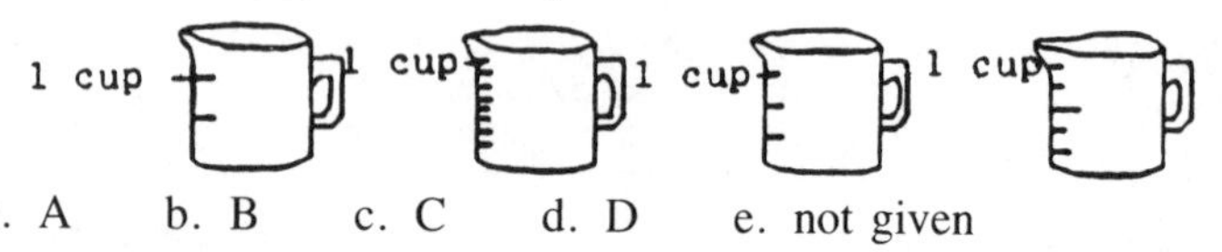

a. A b. B c. C d. D e. not given

Item C–26 (11): .727, Grades 4–8, n = 650.
a) 94 b) 22 c) 473 d) 25 e) 25 na) 8
G4–.693 (15/43) G5–.746 (24/43) G6–.745 (26/60) G7–.734 (32/60) G8–.724 (41/60)

49. What is the numerator of the fraction that tells what part of the picture below is shaded?

a. three b. four c. seven d. three-sevenths
e. not given

Item C–27 (49): .723, Grades 6–8, n = 325.
a) 14 b) 235 c) 28 d) 14 e) 26 na) 8
— — G6–.644 (36/60) G7–.744 (31/60) G8–.788 (31/60)

21. Which picture below shows $\frac{3}{4}$ shaded?

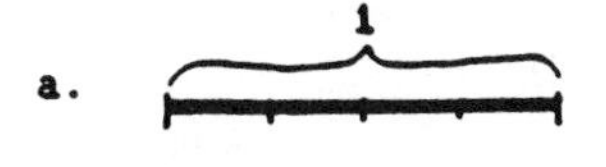

e. not given

Item C–28 (21): .695, Grades 4–8, n = 650.
a) 13 b) 38 c) 18 d) 118 e) 452 na) 10
G4–.503 (28/43) G5–.734 (25/43) G6–.686 (33/60) G7–.785 (28/60) G8–.853 (25/60)

30. What fraction of this circle is shaded?

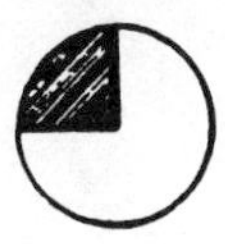

a. one-third b. one-fifth
c. one-fourth d. three e. not given

Item C–29 (30): .683, Grades 4–8, n = 650.
a) 122 b) 40 c) 444 d) 6 e) 29 na) 9
G4–.539 (27/43) G5–.592 (27/43) G6–.728 (28/60) G7–.857 (19/60) G8–.825 (28/60

➢ The following explanation and picture preceded a set of three items on the original CA, and applies to Items C–30, C–43, and C–52 on this reordered version.

In 35–37, tell what ratio is suggested by each picture. Here is a sample problem:

Wheels on one bicycle is 2 to 1.

35. Bottles in one carton is

a. 5 to 1 b. 6 to 1 c. 1 to 6
d. 3 to 3 e. not given

Item C–30 (35): .675, Grades 4–8, n = 650.
a) 15 b) 439 c) 46 d) 85 e) 48 na) 17
G4–.472 (31/43) G5–.555 (31/43) G6–.771 (24/60) G7–.836 (21/60) G8–.908 (17/60)

➢ For the following item, see the directions and illustration preceding Item C–18.

45. △ $\frac{1}{4}$ a. 1 b. 2 c. 3 d. 4 e. not given

Item C–31 (45): .670, Grades 6–8, n = 325.
a) 20 b) 18 c) 218 d) 35 e) 27 na) 7
— — G6–.610 (39/60) G7–.704 (35/60) G8–.706 (43/60)

14. Which picture shows $\frac{2}{3}$ shaded?

a. 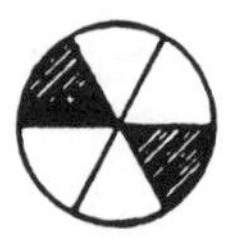

b. 

c. 

d. 

e. not given

Item C–32 (14): .652, Grades 4–8, n = 650.
a) 14 b) 9 c) 424 d) 71 e) 129 na) 2
G4–.546 (26/43) G5–.604 (26/43) G6–.728 (28/60) G7–.734 (32/60) G8–.724 (41/60)

59. What fraction of the whole numbers 1, 2, 3, 4, 5, 6, 7 are odd numbers?
a. Three sevenths b. four-sevenths
c. four ninths d. one-third e. not given

Item C–33 (59): .649, Grades 6–8, n = 325.
a) 44 b) 211 c) 12 d) 14 e) 32 na) 12
— — G6–.584 (42/60) G7–.622 (43/60) G8–.743 (36/60)

54. $2\frac{1}{3}$ = a. $\frac{4}{3}$ b. $\frac{8}{3}$ c. $\frac{7}{3}$ d. $\frac{3}{3}$ e. not given

Item C–34 (54): .643, Grades 6–8, n = 325.
a) 24 b) 15 c) 209 d) 32 e) 27 na) 17
— — G6–.576 (43/60) G7–.510 (48/60) G8–.834 (26/60)

➢ For the following item, see the comments and illustration preceding Item C–6.

25. What is the ratio of shaded to unshaded rectangles?

a. three to seven b. three to four
c. four to three d. seven to three e. not given

Item C–35 (25): .632, Grades 4–8, n = 650.
a) 73 b) 411 c) 121 d) 23 e) 14 na) 7
G4–.576 (25/43) G5–.592 (27/43) G6–.635 (38/60) G7–.704 (35/60) G8–.706 (43/60)

15. How many halves equal one whole?
a. $\frac{2}{2}$ b. 2 c. $1\frac{1}{2}$ d. 1 e. not given

Item C–36 (15): .632, Grades 4–8, n = 650.
a) 137 b) 411 c) 28 d) 39 e) 26 na) 8
G4–.447 (13/43) G5–.641 15/43) G6–.618 (10/60) G7–.816 (14/60) G8–.743 (11/60)

56. $\frac{11}{3}$ = a. $3\frac{3}{2}$ b. $2\frac{3}{2}$ c. $3\frac{2}{3}$ d. $10\frac{1}{3}$ e. not given

Item C–37 (56): .621, Grades 6–8, n = 325.
a) 15 b) 11 c) 202 d) 31 e) 41 na) 25
— — G6–.593 (41/60) G7–.530 (45/60) G8–.733 (38/60)

34. If this is the unit, , then what fraction is shown by this picture?

a. 3 b. $\frac{1}{2}$ c. $2\frac{1}{2}$ d. $\frac{3}{2}$ e. not given

Item C–38 (34): .595, Grades 4–8, n = 650.
a) 15 b) 19 c) 387 d) 40 e) 166 na) 23
G4–.325 (36/43) G5–.543 (32/43) G6–.652 (35/60) G7–.795 (27–60) G8/.834 (26/60)

41. If this is the unit, , then what fraction is shaded in this picture?

a. $2\frac{1}{3}$ b. $2\frac{2}{3}$ c. 3 d. $\frac{3}{8}$ e. not given

Item C–39 (41): .584, Grades 4–8, n = 650.
a) 131 b) 380 c) 8 d) 15 e) 82 na) 33
G4–.380 (33/43) G5–.561 (30/43) G6–.644 (36/60) G7/.693 (37/60) G8–.761 (34/60)

➢ For the following item, see the comments and illustration preceding Item C–6.

23. What is the ratio of circles to triangles?

a. 3 to 2 b. 3 to 5 c. 2 to 3
d. 2 to 5 e. not given

Item C–40 (23): .578, Grades 4–8, n = 650.
a) 196 b) 22 c) 376 d) 29 e) 16 na) 11
G4–.478 (30/43) G5–.524 (33/43) G6–.576 (43/60) G7–.653 (40/60) G8–.743 (36/60)

26. How many thirds equal one whole?
a. 1 b. 2 c. 3 d. 4 e. not given

Item C–41 (26): .569, Grades 4–8, n = 650.
a) 27 b) 53 c) 370 d) 89 e) 88 na) 22
G4–.343 (35/43) G5–.512 (34/43) G6–.669 (34/60) G7–.693 (37/60) G8–.770 (33/60)

40. How many thirteenths equal one whole?
a. $\frac{1}{13}$ b. 1 c. 12 d. 13 e. not given

Item C–42 (40): .564, Grades 4–8, n = 650.
a) 71 b) 48 c) 28 d) 367 e) 94 na) 41
G4–.306 (37/43) G5–.500 (35/43) G6–.720 (31/60) G7–.693 (37/60) G8–.761 (34/60)

➢ For the following item, see the comments and illustration preceding Item C–6.

➢ For the following item, see the comments and sample ratio preceding Item C–30.

36. Toes on two feet is

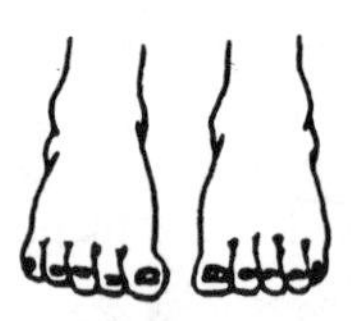

a. ten to two b. two to ten c. five to two
d. two to five e. not given

Item C–43 (36): .563, Grades 4–8, n = 650.
a) 366 b) 68 c) 41 d) 28 e) 130 na) 17
G4–.386 (32/43) G5–.462 (36/43) G6–.601 (40/60) G7–.734 (32/60) G8–.779 (32/60)

42. What fraction of the balls are tennis balls?

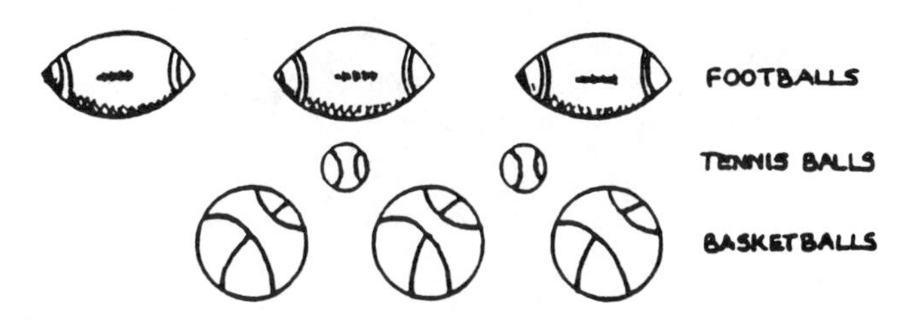

a. $\frac{2}{8}$ b. $\frac{3}{2}$ c. $\frac{2}{6}$ d. $\frac{6}{2}$ e. not given

Item C–44 (42): .561, Grades 4–8, n = 650.
a) 365 b) 29 c) 166 d) 18 e) 54 na) 18
G4–.355 (34/43) G5–.574 (29/43) G6–.694 (32/60) G7–.632 (42/60) G8–.642 (47/60)

➢ For the following item, see the directions and illustration preceding Item C–18.

48. $\frac{1}{4}$ a. 1 b. 2 c. 3 d. 4 e. not given

Item C–45 (48): .553, Grades 6–8, n = 325.
a) 65 b) 32 c) 180 d) 28 e) 12 na) 7
— — G6–.508 (46/60) G7–.530 (45/60) G8–.623 (48/60)

51. Which letter shows the point $\frac{8}{3}$?

a. A b. B c. C d. D e. not given

Item C–46 (51): .553, Grades 6–8, n = 325.
a) 5 b) 18 c) 33 d) 180 e) 82 na) 6
— — G6–.466 (48/60) G7–.530 (45/60) G8–.669 (46/60)

➢ For the following item, see the directions and illustration preceding Item C–18.

46. {OO} $\frac{1}{3}$ a. 1 b. 2 c. 3 d. 4 e. not given

Item C–47 (46): .532, Grades 6–8, n = 325.
a) 39 b) 173 c) 41 d) 30 e) 28 na) 14
— — G6–.516 (45/60) G7–.459 (51/60) G8–.614 (49/60)

> ➢ The following directions preceded a set of two items on the original CA, and applies to Items C–48 and C–50 on this reordered version.

The first picture of 57 and 58 is a fraction of the whole. For each of these problems, tell which answer shows the whole.

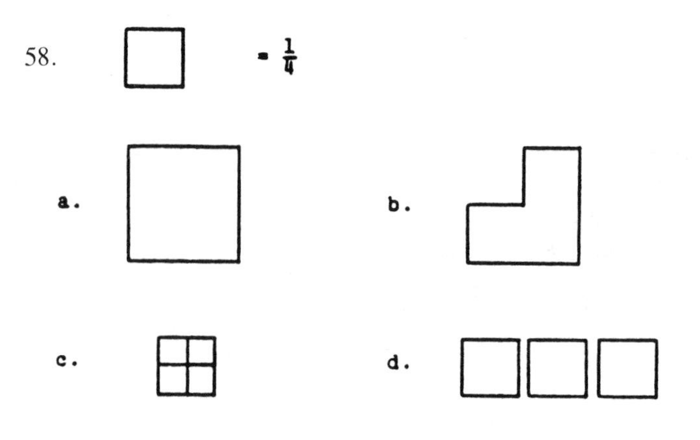

e. not given

Item C–48 (58): .483, Grades 6–8, *n* = 325.
a) 157 b) 62 c) 72 d) 12 e) 9 na) 13
— — G6–.423 (52/60) G7–.479 (50/60) G8–.550 (51/60)

22. This ruler measures inches by

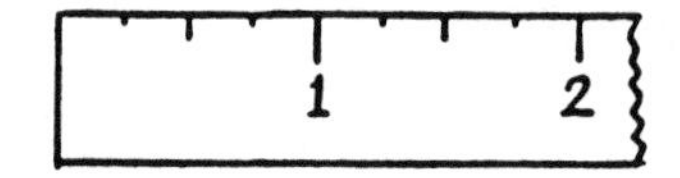

a. wholes, halves, and fourths
b. wholes and halves only
c. wholes, halves, and thirds
d. wholes and fourths only
e. not given

Item C–49 (22): .483, Grades 4–8, *n* = 650.
a) 314 b) 64 c) 99 d) 72 e) 82 na) 19
G4–.233 (40/43) G5–.456 (37/43) G6–.491 (47/60) G7–.653 (40/60) G8–.733 (38/60)

➢ For the following item, see the directions preceding Item C–48.

57. 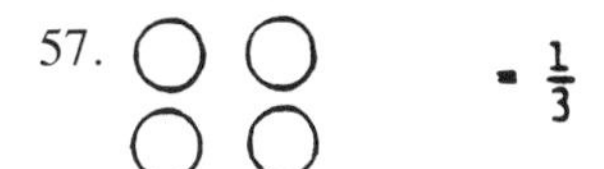

a. 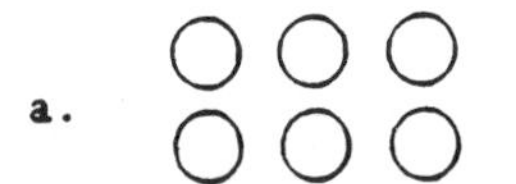

b. 

c. 

d. 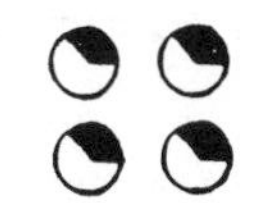

e. not given

Item C–50 (57): .458, Grades 6–8, n = 325.
a) 30 b) 149 c) 42 d) 49 e) 36 na) 19
— — G6–.398 (53/60) G7–.438 (52/60) G8–.541 (52/60)

7. Which picture shows fourths?

a. 

b. 

c. 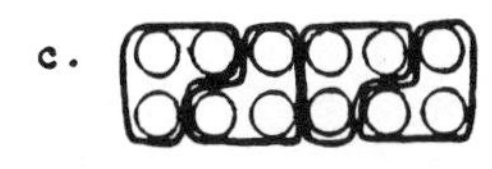

d. 

e. not given

Item C–51 (7): .430, Grades 4–8, n = 650.
a) 15 b) 13 c) 280 d) 299 e) 36 na) 7
G4–.306 (37/43) G5–.425 (38/43) G6–.457 (50/60) G7–.438 (52/60) G8–.487 (50/60)

➢ For the following item, see the comments and illustration preceding Item C–6.

➢ For the following item, see the comments and sample ratio preceding Item C–30.

37. Feet to inches is

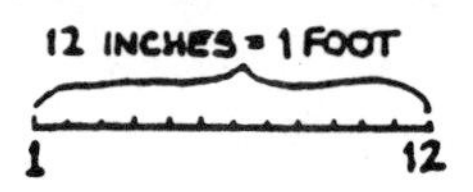

a. 1 to 12 b. 6 to 6 c. 12 to 1
d. 3 to 4 e. not given

Item C–52 (37): .429, Grades 4–8, n = 650.
a) 279 b) 27 c) 290 d) 4 e) 21 na) 29
G4–.484 (29/43) G5–.376 (39/43) G6–.466 (48/60) G7–.346 (55/60) G8–.458 54/60)

12. Which letter is above the point $\frac{1}{2}$?

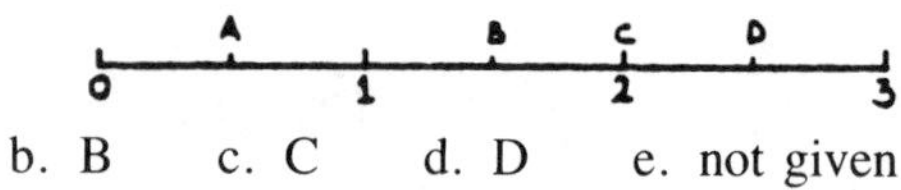

a. A b. B c. C d. D e. not given

Item C–53 (12): .406, Grades 4–8, n = 650.
a) 264 b) 262 c) 50 d) 34 e) 26 na) 12
G4–.208 (41/43) G5–.290 (41/43) G6–.440 (51/60) G7–.571 (44/60) G8–.688 (45/60)

➢ For the following item, see the comments and illustration preceding Item C–6.

38. What is the ratio of shaded to unshaded parts in this picture?

a. 1 to 2 b. 1 to 5 c. 1 to 3
d. 3 to 2 e. not given

Item C–54 (38): .383, Grades 4–8, n = 650.
a) 367 b) 11 c) 8 d) 3 e) 249 na) 12
G4–.282 (39/43) G5–.339 (40/43) G6–.389 (54/60) G7–.489 (49/60) G8–.495 (53/60)

50. What is the denominator of the fraction that tells what part of the picture below is shaded?

a. five-thirds b. five c. three
d. two e. not given

Item C–55 (50): .332, Grades 6–8, n = 325.
a) 8 b) 108 c) 52 d) 125 e) 28 na) 4
— — G6–.305 (56/60) G7–.265 (56/60) G8–.422 (55/60)

53. How long is the snake?

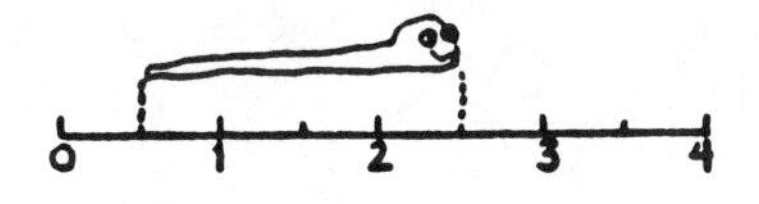

a. $\frac{4}{2}$ b. $\frac{5}{2}$ c. $\frac{6}{2}$ d. $\frac{3}{2}$ e. not given

Item C–56 (53): .286, Grades 6–8, n = 325.
a) 93 b) 44 c) 9 d) 11 e) 158 na) 10
— — G6–.262 (57/60) G7–.173 (59/60) G8–.412 (56/60)

60. If ⊞ = $\frac{1}{3}$, then what fraction of the picture below is shaded?

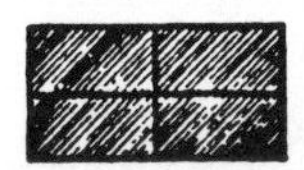
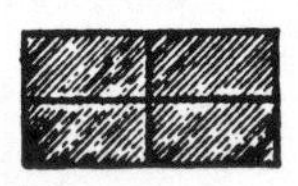
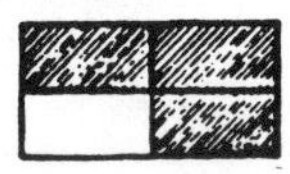

a. three b. two and one-half
c. two and three fourths d. eleven-twelfths
e. not given

Item C–57 (60): .233, Grades 6–8, n = 325.
a) 17 b) 14 c) 129 d) 76 e) 71 na) 18
— — G6–.161 (59/60) G7–.265 (56/60) G8–.284 (58/60)

39. What fraction of the set of objects are triangles?

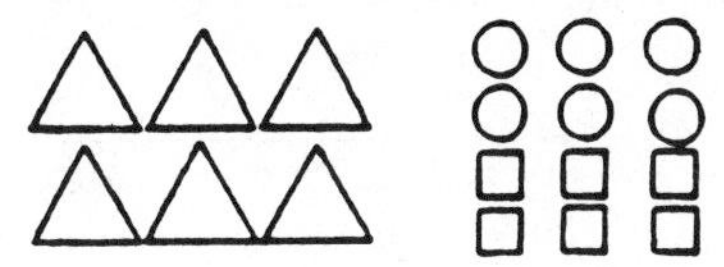

a. $\frac{6}{6}$ b. $\frac{1}{2}$ c. $\frac{1}{3}$ d. $\frac{2}{1}$ e. not given

Item C–58 (39): .215, Grades 4–8, n = 650.
a) 194 b) 39 c) 140 d) 11 e) 235 na) 30
G4–.104 (42/43) G5–.080 (42/43) G6–.279 (56/60) G7–.367 (55/60) G8–.376 (57/60)

55. $3\frac{4}{3}$ = a. $\frac{7}{3}$ b. $4\frac{1}{3}$ c. $\frac{12}{3}$ d. $\frac{16}{3}$ e. not given

Item C–59 (55): .203, Grades 6–8, n = 325.
a) 33 b) 66 c) 30 d) 14 e) 158 na) 24
— — G6–.203 (58/60) G7–.214 (57/60) G8–.192 (60/60)

43. If this is the unit, 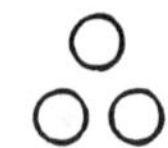 , then what fraction is shaded in the picture below?

a. $\frac{7}{9}$ b. $\frac{7}{3}$ c. $\frac{7}{2}$ d. $\frac{2}{9}$ e. not given

Item C–60 (43): .090, Grades 4–8, n = 650.
a) 235 b) 59 c) 104 d) 47 e) 178 na) 27
G4–.049 (43/43) G5–.049 (43/43) G6–.084 (60/60) G7–.122 (60/60) G8–.192 (60/60)

➢ The following directions were given on the original OA. The information lines for each item follow the same format as the corresponding lines on the CA.

OPERATIONS ASSESSMENT

Directions: Read each question and set of answers carefully. Select the choice that you think answers the question. Mark the appropriate space on your answer sheet. If a question does not have the correct answer given, mark space "e" on your answer sheet. If you do not know how to do a problem, leave the answer space blank.

5. Jeremy's dog had six puppies. He gave away one-half of them. How many puppies did he give away.
   a. 3 b. 2 c. 1 d. 8 e. not given

Item O–1 (5): .876, Grades 4–8, n = 608.
a) 533 b) 28 c) 13 d) 9 e) 17 na) 8
G4–.831 (1/15) G5–.846 (1/28) G6–.845 (2/28) G7–.921 (1/35) G8–.962 (1/35)

7. Three children went to the store. Two-thirds of them rode bicycles. How many did not ride bicycles?
   a. 1 b. 3 c. 2 d. 5 e. not given

Item O–2 (7): .807, Grades 4–8, n = 608.
a) 491 b) 15 c) 73 d) 10 e) 11 na) 8
G4–.688 (2/15) G5–.779 (2/28) G6–.827 (3/28) G7–.882 (2/35) G8–.898 (3/35)

4. Willie had \$1.00. He spent $\frac{1}{4}$ of it on some baseball cards. How much did he spend?
   a. 4¢ b. 25¢ c. 75¢ d. 96¢ e. not given

Item O–3 (4): .703, Grades 4–8, n = 608.
a) 19 b) 428 c) 93 d) 25 e) 24 na) 19
G4–.567 (3/15) G5–.638 (3/28) G6–.781 (5/28) G7–.774 (5/35) G8–.814 (5/35)

2. What is one-half of eighteen?
   a. $18\frac{1}{2}$ b. 9 c. $17\frac{1}{2}$ d. 8 e. not given

Item O–4 (2): .662, Grades 4–8, n = 608.
a) 105 b) 403 c) 21 d) 24 e) 40 na) 15
G4–.415 (6/15) G5–.607 (4/28) G6–.736 (7/28) G7–.823 (4/35) G8–.805 (6/35)

8. Subtract $\frac{5}{6} - \frac{2}{6} =$
   a. $\frac{3}{6}$ b. 3 c. $\frac{7}{6}$ d. $\frac{10}{6}$ e. not given

Item O–5 (8): .659, Grades 4–8, n = 608.
a) 401 b) 131 c) 9 d) 11 e) 44 na) 11
G4–.351 (8/15) G5–.447 (9/28) G6–.890 (1/28) G7–.833 (3/35) G8–.935 (2/35)

➢ The following direction preceded a set of three items on the original OA, and applies to Items O–6, O–21, and O–30 on this reordered version.

In 16–18, find the sums.

16. $\frac{2}{10} + \frac{5}{10} =$ a. $\frac{2}{5}$ b. $\frac{10}{100}$ c. $\frac{10}{10}$ d. $\frac{7}{10}$ e. not given

Item O–6 (16): .645, Grades 5–8, n = 483.
a) 10 b) 24 c) 26 d) 312 e) 104 na) 7
— G5–.380 (10/28) G6–.827 (3/28) G7–.686 (8/35) G8–.824 (4/35)

1. What is 1-fourth plus 1-fourth?
   a. 1-sixteenth b. 2-eighths c. 2-sixteenths
   d. 2-fourths e. not given

Item O–7 (1): .621, Grades 4–8, n = 608.
a) 5 b) 190 c) 6 d) 378 e) 22 na) 7
G4–.479 (4/15) G5–.546 (5/28) G6–.754 (6/28) G7–.686 (8/35) G8–.703 (10/35)

➢ The following direction preceded a set of three items on the original OA, and applies to Items O–8, O–13, and O–18 on this reordered version.

In 11–13, find the products.

13. $\frac{4}{5} \times \frac{4}{7} =$ a. $\frac{8}{12}$ b. $\frac{48}{35}$ c. $\frac{16}{35}$ d. $\frac{8}{35}$ e. not given

Item O–8 (13): .613, Grades 4–8, $n = 608$.
a) 49 b) 44 c) 373 d) 47 e) 63 na) 32
G4–.464 (5/15) G5–.515 (6/28) G6–.709 (9/28) G7–.705 (7/35) G8–.750 (8/35)

22. Find the sum: $2\frac{2}{3} + 3 =$
a. $\frac{8}{9}$ b. $1\frac{2}{3}$ c. $5\frac{2}{3}$ d. $2\frac{5}{3}$ e. not given

Item O–9 (22): .612, Grades 5–8, $n = 483$.
a) 23 b) 16 c) 296 d) 49 e) 72 na) 27
— G5–.503 (7/28) G6–.581 (10/28) G7–.715 (6/35) G8–.712 (9/35)

23. What does this picture show?

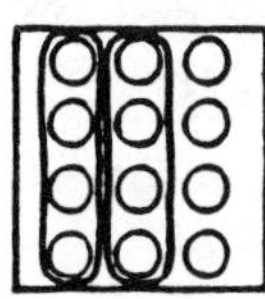

a. $\frac{2}{3}$ of 12 = 8 b. $\frac{2}{4}$ of 12 = 8
c. $\frac{1}{2}$ of 12 = 3 d. 2 to 1 = 8 e. not given

Item O–10 (23): .523, Grades 5–8, $n = 483$.
a) 253 b) 119 c) 20 d) 22 e) 45 na) 24
— G5–.337 (12/28) G6–.572 (11/28) G7–.578 (12/35) G8–.703 (10/35)

26. Sue is $2\frac{1}{2}$ years old. Her brother Tim is $6\frac{1}{2}$ years older than Sue. How old is Tim?
a. 4 b. 8 c. 9 d. $6\frac{1}{2}$ e. not given

Item O–11 (26): .517, Grades 5–8, $n = 483$.
a) 67 b) 68 c) 250 d) 34 e) 43 na) 21
— G5–.337 (12/28) G6–.445 (14/28) G7–.617 (10/35) G8–.768 (7–35)

3. Add: $\frac{1}{3} + \frac{1}{3} =$
a. $\frac{2}{6}$ b. $\frac{2}{3}$ c. $\frac{2}{9}$ d. $\frac{1}{9}$ e. not given

Item O–12 (3): .508, Grades 4–8, $n = 608$.
a) 264 b) 309 c) 6 d) 5 e) 15 na) 9
G4–.287 (9/15) G5–.343 (11/28) G6–.736 (7/28) G7–.607 (11/35) G8–.685 (13/35)

➢ For the following item, see the direction preceding Item O–8.

11. $\frac{1}{2} \times \frac{1}{4} =$ a. $\frac{1}{6}$ b. $\frac{2}{6}$ c. $\frac{2}{8}$ d. $\frac{1}{8}$ e. not given

Item O–13 (11): .478, Grades 4–8, $n = 608$.
a) 40 b) 49 c) 125 d) 291 e) 73 na) 30
G4–.376 (7/15) G5–.472 (8/28) G6–.400 (16/28) G7–.568 (13/35) G8–.601 (17/35)

33. What fraction of this diagram is shaded?

a. $\frac{3}{5}$, since there are 3 shaded parts and 5 unshaded parts.
b. $\frac{4}{8}$, since $\frac{2}{4}$ of $\frac{1}{2}$ plus $\frac{1}{4}$ of $\frac{1}{2} = \frac{4}{8}$.
c. $\frac{5}{8}$, since $\frac{1}{4}$ of $\frac{1}{2}$ is $\frac{1}{8}$, and $\frac{1}{8} + \frac{1}{2} = \frac{5}{8}$.
d. $\frac{3}{8}$, since $\frac{2}{4}$ of $\frac{1}{2}$ plus $\frac{1}{4}$ of $\frac{1}{2} = \frac{3}{8}$.
e. not given

Item O–14 (33): .457, Grades 7–8, n = 210.
a) 28 b) 32 c) 18 d) 96 e) 30 na) 6
— — — G7–.450 (15/35) G8–.462 (24–35)

28. Pam has $4\frac{1}{2}$ cups of flour. If she uses $3\frac{1}{4}$ cups to make a cake, how much flour will she have left?
a. $7\frac{3}{4}$ cups b. $1\frac{1}{4}$ cups c. 1 cup
d. $1\frac{1}{2}$ cups e. not given

Item O–15 (28): .436, Grades 5–8, n = 483.
a) 19 b) 211 c) 51 d) 93 e) 57 na) 52
— G5–.282 (15/28) G6–.354 (20/28) G7–.490 (14/35) G8–.703 (10–35)

31. In a group of thirty-six people, two-ninths of the men have blue eyes. Three-fourths of the people are men. How should you find out how many men have blue eyes?
a. First find $\frac{3}{4}$ of 36, then take $\frac{2}{9}$ of that number.
b. First find $\frac{2}{9}$ of 36, then take $\frac{3}{4}$ of that number.
c. Find $\frac{5}{13}$ of 36.
d. Find $\frac{35}{36}$ of 36.
e. not given

Item O–16 (31): .433, Grades 7–8, n = 210.
a) 91 b) 59 c) 14 d) 6 e) 16 na) 24
— — — G7–.382 (16/35) G8–.481 (21/35)

➢ The following direction preceded a set of two items on the original OA, and applies to Items O–17 and O–29 on this reordered version.

In 24–25, find the products.

24. $3 \times \frac{1}{5} =$
a. $\frac{1}{15}$ b. $\frac{3}{15}$ c. 15 d. $\frac{3}{5}$ e. not given

Item O–17 (24): .389, Grades 5–8, n = 483.
a) 54 b) 125 c) 28 d) 188 e) 56 na) 32
— G5–.171 (19/28) G6–.509 (12/28) G7–.352 (19/35) G8–.629 (15–35)

➢ For the following item, see the direction preceding Item O–8.

12. $4 \times \frac{1}{3} =$ a. $\frac{1}{12}$ b. $\frac{4}{3}$ c. $\frac{4}{12}$ d. $\frac{13}{3}$ e. not given

Item O–18 (12): .348, Grades 4–8, n = 608.
a) 74 b) 212 c) 164 d) 39 e) 78 na) 41
G4–.200 (11/15) G5–.171 (19/28) G6–.481 (13/28) G7–.362 (17/35) G8–.638 (14–35)

20. Gwen did an experiment on vitamins. She used 20 mice. She gave $\frac{3}{5}$ of them special vitamin food. How many mice got special vitamin food?
a. 15 b. 12 c. 4 d. 16 e. not given

Item O–19 (20): .339, Grades 5–8, n = 483.
a) 158 b) 164 c) 42 d) 38 e) 59 na) 22
— G5–.220 (17/28) G6–.400 (16/28) G7–.284 (23/35) G8–.509 (20/35)

6. What does $\frac{3}{4}$ of 12 equal?
a. $\frac{48}{3}$ b. $12\frac{3}{4}$ c. 9 d. 3 e. not given

Item O–20 (6): .330, Grades 4–8, n = 608.
a) 21 b) 146 c) 201 d) 114 e) 82 na) 43
G4–.111 (13/15) G5–.245 (16/28) G6–.390 (18/28) G7–.362 (17/35) G8–.620 (16/35)

➢ For the following item, see the direction preceding Item O–6.

17. $\frac{1}{3} + \frac{1}{2} =$ a. $\frac{5}{6}$ b. $\frac{2}{5}$ c. $\frac{2}{6}$ d. $\frac{1}{6}$ e. not given

Item O–21 (17): .308, Grades 5–8, n = 483.
a) 149 b) 206 c) 67 d) 26 e) 26 na) 9
— G5–.110 (24/28) G6–.381 (19/28) G7–.264 (27/35) G8–.574 (19/35)

21. Find the sum: $\begin{array}{r} \frac{4}{15} \\ + \frac{5}{9} \\ \hline \end{array}$

a. $\frac{9}{24}$ b. $\frac{20}{135}$ c. $\frac{37}{45}$ d. $\frac{14}{15}$ e. not given

Item O–22 (21): .308, Grades 5–8, n = 483.
a) 169 b) 36 c) 149 d) 16 e) 98 na) 14
— G5–.085 (26/28) G6–.327 (22/28) G7–.343 (20/35) G8–.592 (18/35)

32. What fraction of this circle is shaded?

a. $\frac{1}{10} + \frac{1}{2}$ b. $\frac{1}{2} + \frac{1}{5}$ c. $1 + \frac{1}{5}$ d. $\frac{2}{6}$
e. not given

Item O–23 (32): .280, Grades 7–8, n = 210.
a) 59 b) 79 c) 24 d) 19 e) 25 na) 4
— — — G7–.274 (25/35) G8–.287 (30/35)

27.

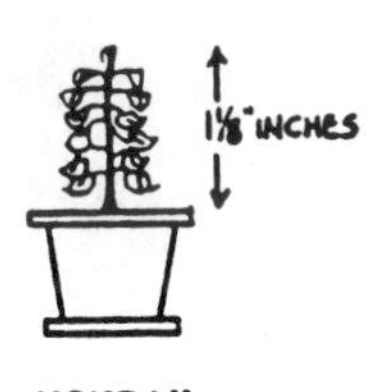

MONDAY

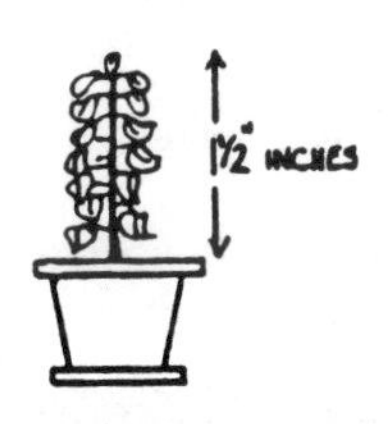

SATURDAY

Which of the following expressions represents how many inches this plant grew from Monday to Saturday?
a. $1\frac{1}{8} + 1\frac{1}{2}$ b. $1\frac{1}{2}$ c. $1\frac{1}{2} - 1\frac{1}{8}$
d. $1\frac{1}{8} - 1\frac{1}{2}$ e. not given

Item O–24 (27): .275, Grades 5–8, n = 483.
a) 115 b) 31 c) 133 d) 93 e) 80 na) 31
— G5–.153 (21/28) G6–.254 (26/28) G7–.313 (21/35) G8–.444 (26/35)

19. Which means the same as "one-third of one-fourth"?
a. $\frac{1}{3} + \frac{1}{4}$ b. $3 + \frac{1}{4}$ c. $\frac{1}{3} \times \frac{1}{4}$ d. $\frac{1}{3} + 14$ e. not given

Item O–25 (19): .269, Grades 5–8, n = 483.
a) 161 b) 27 c) 130 d) 8 e) 129 na) 28
— G5–.184 (18/28) G6–.290 (24/28) G7–.333 (25/35) G8–.314 (29/35)

10. How can you find $\frac{1}{2}$ of $\frac{1}{3}$?
    a. Add the numerators, multiply the denominators to get $\frac{2}{6}$.
    b. Find common denominators and add numerators to get $\frac{5}{6}$.
    c. Find common denominators and multiply numerators to get $\frac{6}{6}$.
    d. Multiply the numerators and add the denominators to get $\frac{1}{5}$.
    e. not given

Item O–26 (10): .264, Grades 4–8, n = 608.
a) 119 b) 111 c) 62 d) 68 e) 161 na) 87
G4–.263 (10/15) G5–.331 (14/28) G6–.281 (25/28) G7–.147 (32/35) G8–.259 (31–35)

15. How can you add three-fourths and one-third?
    a. First add three and one, then four and three to get four-sevenths.
    b. First change to twelfths, then add numerators to get thirteen-twelfths.
    c. First change to twelfths, then add numerators and denominators to get thirteen-twenty-fourths.
    d. First add three and one, then multiply four times three to get four-twelfths or one-third.
    e. not given

Item O–27 (15): .250, Grades 4–8, n = 608.
a) 230 b) 152 c) 52 d) 46 e) 84 na) 44
G4–.064 (14/15) G5–.116 (23/28) G6–.409 (15/28) G7–.284 (23/35) G8–.472 (23–35)

9. Which picture shows one-sixth of twelve?

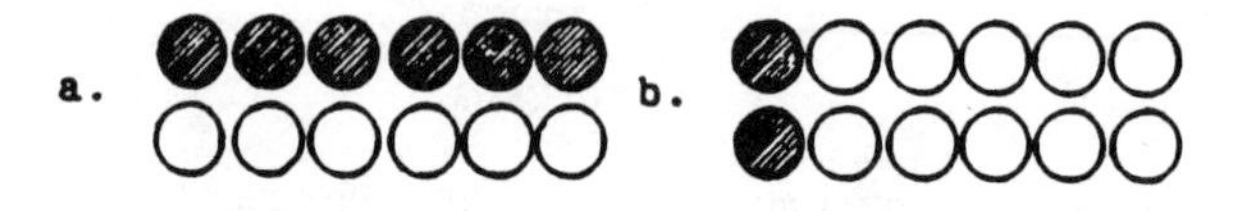

Item O–28 (9): .245, Grades 4–8, n = 608.
a) 247 b) 149 c) 11 d) 72 e) 117 na) 12
G4–.055 (15/15) G5–.153 (21/28) G6–.336 (21/28) G7–.294 (22/35) G8–.462 (24/35)

➢ For the following item, see the direction preceding Item O–17.

25. $9 \times \frac{2}{3} =$
 a. $9\frac{2}{3}$ b. $\frac{2^7}{2}$ c. 6 d. $\frac{18}{27}$ e. not given

Item O–29 (25): .242, Grades 5–8, n = 483.
a) 59 b) 35 c) 117 d) 145 e) 91 na) 35
— G5–.030 (28/28) G6–.318 (23/28) G7–.245 (29/35) G8–.481 (21/35)

➢ For the following item, see the direction preceding Item O–6.

18. $\frac{1}{3} + \frac{1}{6} =$ a. $\frac{1}{18}$ b. $\frac{1}{2}$ c. $\frac{2}{9}$ d. $\frac{2}{18}$ e. not given

Item O–30 (18): .217, Grades 5–8, n = 483.
a) 16 b) 105 c) 235 d) 26 e) 91 na) 10
— G5–.085 (26/28) G6–.236 (27/28) G7–.225 (31/35) G8–.388 (27/35)

14. Which picture shows the result of $\frac{1}{6} + \frac{1}{3}$?

a. 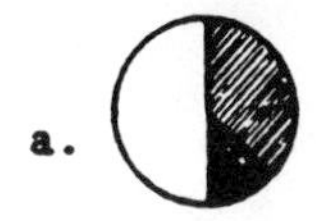 b.  c.  d.  e. not given

Item O–31 (14): .187, Grades 4–8, n = 608.
a) 114 b) 51 c) 232 d) 54 e) 135 na) 22
G4–.119 (12/15) G5–.104 (25/28) G6–.145 (28/28) G7–.254 (28/35) G8–.370 (28/35)

35. Tameka runs $\frac{2}{3}$ of a mile each morning. Today she stopped running after going $\frac{4}{5}$ of the way. How far did she run?
 a. $\frac{2}{15}$ b. $\frac{6}{15}$ or $\frac{2}{5}$ c. $\frac{6}{8}$ or $\frac{3}{4}$ d. $\frac{8}{15}$ e. not given

Item O–32 (35): .180, Grades 7–8, n = 210.
a) 40 b) 32 c) 48 d) 38 e) 28 na) 24
— — — G7–.156 (29/35) G8–.203 (32/35)

29. How many thirds equal $1\frac{1}{2}$?
 a. $4\frac{1}{2}$ b. $5\frac{1}{2}$ c. $3\frac{1}{2}$ d. cannot be done
 e. not given

Item O–33 (29): .138, Grades 7–8, n = 210.
a) 29 b) 10 c) 28 d) 108 e) 27 na) 8
— — — G7–.107 (34/35) G8–.166 (33/35)

30. Andre had $\frac{2}{3}$ of a pie. He ate $\frac{4}{5}$ of that. How much of the pie did he eat?
a. $\frac{2}{15}$ b. $\frac{6}{15}$ or $\frac{2}{5}$ c. $\frac{6}{8}$ or $\frac{3}{4}$
d. $\frac{8}{15}$ e. not given

Item O–34 (30): .128, Grades 7–8, $n = 210$.
a) 56 b) 33 c) 31 d) 27 e) 41 na) 22
— — — G7–.147 (32/35) G8–.111 (34/35)

34. What part of this diagram is shaded?

a. $\frac{2}{6}$ or $\frac{1}{3}$ b. $\frac{2}{4}$ or $\frac{1}{2}$ c. $\frac{2}{12}$ or $\frac{1}{6}$ d. $\frac{3}{8}$
e. not given

Item O–35 (34): .080, Grades 7–8, $n = 210$.
a) 55 b) 33 c) 14 d) 17 e) 73 na) 18
— — — G7–.068 (35/35) G8–.092 (35/35)

Appendix 9.B

Rational Number Characteristics on the CA and OA Instruments

The following lists catalog the item characteristics for the Rational Number Concepts Assessment (CA) and Operations Assessment (OA). For each test, a list of numbered characteristics is given, followed by a list of numbers (an "item vector") given for each of the 95 items (60 on the CA and 35 on the OA). Each of these numbers represents one of the rational number characteristics in the respective item. Each CA item is identified by IC characteristics, and each OA item is identified by 6 characteristics. The item numbering for the lists of vectors corresponds to the ordering on the *original* tests, and thus to the number appearing before each item stem (and the number in parentheses in the information line following) in the reordered versions in Appendix 9.A.

CONCEPTS TEST ITEM CHARACTERISTICS

Translation type:

1. Symbol-picture
2. Picture-symbol
3. Picture-written
4. Written-picture
5. Picture-picture
6. Symbol-symbol
7. Written-symbol
8. Symbol-written
9. (unused code)

Picture types:

10. Circular region representation
11. Polygonal region representation
12. Discrete object representation
13. Number line representation
14. Nonpicture items
15. Volume representation
16. Discrete object–polygonal region translation
17. Polygonal region–discrete object translation
18. Discrete object–discrete object translation
19. Polygonal region–polygonal region translation
20. Number line–polygonal region translation

Imposition of measure on picture:

21. Measure imposed on picture or measure not applicable
22. No measure imposed on picture

Subconstruct:

23. Ratio subconstruct
24. Fraction subconstruct

Numerator characteristics:

25. Unit rational
26. Subitizible numerator (>1)
27. Nonsubitizible numerator
28. Numerator size not applicable

Denominator characteristics:

29. Denominator = 1
30. Subitizible denominator
31. Nonsubitizible denominator (>1)
32. Denominator size not applicable

Rational number size characteristics:

33. Rational = 1/2
34. Rational is less than 1 (<>1/2)
35. Rational greater than 1 in mixed form
36. Improper fraction form
37. Problem is to identify whole
38. Whole number rational

Congruency of parts in picture:

39. Parts congruent or congruency not applicable
40. Parts not congruent

Perceptual distractor characteristics:

41. Discrete object perceptual distractor
42. Polygonal perceptual distractor
43. Distractors not analyzed

Task characteristic:

44. Identify p/q
45. Identify ordinal denominator or size of parts
46. Identify cardinal numerator or denominator

CONCEPTS TESTS: ITEM VECTORS KEYED TO NUMBERED CHARACTERISTICS ABOVE

1.	2	10	21	24	25	30	33	39	43	44
2.	3	11	22	24	26	30	34	39	43	44
3.	2	12	21	24	26	30	34	39	43	44
4.	4	11	21	24	26	30	34	39	43	44
5.	4	12	21	24	26	30	34	39	43	44
6.	1	10	21	24	25	30	33	39	43	44
7.	4	12	21	24	28	30	38	39	41	45
8.	3	13	21	24	25	30	35	39	43	44
9.	5	20	21	24	25	30	33	39	43	44
10.	7	14	21	24	26	30	34	39	43	44
11.	4	15	21	24	28	30	38	39	43	45
12.	1	13	21	24	25	30	33	39	43	44
13.	4	10	21	24	28	31	38	39	43	45
14.	1	10	22	24	26	30	34	39	43	44
15.	7	14	21	24	28	30	38	39	43	45
16.	2	12	21	24	26	31	34	39	43	44
17.	7	14	21	24	26	30	34	39	43	44
18.	8	14	21	24	26	30	34	39	43	44
19.	2	13	21	24	26	30	35	39	43	44
20.	5	16	21	24	26	30	34	39	42	44
21.	1	13	21	24	26	30	34	39	43	44
22.	3	13	21	24	28	30	38	39	43	45
23.	2	12	21	23	26	30	34	39	43	44
24.	4	12	21	23	26	30	34	39	43	44
25.	3	12	21	23	26	30	34	39	43	44
26.	7	14	21	24	28	30	38	39	43	45
27.	1	12	21	24	26	30	34	39	43	44
28.	5	17	21	23	26	30	35	39	43	44
29.	2	10	22	24	25	30	34	39	43	44
30.	3	10	22	24	25	30	34	39	43	44
31.	2	11	21	24	25	30	34	39	42	44
32.	3	13	21	24	25	30	33	39	43	44
33.	2	12	21	24	25	31	34	39	43	44
34.	2	12	21	24	25	30	35	39	43	44
35.	2	12	21	23	27	29	35	39	43	44
36.	3	12	21	23	27	30	35	39	43	44
37.	2	13	21	23	25	31	34	39	43	44
38.	2	10	21	23	25	29	38	39	43	44
39.	2	12	21	24	25	30	34	40	43	44

40.	7	14	21	24	25	31	34	39	43	45
41.	2	11	22	24	25	30	35	39	43	44
42.	2	12	21	24	26	31	34	40	43	44
43.	2	12	21	24	25	30	35	39	43	44
44.	2	10	21	24	25	30	34	39	43	37
45.	2	11	21	24	25	30	34	39	42	37
46.	2	12	21	24	25	30	34	39	41	37
47.	2	11	21	24	25	30	33	39	42	37
48.	2	10	21	24	25	30	34	39	42	37
49.	3	12	21	24	26	31	34	39	41	46
50.	3	11	21	24	26	30	34	39	41	46
51.	1	13	21	24	27	30	36	39	43	44
52.	5	19	21	24	26	30	34	39	43	44
53.	2	13	21	24	25	32	38	39	43	44
54.	6	14	21	24	25	30	35	39	43	44
55.	6	14	21	24	26	30	35	39	43	44
56.	6	14	21	24	27	30	36	39	43	44
57.	5	18	22	24	25	30	34	39	41	37
58.	5	19	22	24	25	30	34	39	42	37
59.	3	12	21	24	26	31	34	39	43	44
60.	3	11	21	24	27	31	34	39	42	44

OPERATIONS TEST ITEM CHARACTERISTICS

Translation type:

1. Written language–written language
2. Written language–symbols
3. Symbols–symbols
4. Symbols and written language–symbols
5. Symbols and written language–symbols and written language
6. Symbols and written language–symbols
7. Word problem with symbols–symbols
8. Word problem with written language–symbols
9. Written language–picture(s)
10. Symbols–symbols and written language
11. Symbols–picture(s)
12. Picture(s)–symbols
13. Word problem with picture(s) and symbols–symbols

Picture type:

14. Nonpictorial item
15. Discrete object representation

16. Circular region representation
17. Continuous real object in picture
18. Polygonal region in picture

Operation:

19. Addition
20. Multiplication or division
21. Subtraction
22. Multiplication and subtraction
23. Division

Word cue:

24. Word cue - ''plus''
25. word cue - ''of''
26. Word cue - ''left''
27. Word cue - ''add''
28. Word cue - ''subtract''
29. Word cue - ''sum''
30. Word cue - ''older than''
31. Word cue - ''product''
32. No word cue or miscellaneous word cue

Response type:

33. Response is translation only
34. Response is a calculation with little or no simplifying
35. Response is a calculation requiring simplification
36. Response identifies a process

Rational size / notation characteristics:

37. Item involves unit fraction(s)
38. Item involves proper fraction(s)
39. Item involves one proper fraction and one whole number
40. Item contains mixed number(s)
41. Non-unit proper fractions need a common denominator
42. Mixed numbers need a common denominator

OPERATIONS TEST: ITEM VECTORS KEYED TO CHARACTERISTICS ABOVE:

1. 5 14 19 24 34 37
2. 2 14 9 25 34 37
3. 3 14 19 27 34 37

4.	6	14	20	25	34	37
5.	8	14	20	25	34	37
6.	6	14	20	25	35	39
7.	8	14	22	25	35	39
8.	3	14	21	28	34	38
9.	9	15	20	25	33	37
10.	5	14	20	25	36	37
11.	3	14	20	31	34	37
12.	3	14	20	31	34	39
13.	3	14	20	31	34	38
14.	11	16	19	32	33	37
15.	1	14	19	27	36	38
16.	3	14	19	29	34	38
17.	3	14	19	29	34	41
18.	3	14	19	29	34	41
19.	2	14	20	25	33	37
20.	6	14	20	25	34	39
21.	3	14	19	29	34	41
22.	3	14	19	29	34	40
23.	12	15	20	25	34	39
24.	3	14	20	31	34	39
25.	3	14	20	31	34	39
26.	7	14	19	30	35	40
27.	13	17	21	32	33	42
28.	6	14	21	26	35	42
29.	6	14	23	32	33	42
30.	6	14	20	25	34	38
31.	8	14	20	25	36	38
32.	12	16	19	32	33	37
33.	12	18	19	32	36	37
34.	12	18	19	32	35	37
35.	6	14	20	25	34	38

Reference Notes

1. Behr, M., Lesh, R., & Post, T. *The role of manipulative aids in the learning of rational numbers.* RISE grant #SED 79–20591, Northern Illinois University.
2. Lesh, R. *Applied problem-solving in middle-school mathematics.* RISE grant #SED 80–17771, Northwestern Universtiy.
3. Behr, M., Lesh, R., & Post, T. *The role of representational systems in the acquisition and use of rational number concepts.* RISE grant #SED 79–20591, Northern Illinois University.

4. Usiskin, Z. *Arithmetic and its applications.* RISE grant #SED 79–19065, The University of Chicago.
5. Landau, M. *The effect of spatial abilities and problem presentation formats on problem solving performance in middle school students.* Doctoral dissertation, in progress. Northwestern University.

References

Cardone, I. P. *Centering/decentering and socio-emotional aspects of small groups: An ecological approach to reciprocal relations.* Unpublished doctoral dissertation, Northwestern University, 1977.

Kieren, T. E. On the mathematical, cognitive, and instructional foundations of rational numbers. In R. A. Lesh (Ed.), *Number and measurement: Papers from a research workshop.* Columbus: ERIC/SMEAC, 1976.

Kieren, T. E., & Southwell, B. The development in children and adolescents of the construct of rational numbers as operators. *The Alberta Journal of Educational Research,* 1979, *25*(4), 234–247.

Landau, M., Hamilton, E., & Hoy, C. *Relationships between process use and content understanding.* Paper presented at the Annual Meeting of the American Educational Research Association, Los Angeles, April 1981.

Lesh, R. Directions for research concerning number and measurement concepts. In R. Lesh (Ed.), *Number and measurement: Papers from a research workshop.* Columbus: ERIC/SMEAC, 1976.

Lesh, R. *Social/affective factors influencing problem solving capabilities.* Paper presented at the Third International Conference for the Psychology of Mathematics Education, Warwick, England, 1979. (a)

Lesh, R. Mathematical learning disabilities: Considerations for identification, diagnosis, and remediation. In R. Lesh, D. Mierkiewicz, & M. G. Kantowski (Eds.), *Applied mathematical problem solving.* Columbus, ERIC/SMEAC, 1979. (b)

Lesh, R. Applied mathematical problem solving. *Educational Studies in Mathematics,* 1981, *12,* 235–264.

Lesh, R. Modeling students' modeling behaviors. In S. Wagner (Ed.), *Proceedings of the Fourth Annual Meeting of the North American Chapter of the International Group for the Psychology of Mathematics Education.* Athens, Georgia: University of Georgia, 1982.

Lesh, R. *Metacognition in mathematical problem solving* (Tech. Rep.). Evanston, Illinois: Mathematics Learning Research Center, Northwestern University, 1983.

Lesh, R., & Hamilton, E. *The rational number testing program.* Paper presented at the annual meeting of the American Educational Research Association, Los Angeles, April, 1981.

Lesh, R., Landau, M., & Hamilton, E. Rational number ideas and the role of representational systems. In R. Karplus (Ed.), *Proceedings of the Fourth International Conference for the Psychology of Mathematics Education.* Berkeley: Lawrence Hall of Science, 1980.

Lesh, R., & Mierkiewicz, D. *Recent research concerning the development of spatial and geometric concepts.* Columbus: ERIC/SMEAC, 1978.

CHAPTER 10

Episodes and Executive Decisions in Mathematical Problem-Solving*

Alan H. Schoenfeld

Introduction

This is a somewhat speculative chapter dealing with managerial or executive decisions in human problem-solving. It presents a framework for analyzing problem-solving protocols at the macroscopic level, with a focus on these strategic, often metacognitive behaviors.

In this section, the two major perspectives underlying the structure of this framework are discussed: the nature of tactical and strategic decision-making and the nature of problem-solving expertise.

On Decision Making

Two qualitatively different kinds of decisions, tactical and strategic, seem to be needed for problem-solving success in broad, semantically rich domains (for example, in mathematics at the college level). The first, tactical decision-making, is the more easily seen of the two and has received the lion's share of attention. Here, *tactics* include most standard procedures for implementation in problem solving. Tactics include all algorithms and most heuristics, both of the Pólya type (e.g., draw a diagram whenever possible; consider special cases) and

*The research reported here was supported in part by the National Science Foundation under RISE Grant SED 79–19049. Any opinions, findings, and conclusions expressed in this report are those of the author and do not necessarily reflect the views of the National Science Foundation.

ISBN No. 0-12-444220-X

of the kinds used in artificial intelligence (means–ends analysis, hill climbing). Tactical decisions are "local." Suppose, for example, that one has decided to calculate the area of a particular region. The choice whether to approach that calculation via trigonometry or via analytic geometry is tactical.

In contrast, *strategic* decisions are those that have a major impact on the direction a solution will take, and on the allocation of one's resources during the problem-solving process. These will also be called *executive* or *managerial* decisions. For example, if one is given 20 minutes to work on a problem, and calculating the area of a region is likely to take 10 minutes, the decision to calculate the area of that region is strategic—regardless of whatever particular method is chosen for performing the calculation. Like a decision during wartime to open a front, this one choice may determine the success or failure of the entire enterprise.

This separation of managerial decisions from implementation decisions has implications for both human and machine problem-solving. Mathematics problem-solving instruction to date has focused largely, and with questionable success, on heuristics or tactics. I propose that much of the lack of success stems from the fact that managerial behaviors have been largely ignored. The protocols discussed below will illustrate that heuristic fluency is of little value if the heuristics are not managed properly. I believe tbat much greater attention will have to be paid to managerial actions (in fact, to metacognitive behavior in general) in classroom instruction if we are to be successful in understanding and teaching problem-solving skills.

On the Nature of Problem-Solving Expertise

Most definitions of *expert* and *novice* are domain specific. An *expert* is someone who knows the domain inside out, whereas a *novice* is someone who does not, and is usually new to it. The standard definition of *novice* is broadly inclusive: it allows expert problem-solvers to serve as novices in a new domain. Thus, professional problem-solvers (say a mathematician who does research in mathematical problem-solving or a psychologist who does computer simulations of problem solving) could, according to this convention, serve as novices in an experiment on kinematics if they had not studied physics for years (see, e.g., Simon and Simon, 1978). Protocols gathered from these problem-solving experts who serve as domain-specific novices might well differ from protocols gathered from naive college freshman—*nonexpert* problem-solving novices. This is a potential source of confusion about the capacities of problem-solving novices.

Further, the domain-specific definition of *expertise* is narrow and exclusive; it requires that experts be *proficient* in the domains in which we study them. Because such experts have a domain at their fingertips, they have ready access to a range of problem-solving schemata and their performance is nearly automatic. Of course, proficiency is often a major component of expert performance, but it is not usually the whole story. According to the domain-specific definition of expertise, the professional mathematician who is an excellent problem-solver but who has not worked geometry problems for 15 years no longer qualifies as an expert when working geometry problems. Yet, in Protocol 4 (Appendix 10.D) we see how one such person, who had less factual and procedural geometric knowledge at his disposal than did students in a problem-solving course, managed to solve a problem that none of them could. The expert's success in Protocol 4 is strategic rather than tactical: avoiding wild-goose chases and marshalling his resources well account for his success. Such behavior will not appear when he is solving routine (for him) problems. It is, however, an important component of his expertise, one that seems to be lacking in nonexpert novices. Such behavior becomes evident only when we examine the performance of nonproficient experts and has, thus, received very little attention. It is a major focus of this chapter.

A framework is described for examining, at the macroscopic level, a broad spectrum of problem-solving protocols. Protocols are parsed into major *episodes*. These are periods of time during which the problem solver(s) is engaged in a single set of like actions, such as planning or exploration. It is precisely between such episodes that the managerial decisions that can make or break a solution are often made, or not made. This chapter focuses on decision making at these points, and on the impact of such decisions—or their absence—on problem-solving performance. The quality and success of problem-solving endeavors is shown to correspond closely (in human problem-solving) to the presence, and vigilance, of managerial decision-making.

A Discussion of Antecedents

By definition, *protocol coding schemes* are concerned with producing objective records or traces of a sequence of overt actions taken by individuals in the process of solving problems. In mathematics education research, the coded protocol is generally subjected to a qualitative analysis; often correlations are sought between certain types of behavior (e.g., the presence of goal-oriented heuristics) and problem-solving success. In artificial intelligence, the goal is often to write a program that will simulate a given protocol or the idealized behavior culled from a variety of protocols. In both cases, the level of analysis is microscopic.

My goal here is to indicate that in many cases the microscopic-level analysis is inappropriate. In analyzing human problem-solving, exclusive attention to the microscopic level may cause one to miss the forest for the trees; if the wrong strategic decisions are made, tactical ones are virtually irrelevant. In artificial intelligence, great progress has been made at the tactical level. It is not at all clear, however, that tactical decision-makers (as embodied in production systems, for example) will serve well for making managerial decisions. I believe that we may wish to think of executive decisions as belonging to a qualitatively different category than tactical ones, and we may want to deal with these higher level strategies separately.

(Note: What follows is an opinionated discussion of the recent literature, which depends heavily on the distinction between tactical and strategic, or managerial, decisions. These distinctions may be much clearer after the reader has considered the examples discussed in the next section. Thus the reader may wish to skip ahead to that section, and later consider the comments made here in the light of those examples.)

The following description, taken from Lucas, Branca, Goldberg, Kantowski, Kellogg, and Smith (1979, p. 354) is typical of the efforts of mathematics educators to deal with problem-solving protocols:

> The authors came to agreement on the definitions for a set of constructs which were to represent observable, disjoint problem solving behaviors and related phenomena. . . . Each event was assigned a symbol, and the collection of events which comprised a problem-solving sequence of processes was recorded in a horizontal string of symbols corresponding to the chronological order of appearance during the actual problem solution. In this manner a researcher could listen to a tape of a problem solution (in conjunction with observing written work, interviewer notes, and/or a verbatim transcript) and produce a string of symbols which represented the composite perception of the solution process. Conversely, an examination of the given string of symbols could be used to provide a reasonably clear picture of what had happened during a problem-solving episode.

That particular coding scheme included a two-page "dictionary" of processes that were assigned coding symbols. All behavior was "required to be explicit; otherwise it is not coded" (p. 359). As an example of the coding, the following sequence was coded as R, R, L_8 P_i D_{a5} 4): "The problem solver reads the problem, hesitates, rereads part of the problem, says the problem resembles another problem and he will try to use the same method, then deduces correctly a piece of information from one of the given data" (p. 361).

In part because of the cumbersome nature of such systems and the wealth of symbols that must be dealt with once the processes are coded, other researchers have opted to focus on more restricted subsets of behaviors. Kulm's recent NSF (National Science Foundation)-supported work uses a revised and more condensed process code dictionary (Kulm, Campbell, Frank, & Talsma, 1981). Kantowski's recent work (Note 1) includes a "coding scheme for heuristic processes of interest" that focuses on five heuristic processes related to planning, four related to memory for similar problems, and seven related to looking back.

Researchers explore the frequency of such processes in relation to problem-solving performance.

So far as I know, there are no systems for protocol analysis that focus in any substantive way on strategic decisions. At the microscopic level, state–space descriptions deal with the selection of "good" solution paths and the consequences of poor selection. But there has not been, at the global level, an adequate framework for dealing with decisions that ought to have been considered, but were not. (See Protocol 1 [Appendix 10.A] for a wild-goose chase and its consequences.) For the most part, discussions in the literature of strategic decision-making during problem solving are weak. Pólya, for example (1965, p. 96) offers "Rules of Preference" for choosing among options in a problem-solving task. These include injunctions such as "The less difficult precedes the more difficult" and "Formerly solved problems having the same kind of unknown as the present problem precede other formerly solved problems." My own attempts (Schoenfeld, 1979; 1980) at capturing a managerial strategy in flow-chart form for students' implementation were somewhat impoverished: the flow chart in effect presented a default strategy. All other factors being equal—meaning that the problem solver had exhausted the lines of attack that had appeared fruitful (his "productions?") and had no strong leads to follow—it was considered reasonable to try the heuristic suggestions in this managerial strategy, roughly in the order suggested by the flow chart. This approach bypassed the tough questions. Issues such as: How does one decide what to pursue? For how long? How does one evaluate progress toward a solution? When should the manager interfere? were discussed in class, but were not formally a part of the strategy. While the importance of executive issues was suggested, there was no systematic and rigorous framework offered for examining these questions.

As a result of (*a*) the narrowness of the problem domains in which artificial intelligence has (until recently) successfully operated, and (*b*) the definitions of domain-specific expertise discussed in the first section, the artificial-intelligence community has paid even less attention to exeucitive strategies than has the mathematics-education community. Again, the questions are not new: the "considerations at a position in problem space" listed by Allen Newell (1966: Figure 5) are quite similar to those we pose below. But

> "Select new operator:
> Has it been used before?
> Is it desirable: Will it lead to progress?
> Is it feasible: Will it work in the present situation if applied?"

takes on very different shades of meaning at the strategic rather than the tactical level. In a typical domain-specific expert approach to a problem, the state space and the operators available are all well specified; the problem is in the selection of operators. But in Protocol 4, part of the problem space was altered significantly during the solution attempt—as was the pool of available operators. Thus the usable solution space is dynamic, and is influenced by metacognitive decision-making (e.g., choices of new perspectives are an important component of problem-solving skills).

Recent advances in programming approaches allow for rather clever decision-making. There are computationally efficient means of keeping track of and sorting through productions in a production system for relevancy, and there are conflict-resolution systems (McDermott & Forgy, 1978) for selecting among productions when the conditions for more than one of them have been satisfied. Such structures prohibit productions from executing more than once on the same data. This prevents the kind of endless repetitions all too common in students, and it forces, if necessary, the examination of all available information. Because preference is given to productions whose conditions are satisfied by elements most recently placed in working memory, there is a ''natural'' continuity to the sequence of operations. Other rules for selection (e.g., specificity precedes generality) provide plausible means of selecting tactics in relatively narrow domains. However, these still seem to be *tactical* decisions. Similar comments apply to the *adaptive* or *self-modifying* production systems described by Anzai and Simon (1979), Neves (1978), and Neches (Note 2). Although the learning principles they exemplify may be general, the embodiments of those principles in those papers are at the tactical level. Simon (1980) argues that ''effective professional education calls for attention to both subject matter knowledge and general skills'' (p. 86). This chapter focuses on the latter.

An Informal Analysis of Two Protocols

The artificial intelligence literature is filled with beautiful protocols. Unfortunately, those generated by my students (and to some extent by my colleagues) in the process of grappling with relatively unfamiliar problems have been, on the whole, rather unesthetic. This section considers two such protocols, each generated by a pair of students. (Following a suggestion from John Seely Brown, I have students work on problems in pairs. Although the question, ''Why did you do that?'' may be terribly intimidating coming from me and is likely to alter the solution path, the question, ''Why should we do that?'' from a fellow student working on a problem is not. Dialogue between students often serves to make managerial decisions overt, whereas such decisions are rarely overt in single-student protocols). An informal analysis, focusing on the importance of managerial decisions, follows. The formal analytic structure is given in the next section.

Protocols 1 and 2 are given in Appendix 10.A and 10.B, respectively. The students were asked to work on the problem together, out loud, as a collaborative effort. They were not to go out of their way to explain things for the tape, if that interfered with their problem solving; their interactions, if truly collaborative, would provide the needed information. (See Ericsson and Simon [1980] for a discussion of instructions for think-aloud experiments.) The students were undergraduates at a liberal arts college. Students A and K (Protocol 1) had 3 and 1 semesters of college mathematics (calculus), respectively. Students D and B (Protocol 2) each had 3 semesters of college mathematics. It should be recalled that such students, by most standards, are successful problem-solvers: the unsuccessful ones had long since stopped taking mathematics courses. Both protocols are of the same problem:

> Three points are chosen on the circumference of a circle of radius R, and the triangle determined by them is drawn. What choice of points results in the triangle with the largest possible area? Justify your answer as best you can.

If Protocol 1 makes for confused reading, the tape it was taken from makes for even more pained viewing. I would summarize the problem-solving session as follows: The students read and understood the problem, and then quickly conjectured that the answer was the equilateral triangle. They impetuously decided to calculate the area of the triangle, and spent the next 20 minutes doing so. These calculations of the area were occasionally punctuated by suggestions that might have salvaged the solution, but in each case the suggestions were quickly dropped and the students returned to their relentless pursuit of the worthless calculation. Neither student could tell me, after the cassette ran out of tape, what good it would do them to know the area of the equilateral triangle. Observe the following:

1. The single most important event in the 20-minute problem-solving session, upon which the success or failure of the entire endeavor rested, was conspicuous by its absence: the students did not assess the potential utility of their planned action, calculating the area of the equilateral triangle. In consequence, the entire session was spent on a wild-goose chase.
2. Inadequate consideration was given to the utility of potential alternatives that arose (and then submerged) during the problem-solving process. Any of these—the related problem of maximizing a rectangle in a circle (Item 28), the potential aplication of the calculus (Item 52) for what can indeed be considered a max–min problem, the qualitative varying of triangle shape (Item 68)—might have, if pursued, led to progress. Instead, the alternatives simply faded out of the picture. (See, for example Items 27–31.)
3. Progress is never monitored or (re)assessed, so there is no reliable means of terminating wild-goose chases once they have begun. (This contrasts strongly with an expert protocol, in which the problem solver interrupted the implementa-

tion of an outlined solution with "this is too complicated. I know the problem shouldn't be this hard.")

Now, how does one code such a protocol? First, we should observe that matters of detail (such as whether or not the students will accurately remember the formula for the area of an equilateral triangle, Items 73–75) are virtually irrelevant. To return to the military analogy in the opening section: If it was a major strategic mistake to open a second front in a war, the details of how a hill was taken in a minor skirmish on that front are of marginal interest.

A second and more crucial point is that the *overt* actions taken by the problem solvers in that protocol are, in a sense, of minor import. The problem-solving effort was a failure because of the *absence* of assessments and strategic decisions. Any framework that will make sense of that protocol must go beyond simply recording what did happen; it should suggest when strategic decisions ought to have been made, and allow one to interpret success or failure in the light of whether, and how well, such decisions were made.

If Protocol 1 stands as evidence of the damage that can be caused by a manager in absentia, Protocol 2 provides evidence of the catastrophic effects of bad management. The processes in this tape were not muddled, as in Protocol 1; the decisions were overt and clear. The next paragraph summarizes the essential occurrences in the tape. The numbered comments refer to the commentary that follows.

D and *B* quickly conjecture that the solution is the equilateral triangle, and look for ways to show it. *D*, apparently wishing to exploit symmetry in some way, suggests that they examine triangles in a semicircle with one side as diameter. They find the optimum under these constraints, and reject it "by eye" as inferior to the equilateral (Comment 1). Still focusing on symmetry, they decide (Comment 2) to maximize the area of a right triangle in a semicricle, where the right angle lies on the diameter. This (serendipitously correct) decision reduces the original problem to a one-variable calculus problem (Comment 3), which *B* proceeds to work on. Twelve minutes later the attempt is abandoned (Comment 4) and the solution process degenerates into an aimless series of explorations, most of which serve to rehash the previous work (Comment 5).

Comment 1. Rejecting the alternative is quite reasonable, as are their actions in analyzing the problem up to this point. However, this blanket rejection may have cost them a great deal. The variational argument they used to find the isosceles right triangle (holding the base fixed and observing that the area is largest when the triangle is isosceles) is perfectly general and can be used to solve the original problem as stated. But the students simply turn away from their unsuccessful attempt, without asking if they could learn from it. In doing so, they may have thrown out the baby with the bath water.

Comment 2. This decision, which affects the direction of the solution for more than 60% of the allotted time, is made in a remarkably casual way (Items 24–27):

> *D*: (after one attempt at symmetry has failed) You want to make it perfectly symmetrical, but we can, if we maximize this area, just flip it over, if we assume that it is going to be symmetrical.
> *B*: Yea, it is symmetrical.

This assumption is not at all justified (they are assuming part of what they are to prove). The students have changed the problem and proceed, without apparent concern, to work on the altered version.

Comment 3. *B*'s tactical work here is quite decent, as is much of both students' tactical work throughout the solution process. The decision to scale down the problem to the unit circle (Item 37) is just one example of their proficiency. There is awareness of, and access to, a variety of heuristics and algorithmic techniques during the solution. Unfortunately, *B* lost a minus sign during this particular calculation, which gave him a physically impossible answer. He was aware of it; local assessment worked well. However, global assessment (see Comments 4 and 5) did not.

Comment 4. This decision to abandon the analytic approach is just as astonishing in the way it takes place (Items 74 and 75) as the decision to undertake it:

> *D*: Well, let's leave the numbers for a while and see if we can do it geometrically.
> *B*: Yea, you're probably right.

Given that more than 60% of the solution has been devoted to that approach (and that correcting a minor mistake would salvage the entire operation), this casual dismissal of their previous efforts has rather serious consequences.

Comment 5. There were a number of clever ideas in the earlier attempts made by *D* and *B*. Had there been an effort at a careful review of those attempts, something might have been salvaged. Instead, there was simply a "once over lightly" of the previous work that added nothing to what they had already done.

A framework for focusing on the managerial decisions in such protocols is discussed in the next section.

Framework for the Macroscopic Analysis of Problem-Solving Protocols

The two protocols discussed in the preceding section raise the major questions I address here. I believe that decisions at the managerial level may make or break a problem-solving attempt, and that (at least in the case of poor managerial decisions) these may render irrelevant any subsequent tactical (i.e., implementation) decisions. Thus we focus on behavior at the macroscopic level.

Protocol 1, which is typical of students' problem-solving, illustrates one of the major difficulties in dealing with managerial decisions: the absence of intelligent management may doom problem-solving attempts to failure. Yet all extant schemes focus on what is overtly present, ignoring the crucial decisions that might (and should) have taken place. Protocol 2 is, in a sense, easier to deal with. The decisions were overt, though poor. This protocol serves to indicate that decision making means more than simply choosing solution paths: it incorporates local and global assessments of progress, and tries to salvage the valuable elements of ultimately flawed approaches. This section offers a scheme for parsing protocols that tries to address these issues.

There are both objective and subjective components to the framework for analyzing protocols. The objective part consists of identifying, in the protocol, the loci of potential managerial decisions. The subjective part consists of characterizing the nature of the decision-making process at these managerial decision-points and describing the impact of those decisions (or their absence) on the overall problem-solving process.

By definition, managerial or strategic action is appropriate whenever a large amount of tactical resources are about to be expended. This provides the basic idea for parsing the protocols: to partition a protocol into macroscopic chunks of consistent behavior (*episodes*). Then the points between episodes—at which the direction or nature of the problem solution changes significantly—are the managerial decision-points at which, at minimum, managerial action ought to have been considered.

In addition to these junctures between episodes, there are two other loci for managerial action: at the arrival of new information or the suggestion of new tactics, and at the point at which a series of tactical failures indicates that strategic review might be appropriate. The loci that deal with new information are well defined and easy to identify. Observe that this kind of decision point can occur in the middle of an episode: new information may be ignored or dismissed (at least temporarily), and the problem solver may continue working along previously established lines. The tactical-failure situation is more difficult and calls for subjective judgment; I have no easy way of dealing with this at present. At some point when implementation bogs down, or when the problem-solving process degenerates into more or less unstructured explorations, it is time for an "executive review." It is clear from the protocols I have taken that experts have "monitors" that call for such review, and that novices often lack them. We return to this point in the section on subjective analysis.

Figures 10.1 and 10.2 represent a parsing of Protocols 1 and 2, respectively, into episodes (see Appendixes 10.A and 10.B). New-information points within episodes are indicated.

Detailed analyses of Figures 10.1 and 10.2 will not be given because Protocols 1 and 2 have been discussed at some length. (Observe, however, how Figures

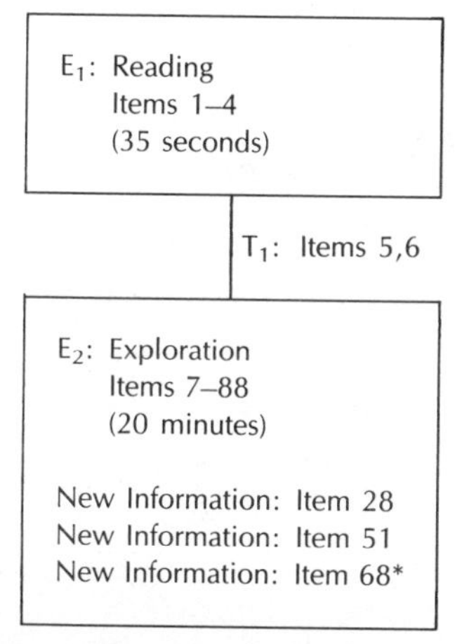

Figure 10.1 A Parsing of Protocol 1.

NOTE: From the written protocol it might appear that Item 68 begins a new episode. In fact, the students had lost virtually all their energy by that point, and were merely doodling; they returned (after the tape clicked off) to musing about the equilateral triangle. Thus Items 6–88 are considered to be one episode.

10.1 and 10.2 reflect the issues singled out for discussion above.) A third protocol is analyzed in detail.

Both parsing into episodes and delineating new-information points turn out to be (more or less) objective decisions. In fact, the parsing of all three protocols used in this chapter was derived, in consensus, by three undergraduates who followed my instructions but arrived at their characterizations of the protocols in my absence. Reliability in parsing protocols is quite high. (This does not, however, obviate the need for an appropriate formalism; see the final commentary.)

Subjectivity lurks around the corner, however. It is, in fact, already present in the labeling of the episodes given in Figures 10.1 and 10.2. This labeling was essential. The potential for combinatorial explosion in characterizing managerial behaviors is enormous. Managerial behaviors include selecting perspectives and frameworks for a problem; deciding at branch points which direction a solution should take; deciding whether, in the light of new information, a path already embarked upon should be abandoned; deciding what (if anything) should be salvaged from attempts that are abondoned or paths not taken; monitoring tactical implementation against a template of expectations for signs that intervention might be appropriate; and much, much more. My early attempts at analyses of managerial behavior called for examining protocols at all managerial decision-points and evaluating at each one a series of questions encompassing the issues just mentioned. This approach, although comprehensive, was completely unwieldy. For example, questions about the assessment of state when (*a*) one has

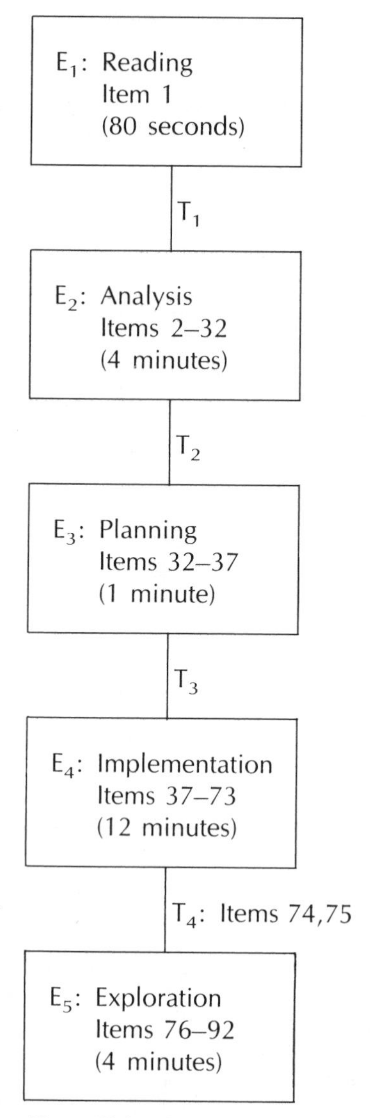

Figure 10.2 A Parsing of Protocol 2.

just read the problem, (*b*) one is stuck, and (*c*) the solution has been obtained, are almost mutually exclusive. Thus at any decision point, 90% of the questions that might be asked were irrevelant. The framework described below provides a workable compromise.

Any episode is characterized as one of the following: Reading, Analysis, Planning, Implementation (or Planning–Implementation if the two are linked),

Exploration, Verification, or Transition. Specific note is also taken of any new information or local assessments that appear during the solution. What follows is the heart of the analytic framework. There is a brief description of the nature of each type of episode once it has been labeled. The parsing, plus the answers to the questions, provide the characterization of the protocol.

Admittedly, the questions are heterogeneous. Some can be answered objectively at the point in the protocol at which they are asked, some in the light of later evidence; some call for inferences or judgments about problem-solving behavior. Further, some ask about the reasonableness of certain behavior. Asking questions in this way, of course, begs the significant question: What is a model of "reasonable" behavior? The creation of such models is the crucial long-term question, and there is no attempt to finesse it here. At present, however, the immediate goal is to deal with the notion subjectively, the better to understand managerial behaviors so that those models can be created. Though highly subjective, these assessments can be made reliably: agreement between my ratings and the consensus scorings of my students was quite high. To quote Mr. Justice Stewart in a 1964 decision of the United States Supreme Court, "I shall not today attempt to further define the kind of materials I understand to be embraced within that shorthand definition; . . . But I know it when I see it."

Episodes and the Associated Questions

READING

The reading episode begins when a subject starts to read the problem statement aloud. It includes the ingestion of the problem conditions, and continues through any silence that may follow the reading—silence that may indicate contemplation of the problem statement, the (nonvocal) rereading of the problem, or blank thoughts. It continues as well through vocal rereadings and verbalizations of parts of the problem statement (observe that in Protocol 1, reading included Items 1–4).

The Reading questions are

1. Have all the conditions of the problem been noted? (Explicitly or implicitly?)
2. Has the goal state been correctly noted? (Again, explicitly or implicitly?)
3. Is there an assessment of the current state of the problem solver's knowledge relative to the problem-solving task (see Transition)?

ANALYSIS

If there is no apparent way to proceed after the problem has been read (i.e., a solution is not *schema driven*), the next (ideal) phase of a problem solution is analysis. An attempt is made to understand a problem fully, to select an appropriate perspective and reformulate the problem in those terms, and to introduce for consideration whatever principles or mechanisms might be appropriate. The problem may be simplified or reformulated. (Often analysis leads directly into plan development, in which case it serves as a transition. Of course, this episode may be bypassed completely.)

Analysis questions are

1. What choice of perspective is made? Is the choice made explicitly, or by default?
2. Are the actions driven by the conditions of the problem? (working forward.)
3. Are the actions driven by the goals of the problem? (working backward.)
4. Is a relationship between conditions and goals sought?
5. Is the episode, as a whole, coherent? In sum (considering Questions 1–4), are the actions reasonable? (Comments?)

EXPLORATION

Both its structure and content serve to distinguish exploration from analysis. Analysis is generally well structured, sticking rather closely to the conditions or goals of the problem. Exploration, on the other hand, is less well structured and is further removed from the original problem. It is a broad tour through the problem space, a search for relevant information that can be incorporated into the analysis–plan–implementation sequence. (One may well return to analysis with new information gleaned during exploration.)

In the exploration phase of problem solving one may find a variety of problem-solving heuristics, the examination of related problems, the use of analogies, and so forth. Ideally, exploration is not without structure: there is a loose metric on the problem space, the perceived distance of objects under consideration from the original problem, that should serve to select items for consideration. Precisely because exploration is weakly structured, both local and global assessments are critical here (see Transition as well). A wild-goose chase, unchecked, can lead to disaster; but so can the dismissal of a promising alternative.

If new information arises during exploration, but is not used, or the examination of it is tentative, "fading in and fading out," the coding scheme calls for delineating new information within the episode. If, however, the problem solver decides to abandon one approach and start another, the coding scheme calls for closing the first episode, denoting (and examining) the transition, and opening another exploration episode.

Exploration questions are

1. Is the episode condition driven? Goal driven?
2. Is the action directed or focused? Is it purposeful?

3. Is there any monitoring of progress? What are the consequences for the solution of the presence or absence of such monitoring?

NEW INFORMATION AND LOCAL ASSESSMENTS

New-information points include any items at which a previously unnoticed piece of information is obtained or recognized. They also include the mention of potentially valuable heuristics (new processes). *Local assessment* is an evaluation of the current state of the solution at a microscopic level.

New Information and Local Assessment questions are

1. Does the problem solver assess the current state of his knowledge? (Was it appropriate?)
2. Does the problem solver assess the relevancy or utility of the new information? (Was it appropriate?)
3. What are the consequences for the solution of the actions (or inactions) described in the sections on Reading and Analysis above?

PLANNING–IMPLEMENTATION

Because the emphasis here is on managerial questions, detailed issues regarding plan formation are not addressed. The primary questions of concern here deal with whether or not the plan is well structured, whether the implementation of the plan is orderly, and whether there is monitoring or assessment of the process on the part of the problem solver(s), with feedback to planning and assessment at local and/or global levels. Many of these judgments are subjective. For example, the absence of any overt planning acts does not necessarily indicate the absence of a plan. In fact, protocols of schema-driven solutions often proceed directly from the reading episode into the coherent and well-structured implementation of a nonverbalized plan. Thus the latitude of the questions below: the scheme should apply to a range of circumstances, from schema-driven solutions to those in which the subject happens upon an appropriate plan by design or by accident.

Planning–Implementation questions are

1. Is there evidence of planning at all? Is the planning overt or must the presence of a plan be inferred from the purposefulness of the subject's behavior?
2. Is the plan relevant to the problem solution? Is it appropriate? Is it well structured?
3. Does the subject assess the quality of the plan as to relevance, appropriateness, or structure? If so, how do those assessments compare with the judgments in Question 2?
4. Does implementation follow the plan in a structured way?

5. Is there an assessment of the implementation (especially if things go wrong) at the local or global level?
6. What are the consequences for the solution of assessments if they occur, or if they do not?

VERIFICATION

The nature of the episode itself is obvious. Verification questions are

1. Does the problem solver review the solution?
2. Is the solution tested in any way? If so, how?
3. Is there any assessment of the solution: either an evaluation of the process or assessment of confidence in the result?

TRANSITION

The juncture between episodes is, in most cases, where managerial decisions (or their absence) will make or break a solution. Observe, however, that the presence or absence of assessment or other overt managerial behavior cannot necessarily be taken as either good or bad for a solution. In an expert's solution of a routine problem, for example, the only actions one sees may be reading and implementation. This explains, in part, the contorted and subjective nature of what follows.

Transition questions are

1. Is there an assessment of the current solution state? Is there an attempt to salvage or store things that might be valuable in it?
2. What are the local and global effects on the solution of the presence or absence of assessment in Question 1? Was the action there appropriate or necessary?
3. Is there an assessment of the short and/or long-term effects on the solution of the new direction, or does the subject simply jump into the new approach?
4. What are the local and global effects on the solution of the presence or absence of assessment in Question 3? Was the action there appropriate or necessary?

The Full Analysis of a Protocol

Appendix 10.C presents the full protocol of two students working on the following problem:

Consider the set of all triangles whose perimeter is a fixed number, P. Of these, which has the largest area? Justify your answer as best you can.

Student K is the same student who appeared in Protocol 1. Student D (not the same as Student D in Protocol 2) was a freshman with one semester of calculus behind him. This protocol was taken at the end of my problem-solving course, whereas Protocols 1 and 2 were taken at the beginning.

The parsing of Protocol 3 is given in Figure 10.3. The analysis given below follows that parsing.

EPISODE 1 (READING, ITEMS 1, 2)

The conditions were noted, explicitly.
The goal state was noted, but somewhat carelessly (Items 10, 11).
There were no assessments, simply a jump into exploration.

TRANSITION 1 (NULL)

Transition questions 1–4. There were no serious assessments of either current knowledge or of directions to come. These might have been costly, but were not—assessments did come in Episode 2.

EPISODE 2 (EXPLORATION, ITEMS 3–17)

The explorations seemed vaguely goal driven.
The actions seemed unfocused.
There was monitoring, at Items 4–17. This grounded the explorations, and led into Transition 2.

TRANSITION 2 (ITEMS 17–19)

Transition question 1–4. Assessments were made both of what the students knew and of the utility of the conjecture they made. The result was the establishment of a major direction: try to prove that the equilateral triangle has the desired property, and to create a plan (Episode 3). (Note: If this seems inconsequential, contrast this behavior with the transition (T1) in Protocol 1. The lack of assessment there, in virtually identical circumstances, sent the students on a 20-minute wild-goose chase!)

EPISODE 3 (PLAN, ITEM 20)

The plan is overt.
It is relevant and well structured. As to appropriateness and assessment, see the discussion of T3.

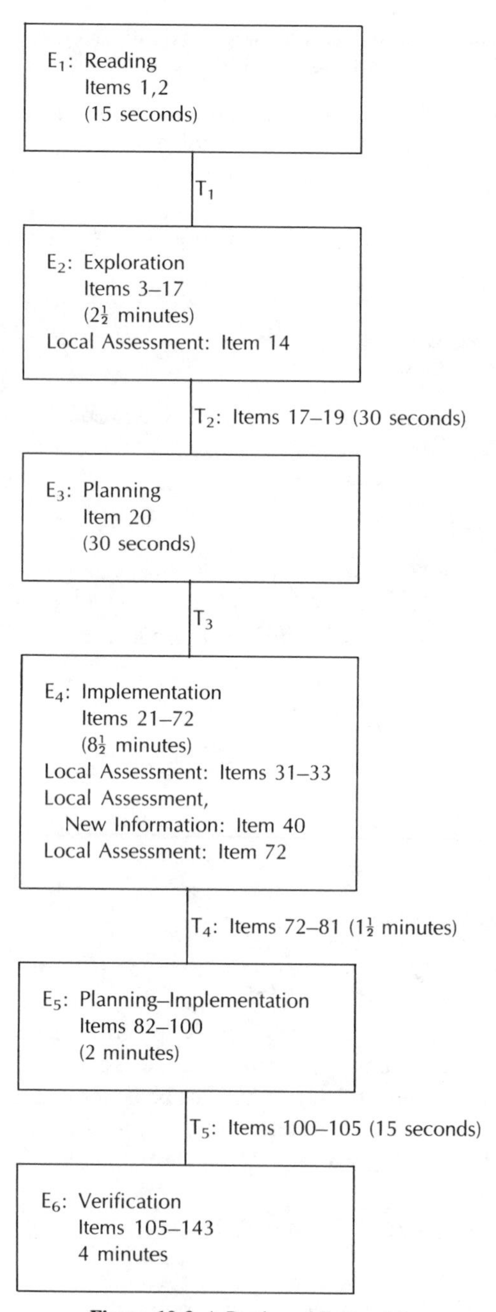

Figure 10.3 A Parsing of Protocol 3.

TRANSITION 3 (NULL)

Transition questions 1 and 2. There was little of value preceding the plan in Item 20; the questions are moot.

Transition question 3. There was no assessment of the plan; there was immediate implementation.

Transition question 4. The plan was relevant but only dealt with half of the problem: showing that the largest isosceles triangle was the equilateral. The other half of the problem is to show that the largest triangle must be isosceles, without which this part of the solution is worthless, a point realized somewhat in Item 72, 8 minutes later. The result was a good deal of wasted effort. The entire solution was not sabotaged, however, because monitoring and feedback mechanisms caused the termination of the implementation episode (see the sequel).

EPISODE 4 (IMPLEMENTATION, ITEMS 21–72)

Implementation followed the lines set out in Episode 3, albeit in somewhat careless form. The conditions were somewhat muddled as the first differentiation was set up. The next two local assessments corrected for that (better late than never).

LOCAL ASSESSMENT (ITEMS 31–33)

Transition questions 1–3. The physically unrealistic answer caused a closer look at the conditions—but not yet a global reassessment (possibly not called for yet).

LOCAL ASSESSMENTS, NEW INFORMATION (ITEM 40)

Transition questions 1–3. The new information here was the realization that one of the problem conditions had been omitted from their implementation (''we don't set any conditions—we're leaving *P* out of that''). This sent them back to the original plan, without global assessment. The cost: squandered energy until Item 72.

LOCAL–GLOBAL ASSESSMENT (ITEM 72)

This closes Episode 4. See Transition 4.

TRANSITION 4 (ITEMS 72–81)

Transition questions 1 and 2. The previous episode was abandoned, reasonably. The goal of that episode, "show it's the equilateral," remained. This, too, was reasonable.

Transition questions 3 and 4. They ease into Episode 5 in Item 82. (It is difficult to say how reasonable this is. Had they chosen something that did not work, it might have been considered meandering. But what they chose did work.)

EPISODE 5 (PLAN–IMPLEMENTATION, ITEMS 82–100)

Planning–implementation questions 1 and 2. "Set our base equal to something" is an obviously relevant heuristic.

Planning–implementation question 3. They plunge ahead as usual.

Planning–implementation question 4. The variational argument evolved in a seemingly natural way.

Planning–implementation question 5. There was local assessment (Item 95). That led to a rehearsal of the subargument (Item 96), from which *D* apparently "saw" the rest of the solution. Further (Item 100), *D* assesses the quality of the solution and his confidence in the result.

TRANSITION 5 (ITEMS 100–105)

Transition questions 1–4. The sequel is most likely the result of a two-person dialectic. It appears that *D* was content with his solution (perhaps prematurely), although his clarity in explaining his argument in E6 suggests he may have been justified.

EPISODE 6 (VERIFICATION, ITEMS 105–143)

This is not a verification episode in the usual sense. *K*'s unwillingness to rest until he understood forced *D* into a full rehearsal of the argument and a detailed explanation, the result being that they were both content with the (correct) solution.

Some Empirical Results

Protocols 1 and 2 are relatively typical of the dozen protocols taken from pairs of students (six pairs, two problems for each pair) before a month-long, intensive problem-solving course that focused on both tactics (heuristics) and strategies. The first problem was the one discussed in Protocols 1 and 2, to find the largest triangle that can be inscribed in a circle. The second problem was a geometric construction:

> You are given two intersecting straight lines and a point marked on one of them, as in the figure below. Show how to construct, using a straightedge and compass, a circle that is tangent to both lines and has the point P as its point of tangency to one of the lines.

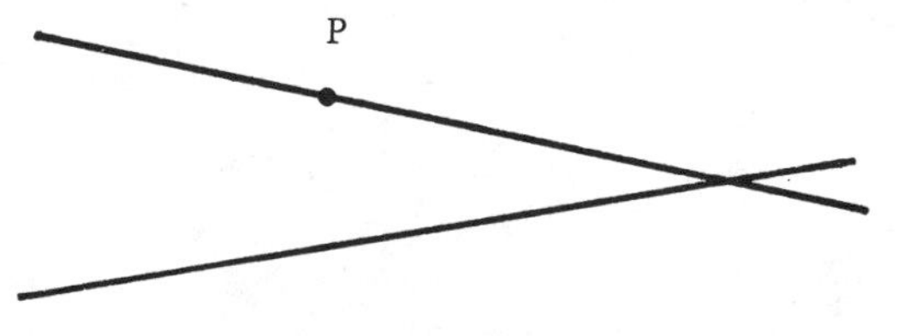

Figure 10.4

Brief "snapshots" of a few representative pretest protocols are given below. These are too condensed to be useful for model building, but they serve to demonstrate again the critical importance of managerial or strategic decision-making. They also stand in (partial) contrast to the students' posttest behavior and (stark) contrast to some expert behavior. The diagrams that represent our episode analyses are here condensed into a sequential list of episode titles, with transitions deleted if there were none. Thus Figure 10.1 is rendered as (Reading/T1/Exploration), and so forth.

ET & DR, PROBLEM 1. (READING/T1/EXPLORATION)

After a brief mention of "max–min" problems, and a brief caveat ("But will it apply for all cases? I don't know if we can check it afterward") in transition, they set off to calculate the area of the equilateral triangle. So much for the next 15 minutes; in spite of some local assessments ("this isn't getting us anywhere") they continue those explorations. Result: all wasted effort.

ET & DR, PROBLEM 2 (READING/EXPLORATION)

In the initial explorations a series of sketches contains all the vital information they need to solve the problem, but they (without any attempt at review or assessment) overlook it. The solution attempt is undirected and rambling. Possibly because they feel the need to do something, they try their hand at an actual construction—already shown to be incorrect by their sketches—and are stymied when it does not work. Overall: lost opportunities, unfocused work, wasted effort.

Note: ET and DR are both bright; both had just completed the first semester calculus course with A's.)

DK & BM, PROBLEM 2
(READ/ANALYZE/T1/EXPLORE/ANALYZE (SOLVE)/VERIFY)

Analysis is extended and coherent, but followed by a poor transition into an inappropriate construction that deflects the students off track for $3\frac{1}{2}$ minutes. When this does not work they return to analysis and solve the problem. A detailed verification completes the session. Managerial decisions worked reasonable well here.

BW & SH, PROBLEM 2
(READING/EXPLORATION/T1/EXPLORATION)

A series of intuition-based conjectures led to a series of attempted constructions, the last of which happened to be correct—though neither student had any idea why, and they were content that it "looked right." This was a classic trial-and-error tape, and only because the trial space was small was there a chance that the right solution would be found. There was one weak assessment (after a construction) that constituted T1, but the result was simply a continuation of trial-and-error search.

Impetuous jumps into a particular direction were pretty much the norm in the pretests, and these first approaches were rarely curtailed. (This behavior was so frequent that it earned the name *proof by assumption,* coined by my assistants.) Because there was little assessment and curtailment, little was ever salvaged from an incorrect first attempt, and a solution was often doomed to failure in the first few minutes of exploration.

Protocol 3, which was discussed above, was obtained after the problem-solving course. It is a representative, perhaps slightly better than average, sample of postinstruction performance. What makes this tape "better" than pretest tapes is not that the students solved the problem, for their discovery of the variational argument that solves it may have been serendipitous. However, that they had the time to consider the approach was no accident: they had evaluated and curtailed other possible approaches as they worked on the problem. In general, more evaluation and curtailment occurred on the posttests than on the pretests, and less pursuit of wild-goose chases. In some cases this allowed for a solution, in some not; but at least the students' actions did not preclude the possibility. The following statistic summarizes the difference: 7 of the 12 pretest protocols were of the

Reading/Exploration type; only 2 of the 12 posttest protocols were of that type. Not at all coincidentally, performance improved on a variety of other measures as well (Schoenfeld, 1982). However, the overall quality of the students' managerial monitoring, assessing, and decision making on the posttests was still quite poor.

To indicate the contrast in managerial behaviors between experts and novices, we turn to the protocol of an expert problem-solver working on a geometry problem. The expert, a number theorist, had a broad mathematical background but (as is apparent) had not dealt with geometric problems for a number of years. By some standards, his solution is clumsy and inelegant. (In a department meeting it was held up for ridicule by the colleague who produced Protocol 5.) Precisely because the expert does run into problems, however, we have the opportunity to see the impact of his metacognitive, managerial skills.

The episode analysis of Protocol 4 is given in Figure 10.5. For reasons of space, the full analysis is condensed.

The critical point to observe in this protocol is that a monitor–assessor–manager is always close at hand during the solution. Rarely does more than a minute pass without some clear indication that the entire solution process is being watched and controlled, both at the local and global levels. The initial actions are an attempt to make certain that the given problem is fully understood. By Item 3 there is the awareness that some other information, or observation, will be necessary in order for a solution to be obtained. The actions in Items 4 and 5 are goal driven and, in Item 6, yield the necessary information. This is a (meta-) comment that the first part of the problem will be solved with one construction, which he knows. The plan is stated in Item 9. Implementation is interrupted twice with refinements (Items 15 and 16; Item 18) that again indicate that the subject is on guard for clarifications and simplifications at almost all times. The first part of the problem concludes with a quick but adequate rehearsal of the argument.

Like Part 1, the second part of the solution begins with a qualitative analysis of the problem. In Item 24, there is a comment that "this is going to be interesting" (i.e., difficult). Such a preliminary assessment of difficulty is, I believe, an indication of an important element of experts' metacognitive behavior. Experts seem to judge their work against a template of expectations when solving a problem. These expectations may be major factors in the experts' decisions to pursue or curtail various lines of exploration during the problem-solving process.

The solution of the second part continues, well structured, with a coherent attempt to narrow down the number of cases that must be considered. This is an implementation of "that kind of induction thought" from Item 29. It appears to be a "forward" or "positive" derivation, verifying that all the cases can be done. Yet the phrase "no contradiction" in Item 33 reveals that the problem solver retains an open mind about whether the constructions could actually be

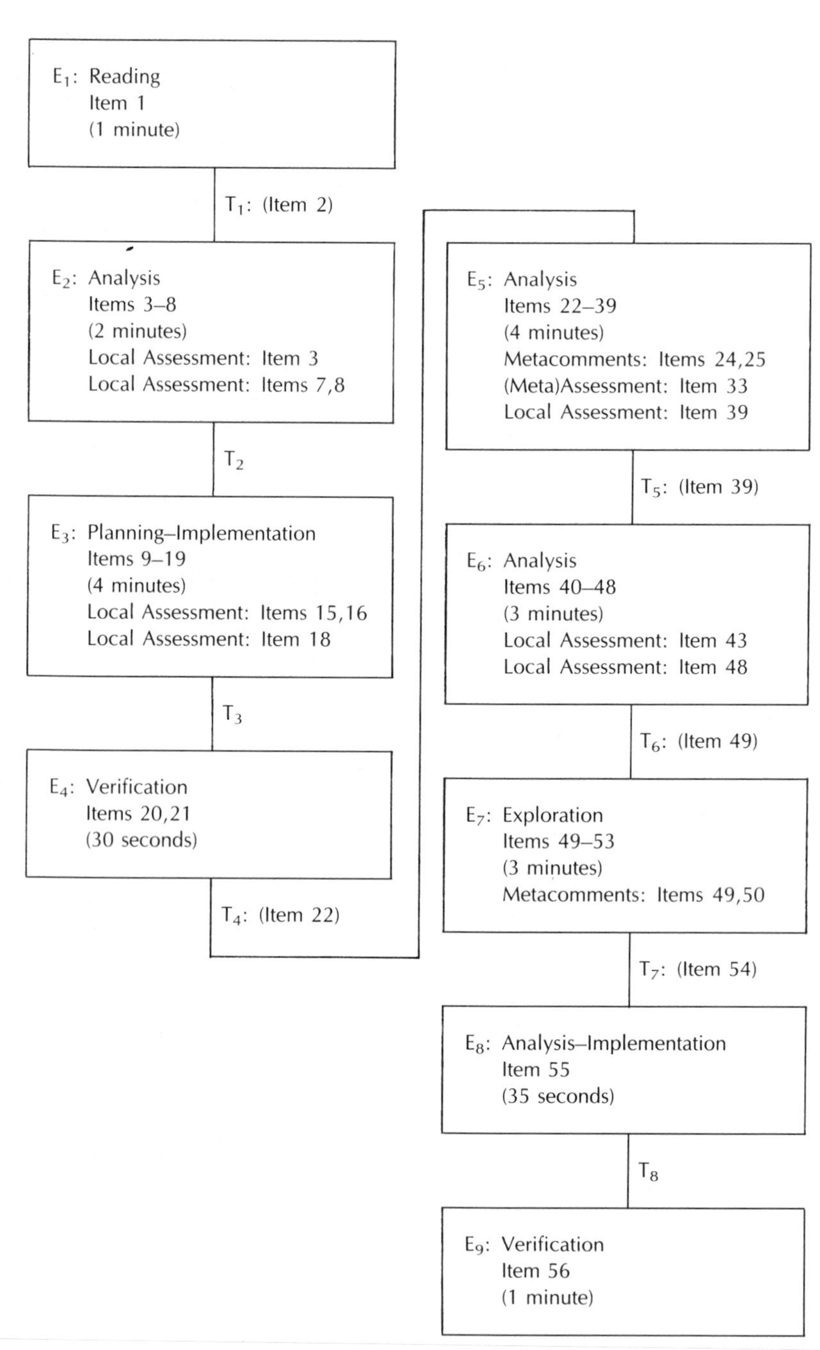

Figure 10.5 A Parsing of Protocol 4.

implemented, and is still probing for trouble spots. The potential for a reversal, using argument by contradiction if he should come to believe one of the constructions impossible, is very close to the surface. This distanced overview, and the maintenance of a somewhat impartial perspective, are confirmed in Item 49.

Assessment is, likewise, always in the immediate vicinity. The comment, "if so this can be done in one shot," in Item 40 indicates not only that solutions are planned ahead, but that the plans are assessed. Even the rather unusual excursion into quadratic extensions (Item 54) is preceded by a comment about "knocking this off with a sledgehammer," and is quickly curtailed.

In sum: this rather clumsy solution (see Protocol 5 in contrast), with its apparent meandering through the solution space, is in reality rather closely controlled. There is constant monitoring of the solution process, both at the tactical and strategic levels. Plans and their implementation are continually assessed and acted upon in accordance with the assessments. Tactical, subject-matter knowledge plays a minor role here. Metacognitive, managerial skills provide the key to success.

Discussion

This chapter deliberately raises many more questions than it can answer. The extended discussions of protocols were designed to make one point absolutely clear: metacognitive or managerial behaviors are of paramount importance in human problem-solving. As Brown observed (1978, p. 82), these types of decisions "are perhaps the crux of intelligent problem solving because the use of an appropriate piece of knowledge . . . at the right time and in the right place is the essence of intelligence." The inverse of this proposition should be given comparable stress: avoiding inappropriate strategies or tactics, at the wrong time or in the wrong place, is an equally strong component of intelligent problem-solving.

To deal coherently with such executive decision-making, one needs a framework for examining, modeling, and judging it. Such a framework must, perforce, be substantially different from extant schemes like those used in mathematics education that focus on overt behaviors at a detailed level (Lucas, *et al.*, 1979; Kantowski, Note 1). As we saw in Protocol 1, the absence of an assessment may doom an entire solution to failure. Schemes that seek only overt behaviors cannot hope to explain that protocol adequately.

An appropriate framework for dealing with such metacognitive behaviors must also differ substantially from those used in artificial intelligence to simulate proficient expert behavior in areas such as physics. Larkin, McDermott, Simon, and Simon (1980) characterize such work as depending on production systems to simulate the pattern recognition that "guide[s] the experts in a fraction of a second to relevant parts of the knowledge store . . . [and] guide[s] a problem's interpretation and solution'' (p. 1336). Although aspects of Protocol 4, such as the recognition of similar triangles (Item 6), are compatible with this perspective, the whole of Protocol 4 is largely complementary to it. At least half of the action in that protocol is metacognitive; it almost seems as if ''manager'' and ''implementer'' work in partnership to solve the problem. And it is precisely when the expert's problem-solving schemata (or *productions*) do not work well that the managerial skills serve to constitute expertise.

The framework presented in this chapter provides a mechanism for focusing directly on certain kinds of managerial decisions. Because a manager ought to be present at major turning points in a problem solution (if only to watch, in case action is necessary), the transition points between episodes are the logical places to look for the presence or absence of such decision making. Here we come to the first serious question: What, precisely, constitutes an *episode*? Although there is reliability among coders in parsing these protocols at the macroscopic level, that begs the question; there is the need for rigorous formalism for characterizing such episodes. Unfortunately, I have not been able to adapt schemata for story understanding or for episodes in memory (see Bobrow & Collins, 1975) to deal with these kinds of macroscopic problem-solving episodes. A formalism needs to be developed in this area.

Questions regarding the characterization and evaluation of the monitoring, assessing, and decision-making processes during problem solving are far more thorny. The role of the monitor was quite clear in Protocol 4; it assured that the solution stayed on track. But how are such decisions made? It is clear from a variety of expert protocols that a priori expectations of problem or subtask difficulty serve as one basis for the decision to intervene. But the nature of the monitoring, the criteria for assessments, what the tolerances are, and how intervention is triggered all remain to be elaborated.

Similarly, assessment is not always desirable or appropriate: in a schema-driven solution, for example, one should simply implement the solution unless or until something untoward occurs. A simple-minded model that looked for assessment at each transition between episodes (and other places) would miss the point entirely: assessment is only valuable some of the time, and we need to know when (and how).

In the long run, we need a detailed model of managerial monitoring and assessment, and of the criteria used for assessment and decision making. This model will enable us to answer questions like those for the transition phase: ''Was the action or inaction appropriate or necessary?'' In the meantime, these questions are not an evasion: they are an attempt to gather data so that the model can be constructed. A further refinement of these questions, and a much more detailed characterization of metacognitive acts in general, are necessary. I hope that this chapter provides a step in that direction.

Appendix 10.A: Protocol 1

1. *K*: (Reads problem) Three points are chosen on the circumference of a circle of radius *R*, and the triangle containing them is drawn. What choice of points results in the triangle with the largest possible area? Justify your answer as best you can.

   You can't have an area larger than the circle. So, you can start by saying that the area is less than $\frac{1}{2}\pi r^2$.

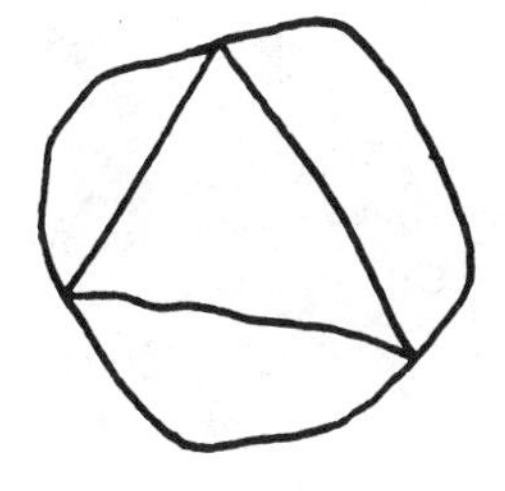

2. *A*: O.K. so we have sort of circle—3 points in front and *r* here and we have let's see—points—
3. *K*: We want the largest one—
4. *K*: We want the largest one—
5. *A*: Right, I think the largest triangle should probably be equilateral. O.K., and the area couldn't be larger than πR^2.
6. *K*: So we have to divide the circumference of the three equal arcs to get this length here. That's true. Right. So, 60–120 are degrees—O.K.—so, let's see, say that it equals *r* over *s*—this radius doesn't help.

7. *A*: Do we have to justify your answer as best as you can? Justify why this triangle—justify why you—O.K. Right.
8. *K*: O.K. Let's somehow take a right triangle and see what we get. We'll get a right angle.
9. *A*: Center of circle of right triangle. Let's just see what a right triangle—is this point in the center? Yep, O.K. Yeah.
10. *K*: This must be the radius and we'll figure out that'll be like that, right?

11. *A*: So the area of this—
12. *K*: is *R*, is *R*—$\frac{1}{2}$ base times height, that's *s* and $2R$, height is *R* so it is $\frac{1}{2}R^2$. It's off by a factor of 2.
13. *A*: O.K. But what we'll need is to say things like—O,K. Let's go back to the angle—probably we can do something with the angle.

14. *K*: Oh, I got it! Here, this is going to be 120—the angle of 120 up here—
15. *A*: Right! Yes, this is 120 and this is 120.
16. *K*: Right!
17. *A*: So—
18. *K*: We have to figure out—
19. *A*: Why do we choose 120—because it is the biggest area—we just give the between the biggest area—120.
20. *K*: Umm. Well—the base and height will be equal at all times.
21. *A*: Base and height—right—
22. *K*: In other words—every right triangle will be the same.
23. *A*: Ah, ah—we have to try to use *R*, too.
24. *K*: Right.
25. *A*: O.K. (seems to reread problem)—justify your answer as best as you can. O.K. (pause)
26. *A*: So—there is the picture again, right? This is—both sides are equal—at this point—equal arc, equal angles—equal sides—this must be the center and this is the radius *R*—this is the radius *R*—
27. *K*: So we have divided a triangle with three equal parts and—
28. *A*: There used to be a problem—I don't know about something being square—the square being the biggest part of the area—do you remember anything about it?
29. *K*: No . . . I agree with you—the largest area . . . of something in a circle, maybe a rectangle, something like that . . .
30. *A*: Oh, well . . . so . . .
31. *K*: Since this is *R*—and this is going to be 120, wouldn't these two be *R* also?
32. *A*: Right.
33. *K*: This is 120.
34. *A*: Ah, ah.
35. *K*: Like a similar triangle—120 and 120 are the same angle—so these two should be *R*.
36. *A*: O.K. Maybe they are.
37. *K*: Why can't they be?

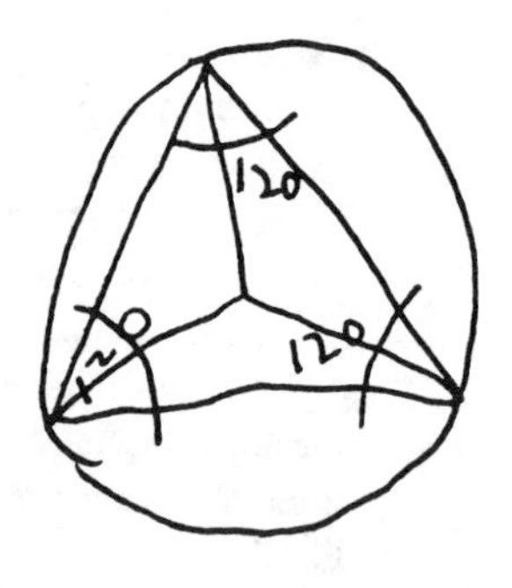

38. *A*: (Mumbles)—
39. *K*: See, look—this is the angle of 120—right?
40. *A*: Right.
41. *K*: And this is an angle of 120. Right? This is like similar triangles—
42. *A*: Wait a second—I think if you—this is true 120 but I don't think this one is—

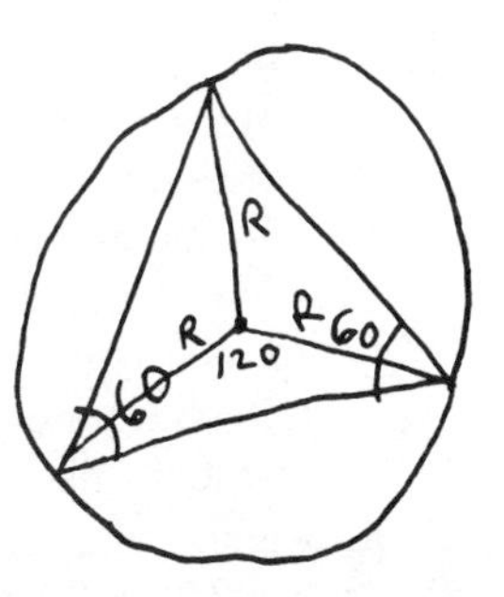

43. *K*: It is an equilateral triangle—that's—
44. *A*: No—it should be a 60.
45. *K*: That's right—it should be a 60. (Mumbles)—that's $\frac{1}{2}$ of it—that's right—$2R$.
46. *A*: What are you trying to read from?
47. *K*: What if we could get one of these sides, we could figure out the whole area.
48. *A*:Ah, ah.
49. *K*: Right?
50. *A*: Presume this to be $\frac{1}{2}$ that side, we've got $\frac{1}{2}$ base times height. We'll get the area—all we have to show is the biggest one.
51. *K*: When we take the formula πR^2 minus $\frac{1}{2}$ base times height and then maximize that—then take the derivative and set it equal to zero, we can get that function—then we can get this in the form of R.
52. *A*: O.K.
53. *K*: Then we can try this as the largest area.
54. *A*: Do you want to get this function, this as a function of R?
55. *K*: Yeah.
56. *A*:We can, I think. So you want this—right?
57. *K*: Well, it is kind of obvious that with H and H you are still going to have an R in it. So you can subtract it.

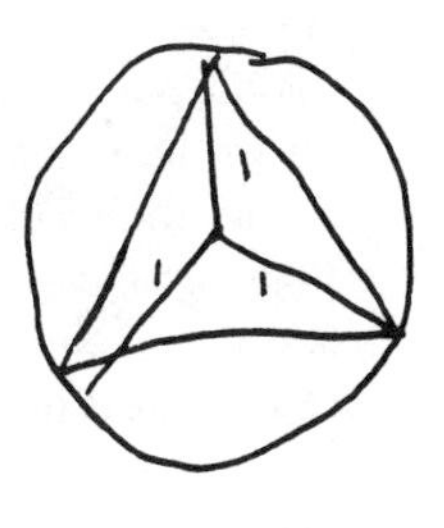

58. *A*: You have H in it. Well we have this one here. (Mumbles)—(repeats the problem). Try this to be $2R$.
59. *K*: No—it can't be. It has to be between R and $2R$.

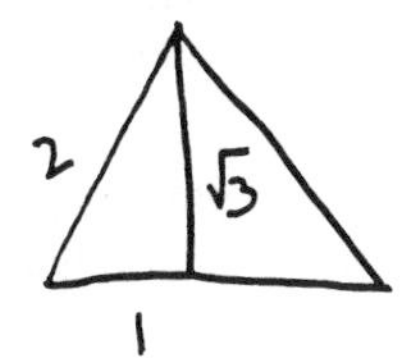

60. *A*: Yeah.
61. *K*: Helps us a lot! Set R equal to 1.
62. *A*: $R = 1$?
63. *K*: Right.

64. *A*: O.K.
65. *K*: That's 1, that 1 that's 1—it'll equal S over R. The area of the triangle is equal with $R = 1$, it's 2.
66. *A*: Well . . . height equals . . .
67. *K*: That's for the sides of the triangle—that's obvious: $R = 1$.[1]
68. *A*: O.K.—divided into equal parts—(lots of mumbling)—This from—well—you know—O.K. If you see we probably try to fix one point and choose the other two—O.K.—we are going to go from something that looks like this all the way down—
69. *K*: Right.
70. *A*: Right. O.K. and here the height is increasing where the base is decreasing.
71. *K*: Right. (Mumbles)
72. *A*: When we reach—O.K.
73. *K*: What is the area, side squared over 4 radical 2 for an equilateral triangle? Is it like that?
74. *A*: You want the area for an equilateral triangle.
75. *K*: The area? I don't know. Something like side squared over radical 2, or something—
76. *A*: If you can probably show . . . at a certain point where we have the equilateral triangle the base and the . . . well . . . you know the product of the base since the base is decreasing and the height is increasing every time we move the line. If you can show a certain point, this product is the maximum—so we have the area is a maximum at that point. So this one is decreasing—and at this point we have R, R, and R.
77. *K*: Ah, ah.
78. *A*: O.K. This is the base—is $2R$—a right angle.
79. *K*: It wouldn't be $2R^2$.
80. *A*: (Mumbles)—One more—I mean—
81. *K*: O.K.
82. *A*: It should be R^2. But base times height—(mumbles)—and this one, say this is $R + x$.
83. *K*: The height equals $R + x$, so the base equals $R - x$.

84. *A*: (Mumbles)—those two things are equal to this—
85. *K*: Right.
86. *A*: All right.
87. *K*: I don't know.
88. *A*: We want this product of *h* as a maximum—as a maximum—and this one . . . I don't know.

Appendix 10.B: Protocol 2

1. *D*: Reads the question.
2. *B*: Do we need calculus for this? So we can minimize, or rather maximize it.
3. *D*: My guess would be more like—(mumbling)—my basic hunch would be that it would be—
4. *B*: An equilateral—
5. *D*: 60, 60, 60.
6. *B*: Yeah.
7. *D*: So what choice of points has to be where on the triangle—these points are gonna be.
8. *B*: Try doing it with calculus—see if you can—just draw the circle—see what we'll do is figure out the right triangle—
9. *D*:Yeah, or why don't we find—or why don't we know the—some way to break this problem down into—like what would a triangle be for half the circle?
10. *B*: 60 degrees here?
11. *D*: Why don't we, why don't we say that—O.K.—why don't we find the largest triangle with base—one of the diameters, O.K.
12. *B*: Base as one of the diameters?
13. *D*: Yeah.
14. *B*: O.K. That would be just a family of right triangles—that go like this.
15. *D*: And they're all the same area?
16. *B*: No, no they're not all the same area—the biggest area would be in one like that. See if

we could figure out—make it into sort of like a—if we could do it with calculus and I know there is a way. I just don't remember how to do it.

17. *D*: I have a feeling we wouldn't need the calculus. So this area then this is *R* and this would be—*R* squared—that would be the area of this—so then the distance here has got to be—45 degrees—

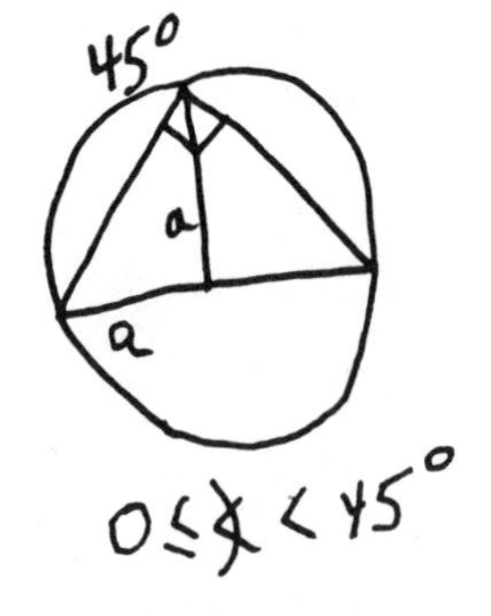

18. *B*: Right—that's got to be 45 degrees because they are the same. That's *A*—*A* over square root of 2—right?
19. *D*: Umma.
20. *B*: If that's radius—*A*—and this is *A*, too, so that would be A^2, that would be R^2, wouldn't it?
21. *D*: Right.
22. *B*: But I think this would be bigger.
23. *D*: Oh, of course it would be bigger—I was just wondering if . . . (pause)
24. *D*: Well we can't build a diamond—so we can't build a diamond that would go like that, obviously you want to make it perfectly symmetrical, but we can, if we maximize this area, and just flip it over, if we can assume that it is going to be symmetrical.

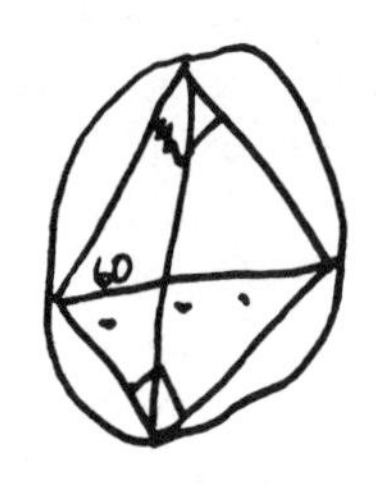

25. *B*: Yeah, it is symmetrical.
26. *D*: And if we can find the best area—
27. *B*: You mean the best—cut it in half in a semicircle.
28. *D*: Right. And if we can find the best area of—

29. *B*: And triangle that fits in a semicircle—well it shouldn't be a semi-
30. *D*: No it's a semicircle.
31. *B*: Largest triangle that fits in there?
32. *D*: Yeah, but it would have to be—if it is going to be symmetrical though, then you know this line has to be flat—it is going to have to form a right angle. So all we really have to do is form a right angle. So all we really have to do is find the largest area of a right triangle—inscribed in a semicircle.

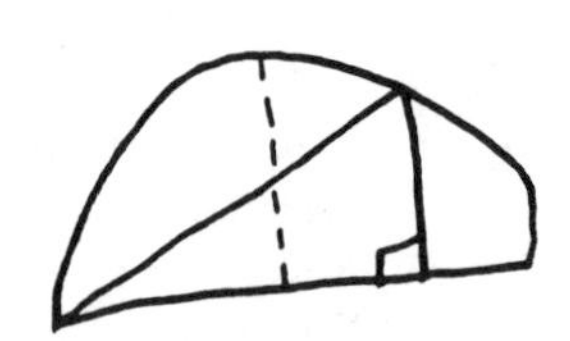

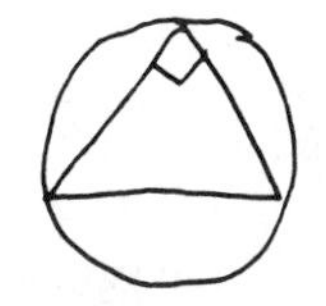

33. *B*: Largest area of a right triangle. Yea, but obviously it is this one which is wrong.
34. *D*: No—No—
35. *B*: One like this.
36. *D*: Yeah with that angle, right.
37. *B*: O.K.—how we go about doing that? Hey, like we can—use the unit circle, right?
38. *D*: Umma.

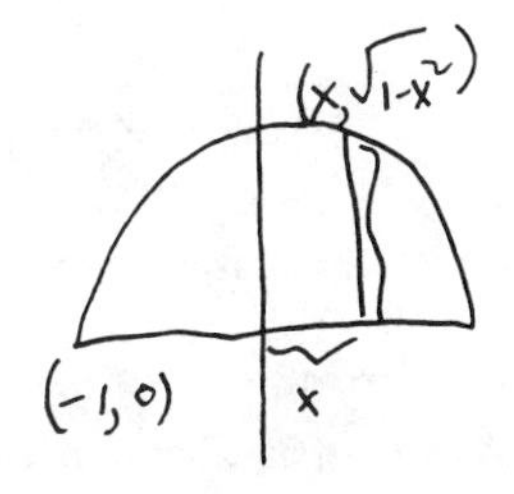

39. *B*: So that means—this is $\sqrt{1 - x^2}$--this point right here—will be $\sqrt{1 - x^2}$, O.K. this squared—(mumbling)—I'll just put some points down to see if . . . pick an arbitrary—
40. *D*: Yeah, yeah, just to find this point—
41. *B*: All right, this is 1. Now I've got to find that point—O.K. What is the area of this—this is the distance right here times that distance, right? Product of those distance—area equals from this distance would be this, would be x value which would be $x - 1$ or $x + 1$? O.K., it's $x + 1$, this distance right here times this distance right there which would be the y coordinate which is x^2. Want to take the derivative of that—to the x—(mumbling).

$$A = \frac{1}{2}(x+1)\sqrt{1-x^2}$$

$$\frac{dA}{dx} = \frac{(x+1)(\frac{1}{2})(-2x)}{\sqrt{1-x^2}} + \sqrt{1-x^2}$$

$$\frac{1}{\sqrt{1-x^2}} = \frac{\sqrt{1-x^2}}{1}$$

42. *D*: O,K.
43. *B*: Times $(2 - x)$. Did I have, oh, the 2 is crossed out so I just have an $-x$—or, that was over the square root of $(1 - x^2)$, plus all this stuff. And set that equal to zero and you get that—oh, this is just one, isn't it—this is just one—so one of that, plus that equals zero, right?
44. *D*: I think we're getting a little lost here—I am not sure. Well, you go ahead with that—
45. *B*: Well, I'll just think about it, as it is just mechanical. There is a minus in here, isn't there? (Mumbling)—O.K. x equals the $\sqrt{2}$ and what was this distance, we said? That was x—so that means it would be $\sqrt{2} + 1$—that's impossible.
46. *D*: Times *R*.

47. *B*: If x equals plus or minus the $\sqrt{2}$—
48. *D*: Umma—
49. *B*: This y thing would be $1 - x^2$, right?
50. *D*: This is just the distance—therefore, this right here has to be $\sqrt{2}$. Guess your calculations are all right.
51. *B*: Yeah, if I got x equals $\sqrt{2}$—we've got a semicircle here, right? O.K.—and I have the points—right, it's a unit circle and I said that $x^2 + y^2 = 1$, so $y = \sqrt{1 - x^2}$. O.K.? And—(pause)—the x can't equal the square of the two because it would be out there. I know this has to be right but—
52. *D*: But all kinds of—let's see—well we know already, O.K. that the triangle is not 45, 45, because that would make it too small, O.K.?
53. *B*: Um—
54. *D*: So we know this angle is greater than zero and less than 90 degrees—
55. *B*: I just want to make sure I didn't—so this is $x + 1, x + 1\ n$ . . . and cross multiply to set $1 - x^2 = 1$ which means $x = \sqrt{2}$.
56. *D*: No, it has to be a 60, 60, 60—right triangle—no I am sorry not a right triangle—has to be a 60, 60, 60 triangle—because no matter where you move these vertices, it has to be a 60, 60, 60 triangle—because no matter where you move these vertices—
57. *B*: O.K.
58. *D*: —you are going to add area to this—like the—(mumbling)—you are going to add area to this.
59. *B*: All right, O.K. I understand, but I don't understand why it didn't work for this. I mean that . . . is there no solution for this equation?
60. *D*: I don't know—are you sure what you are looking for in that one?
61. *B*: Yeah, I marked off these and I just wanted to mark off these dimensions.
62. *D*: O.K. what were you looking for? the length of this?

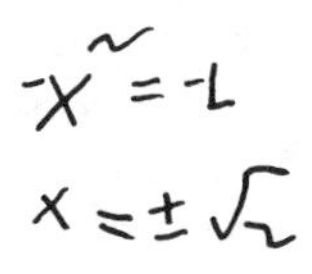

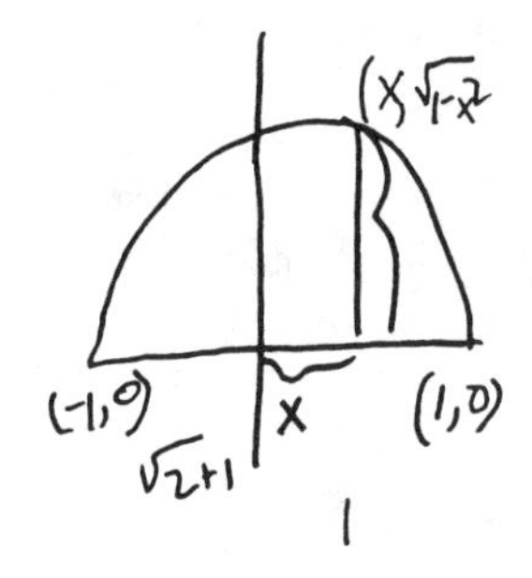

63. *B*: I was just looking for the maximum area of this—I said $A = (x + 1)$ times $\sqrt{1 - x^2}$. That's this height which is $\sqrt{1 - x^2}$ squared. This is the right circle. That's this distance right here—this minus the x value that I used—x value that is just x. O.K.—cause it is all in terms of x—x minus the x value here, which is $x - 1$, which $x + 1$—so area—ah shoot—I should have put $\frac{1}{2}$ that is well,—(mumbling)—I'll get it. That should be $\frac{1}{2}$ there, but I don't think that makes any difference—so that's all in terms of 1.
64. *D*: So—if—
65. *B*: Oh, wait a minute there's a difference—so one for two is $\frac{1}{2}$ the first part—
66. *D*: So if you find the maximum area equal to—
67. *B*: It doesn't make any difference—it is just a factor of $\frac{1}{2}$ here—because the area equals $\frac{1}{2}$ that.
68. *D*: No—what's the next move?
69. *B*: See I get x—see I get a value of x with a plus or minus $\sqrt{2}$, right?
70. *D*: Umma.
71. *B*: If I plug x back into this I get $\sqrt{2} + 1$, right? Then I plug x back into there and I get $(1 - (\sqrt{2})^2)$ which is -1 which doesn't work.
72. *D*: Umma.
73. *B*: Which doesn't seem right. Plus r^2--(mumbling)—Let me just check my derivative over again. Now I know my mistake—hold it. I added this x—it's supposed to be times so we've still got a chance. So let me go from there. It is just a derivative mistake. Let me see it will be (1 − x squared)—no it will be—$(-x + 1)$. This might work—if it does—we solve that and cross out this minus 1. That means $x + 1 + x^2 - 1$, that makes $x^2 + x$—cross this out—(mumbling)—all right? It still doesn't work.
74. *D*: Well, let's leave the numbers for a while and see if we can do this geometrically.
75. *B*: Yeah, you're probably right.

76. *D*: Well, we know that these two are some kind of symmetry.
77. *B*: Yeah.
78. *D*: I still say we should try—yeah—what we were doing before—just try to fix two of the points and let the third one wander around.
79. *B*: Yeah, we were going to fix them—yeah, I know what happens if you fix them on the diameter—then you have a family of right triangles.
80. *D*: Those the maximums.
81. *B*: Well, I don't see how—where are you going to fix the two points?
82. *D*: Well, you just fix them on any diameter. You find the largest triangle.

83. *B*: That would—obviously that would be the 45, 45 triangle if you fix them on the diameter. If you fix them on any chord.
84. *D*: Yeah, why though. Well, we know that if we put two of the points too close together—O.K.—O.K.—no matter where we put the third point—
85. *B*: Yeah.
86. *D*: —It's going to be too small. O.K. If we put them too far apart—O.K.—no matter where we put the third point, we are only using half a triangle.
87. *B*: O.K.
88. *D*: So it's got to be—O.K. So—two of the points, at least well, matter of fact if you've got three points, each two of the points have to be between zero and $\frac{1}{2}$ of the circle distance away from each other.
89. *B*: O.K.
90. *D*: See how I got that? O.K. so therefore each two of the *p* points has to be like that—so how can we construct a circle that's like that? O.K. so we stick one point here—arbitrarily—so now the second point has to be somewhere O.K.—within—O.K. in other words, it can't be right here—it can't be right here—it can be anywhere else. We've

got to place it so that the third points is going to be within half—

91. *B*: Half of what—I don't get you there.

92. *D*: O.K. Now wait a minute—let's see. You know when I said that (pause). O.K. In other words the relationship between every pair of the three points.

At this point the interviewer (*I*) terminated the session and asked the students to sum up what they had done. *B* focused on the algebraic computations he had done in trying to differentiate $(1 + x)$ times $\sqrt{1 - x^2}$. The following dialogue ensued:

I: So what do you wind up doing, when you do that? You wind up finding the area of the largest right triangle that can be inscribed in a semicircle.

D: We determined that.

I: My question is: how does that relate to the original problem?

B: Well. . . .

Appendix 10.C: Protocol 3

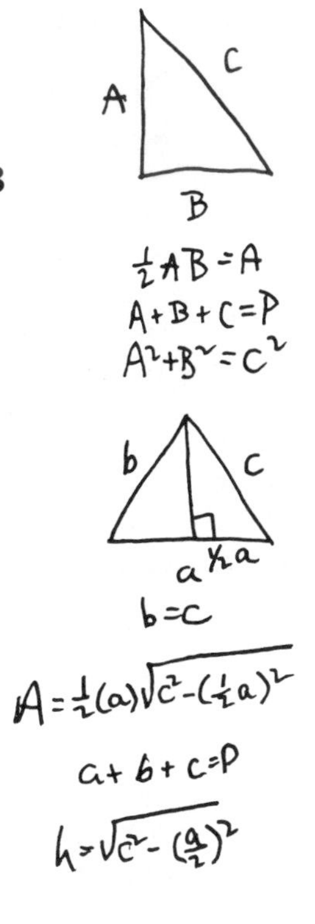

1. *K*: (Reads problem.) Consider the set of all triangles whose perimeter is a fixed number, *P*. Of these, which has the largest area? Justify your assertion as best you can. All right now what do we do?
2. *D*: We got a triangle—well we know we label sides *A*, *B*, and *C*.
3. *K*: Right. I'll make it a right triangle—all right—*A*, *B*, *C* and the relationship such as that $\frac{1}{2}AB$ = Area and $A + B + C = P$ and $A^2 + B^2 = C^2$ and somehow you've got an area of one of these in the perimeter.
4. *D*: Yeah, except for somehow—I mean I don't really know—but I doubt that's the triangle of minimum area—well, O.K. we'll try it.
5. *K*: Largest area. Well, it is the only way we can figure out the area.

6. *D*: All right.
7. *K*: But for an isosceles we can do almost the same thing. This is $\frac{1}{2}A$ so that we know that the area is $A/2$ times $\sqrt{C^2 - (A/2)^2}$.The perimeter $= A + B + C$ and the height equals $\sqrt{C^2 - (A/2)^2}$.
8. *D*: All right.
9. *K*: Now what do we do. We've got to figure out the largest area.
10. *D*: Isn't it the minimum?
11. *D*: The largest area.
12. *D*: So actually if we can get *A*—we have to get everything in terms of one variable and take the derivative, right? Basically?
13. *K*: Yeah, well—
14. *D*: Well, I still don't know if we should do—I mean we can find an area for this and can find an area for that, granted, but if we ever come to a problem like this—I mean we don't know—we have no idea as of yet with a given perimeter what's going to be that.
15. *K*: Right.
16. *D*: So, there—I mean—you can do that again but then what do you do?
17. *K*: Then we're stuck, right? Usually, you know, you could probably take a guess as to what kind of triangle it would be—like you could say it is a right triangle or an isosceles—I think it is an equilateral, but I don't know how to prove it.
18. *D*: Umma.
19. *K*: So we have to figure out some way to try to prove that.
20. *D*: All right, a good guess is that it is an equilateral, then why don't we try an isosceles and if we can find that these two sides have to be equal to form the maximum area, then we can find that—then we should be able to prove that side also has to be equal.
21. *K*: O.K. so *B* will be equal to *C*, so the perimeter $P = A + 2B$, or $A + 2C = P$.
22. *D*: All right.

23. *K*: Ummmm.
24. *D*: See what we've got.
25. *K*: Fix *A* as a constant then we can do this, solve that for *C*.
26. *D*: All right.
27. *K*: For a maximum area we've got $\frac{1}{2}$, let's say $A = 1$; $\sqrt{C^2 - \frac{1}{4}}$. Right? Maximum area: $\frac{1}{2}$ $\sqrt{C^2 - (\frac{1}{4})^{\frac{1}{2}}} = 0$.
28. *D*: C squared minus what?
29. *K*: $(\frac{1}{2})^2$, yeah, $(\frac{1}{2})^2$. $\frac{A}{2}$, where $A = 1$. O.K.?
30. *D*: Ah, ah.
31. *K*: (Mumbling)—this is $\frac{1}{4}$ $(C^2 - \frac{1}{4})^{-\frac{1}{2}}$. $2C$, so we know that $2C$ has to $= 0$ and $C = 0$ and we are stuck!

$$1/2\sqrt{c^2 - 1/4}$$
$$1/2\,(c^2 - 1/4)^{1/2} = 0$$
$$1/8\,(c^2 - 1/4)^{-1/2}(2c) = 0$$
$$2c = 0$$
$$c = 0$$

32. *D*: We should have taken a derivative in it and everything you think?
33. *K*: Yeah, that's the derivative of that. So does it help us? My calculus doesn't seem to work anymore.
34. *D*: The thing is—pause—you are letting *C* be the variable, holding *A* constant. so what was your formula—$\frac{1}{2}$ base times square root.
35. *K*: The base *A* times the square root times the height which is a right triangle to an isosceles which is—so it is $C^2 - (\frac{A}{2})^2$ which would give you this height.
36. *D*: *A* to the $\frac{2}{4}$, no, *A* to the $\frac{2}{2}$, no, $(A/2)$ squared.
37. *K*: How about $P = \ldots$ no, $C = P - (A/2)^2$? Should we try that—
38. *D*: No, see part of the thing is, I think that for here we're just saying we have a triangle, an isosceles triangle, what is going to be the largest area? Largest area.
39. *K*: Largest area—set its derivative equal to 0.
40. *D*: All right. Well the largest area or the smallest area—I mean—what's going to happen is you have a base and it's going to go down like that—I mean—we don't set any conditions—we're leaving *P* out of that.
41. *K*: Ah, ah.
42. *D*: That's absolutely what we have to stick in.

43. *K*: We've got *C* and $P - A/2$.
44. *S*: $P - A/2$.
45. *K*: Formula—isosceles.
46. *D*: $A + 2B = P$—all right?
47. *K*: Shall we try that—(mumbling). $-A$ over 2—we've got to have a minus $\frac{1}{4}$ *PA*—
48. *D*: Well, then you can put *A* back in—then you can have everything in terms of *A*, right? Using this formula, we have the area and we have a—
49. *K*: All right—*P*—so that's

$$\frac{A}{2}\left(\frac{P^2 - 2A + A^2}{4} - \frac{A^2}{4}\right)^{\frac{1}{2}}$$

and that's $\quad \frac{A}{2}\left(\frac{P^2 - 2A}{4}\right)^{\frac{1}{2}}$

. . . (mumbling and figuring)

50. *D*: Wait a minute—you just took the derivative of this right here?
51. *K*: This times the derivative of this plus this times the derivative of this.
52. *D*: Oh.
53. *K*: Mumbling and figuring . . .

$$\frac{A}{4}\left(\frac{P^2 - 2A}{4}\right)^{\frac{1}{2}} 2(P - 2) + \left(\frac{P^2 - 2A}{4}\right)^{\frac{1}{2}}$$

$$\tfrac{1}{2} = 0 \ldots 50\, \frac{2AP - 2A}{4} + \frac{P^2 - 2A}{8} = 0.$$

54. *D*: So can we get *A* in terms of *P*?
55. *K*: P^2--
56. *D*: $8P^2 - 8P^2$ bring the P^2 on this side and multiply it by 8 and we'll have a qudratic in terms—no we won't—then we can just have *A* we can factor out in the equation—you see.
57. *K*: O.K. $P^2 = P^2$
58. *D*: $-8P^2$--oh, are we going to bring everything else to the other side?
59. *K*: Yeah, $2A - 4A - 4AP \times B$—No—

$$\frac{A}{2}\left(\frac{(P-a)^2}{2} - \frac{a^2}{4}\right)^{1/2}$$

$$\frac{a}{2}\left(\frac{P^2-2a-a^2}{4} - \frac{a}{4}\right)^{1/2}$$

$$\frac{a}{2}\left(\frac{P^2-2a}{4}\right)^{-1/2}(2P-2)+$$

$$\left(\frac{P^2-2a}{4}\right)^{1/2}(1/2)=0$$

$$\frac{2ap-2a}{4} + \frac{P^2-2a}{8} = 0$$

$$8 \quad P^2$$

60. *D*: That's not right. Well, the—we can just multiply—
61. *K*: P^2 = all this.
62. *D*: Right.
63. *K*: $P^2 - 4AP$—this isn't getting us anywhere.
64. *D*: P^2 equals—factor out the *A*—then we can get *A* in terms of *P*.
65. *K*: $P^2 = 2A$—so you've got

$$\frac{P^2}{6 + 4P}$$

66. *D*: So if we have an isosceles triangle and *A* is equal to—
67. *K*: be equal to that—
68. *D*: And if *A* has to be equal to that and *B* and *C* are equal—
69. *K*: Sp. *B* =—(whistles)
70. *D*: $B = P -$ that.
71. *D*: $2B = P - A$ over 2.
72. *D*: No we aren't getting anything here—we're just getting—thing is that we assumed *B* to be equal to *C* so of course, I mean—that doesn't't—we want to find out if *B* is going to be equal to *C* and we have a certain base—let's start all over, and forget about this. All right, another triangle. Certain altitude.
73. *K*: Well, let's try to assume that it is an equilateral.
74. *D*: All right.
75. *K*: Sides—(mumbling)—perimeter equals 3*S*, right?
76. *D*: Yeah, but wait a minute—that's still not going to really help us—what are we going to do—simply assume that it is an equilateral. We're just going to get that it is an equilateral if we assume that.
77. *D*: True.
78. *D*: We want to prove that it is an equilateral if we think it is. If we want to do anything we can—
79. *K*: Yeah, how do you prove it?

$p^2 = 2a + 4a - 4ap$

$p^2 = 2a(3 + 2p)$

$\frac{p^2}{3+2p} = 2a$

$a = \frac{p^2}{6+4p}$

80. *D*: Well, we can make up a perimeter—we don't need a perimeter *P*, do we? So,—
81. *K*: Where are you going to get area formula in the form of *P*?
82. *D*: We want to maximize the area so that we can prove—O.K. we have the given base—we'll set our base equal to something.
83. *K*: Yeah, (mumbling) *P*, or something—I don't know.
84. *D*: Then the other two sides have to add up to *P*.
85. *K*: We—how about we say—let's start with an equilateral just for the hell of it—see what happens. You get $\frac{1}{3}P$, $\frac{1}{3}P$ and $\frac{1}{3}P$. And this is $\frac{1}{9} - \frac{1}{36}$ which is the height—

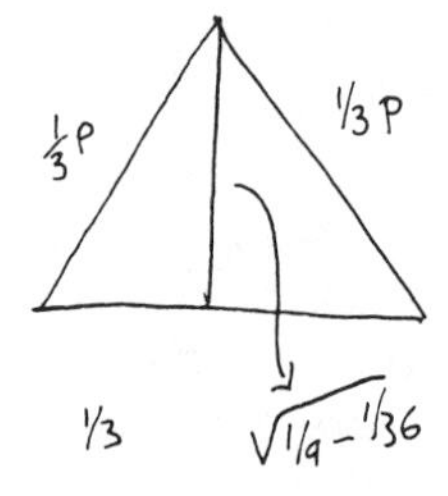

86. *D*: Now the thing we want to do is say—O.K. if we shorten this side at all and then what's going to happen to the height—if we leave this the same.
87. *K*: We can't shorten it.
88. *D*: And we shorten this side—sure we can—
89. *K*: Well—
90. *D*: We can have a—this equal to $\frac{1}{3}$ and then a—this equal to—well you're going to have—I mean—

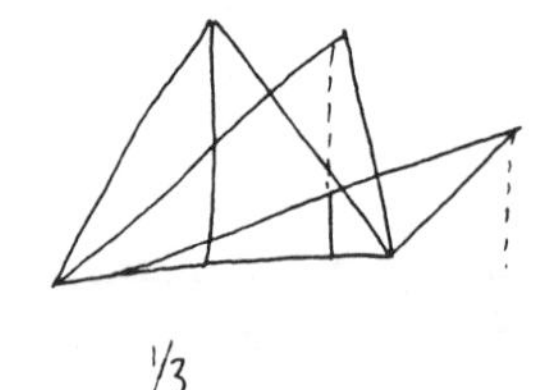

91. *K*: Aha.
92. *D*: This is going to get longer like that. Now we can see from this that all that is going to happen is that the base is going to get shorter so we know from that as far as leaving the base constant goes if we move—if we shorten this side then it is going to—somehow the point's going to go down in either direction.
93. *K*: Semicircle.
94. *D*: Right. That proves that we have to have an equilateral.
95. *K*: No, it proves an isosceles.
96. *D*: No, isosceles, I mean. all right from that if we set—we know that those two have to be equal so if we set his base equal to anything—it doesn't have equal to be $\frac{1}{3}P$—we

can also show that if goes down—the area is going to get smaller, so the constant base then the height is going to get shorter and shorter and is getting smaller and smaller actually.

97. *K*: O.K.
98. *D*: In this case if it goes down to his side, we're going to have again a smaller angle here, shorter base here—and (noise).
99. *K*: So we get—so we know it is an equilateral—well prove it.
100. *D*: I don't know that's not a rigorous proof, but it is proof—good enough for me.
101. *D*: Proves that an equilateral has the largest area.
102. *D*: Oh, we're talking about the largest area.
103. *K*: Yeah.
104. *D*: Oh, we just did.
105. *K*: We have to prove it has fixed number *P*—perimeter.
106. *D*: Well we already—we assumed that we have a fixed *P*, all right? I mean this is a proof as far as I—
107. *K*: Well, we've shown that an equilateral has the largest area. We haven't shown that if you have a certain set perimeter, let's say a right triangle, with a perimeter which is the same—we will not have a larger area.

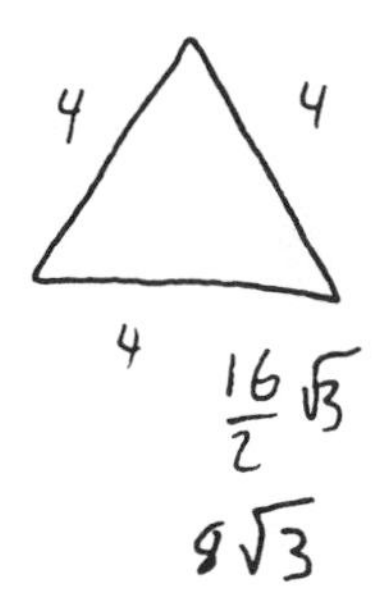

108. *D*: No, but we have because we have shown with the set perimeter—O.K. we know that—
109. *K*: Well what if we have 3, 4, 5 with an equilateral being 4, 4, 4—
110. *D*: 3, 4, 5 is what? (Mumbling.)
111. *K*: 12. So this area will be 6 and this area will be side squared 16.—O.K. that will have the largest area.
112. *D*: What's that, 1.7?
113. *K*: Yeah, 11 is still greater than 6 and that's greater than 1.

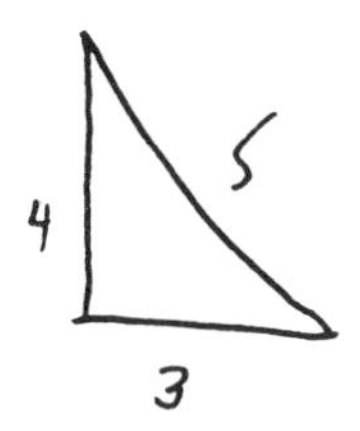

114. *D*: Oh, yeah, that's right. Yeah, but the thing is if we have a fixed dimension, we already

showed that, O.K. what is going to happen is as this side gets longer—say we use 4 as a base here, so then what's going to happen—well say we use 3 as a base, just so we won't have an equilateral when we are done—what's going to happen as 4 gets longer and 5 gets shorter—it's going to go upwards. The optimum area—the maximum area is going to be right there. Because you've got—

115. *D*: Right.
116. *D*: This angle and that height. If you make this angle any less—maybe let me draw a picture.
117. *K*: I can understand that—this will give us largest area, but how can we prove this bottom is one-quarter—$\frac{1}{3}$ the area of the perimeter?
118. *D*: Well, remember all the problems we've done where we say—O.K. let me just start from here once more—so that we have 3, 4, 5—is that what you have—because that's going to be 5. Wasn't a very good 3, 4, 5 anyway. So you start out with 3, 4, 5—all right, we pick the 3 has the base, right?
119. *D*: Aha.
120. *D*: All right, it's 5—(mumbling)—if we have 3 as the base—as this is a little bit off an isosceles, but if we draw an isosceles as 3 as the base—O.K. we've got a right angle—that's got to be the maximum—(mumbling)—(height?) because if it goes any—
121. *K*: Right.
122. *D*: Over this way, it is going to go down.
123. *K*: O.K.
124. *D*: All right, so remember the argument we've used—well if we—
125. *K*: Yeah, I can show that, but what you're not showing is—what you're not proving is that—
126. *D*: That it has to be an equilateral?

127. *D*: Right. But you're not showing that this side is $\frac{1}{3}$ the perimeter.
128. *D*: Right. I'm showing—first of all it has to be an isosceles. Right.
129. *K*: Right.
130. *D*: It has to be an isosceles—that means that we've got these three sides and those two are equal—right?
131. *K*: Umma.
132. *D*: Right—so now I pick this side as my base—I already picked—if that side is my base then the maximum area would have to have an isosceles—so I turn around—this side is my—
133. *K*: That I understand as proof, but you're not showing me that this is $\frac{1}{3}$ the perimeter—(mumbling).
134. *D*: If we have an isoceles triangle—if we have an equilateral triangle—then each side has to be $\frac{1}{3}$ the permieter—that's the whole thing about an equilateral triangle.
135. *K*: I know—O.K.
136. *D*: First we know it must be an isosceles, right?
137. *K*: Umma.
138. *D*: O.K.
139. *K*: I understand this.
140. *D*: If it is an isosceles, it must be an equilateral, right.?
141. *K*: All right.
142. *D*: And if it must be an equilateral—all three sides must be equal and if the perimeter is P, all three sides must be $\frac{1}{3}P$.
143. *K*: O.K. I've got it.

Appendix 10.D: Protocol 4

1. (Reads problem) You are given a fixed triangle T with base B. Show that it is always possible to construct, with ruler and compass, a straight line

parallel to B such that that line divides T into two parts of equal area. Can you similarly divide T into five parts of equal area?

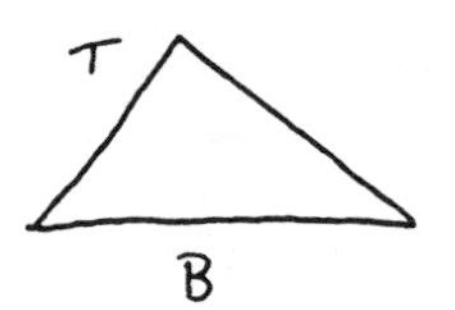

2. Hmm. I don't know exactly where to start.
3. Well I know that the . . . there's a line in there somewhere. Let me see how I'm going to do it. It's just a fixed triangle. Got to be some information missing here. T with base B. Got to do a parallel line. Hmmm.

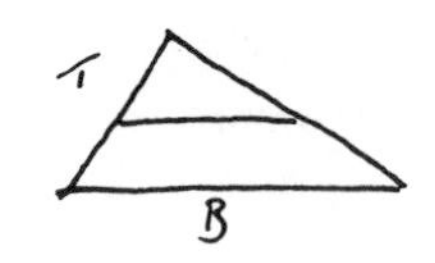

4. It said the line divides T into two parts of equal area. Hmmm. Well, I guess I have to get a handle on area measurement here. So, what I want to do . . . is to construct a line . . . such that I know the relationship of the base . . . of the little triangle to the big one.
5. Now let's see. Let's assume I just draw a parallel line that looks about right, and it will have base little b.
6. Now, those triangles are *similar*.
7. Yeah, all right then I have an altitude for the big triangle and an altitude for the little triangle so I have little a is to big A as little b is to big B. So what I want to have happen is $\frac{1}{2}\,ba = \frac{1}{2}AB - \frac{1}{2}ba$. Isn't that what I want?
8. Right! In other words I want $ab = \frac{1}{2}AB$. Which is $\frac{1}{4}$ of A times . . . (mumbles; confused) . . . $\frac{1}{\sqrt{2}}$ times A times $\frac{1}{\sqrt{2}}$ times B.

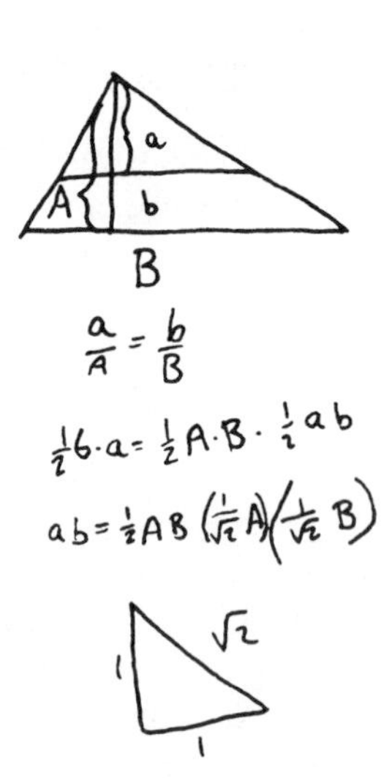

9. So if I can construct the $\sqrt{2}$, which I can! Then I should be able to draw this line . . . through a point which intersects an altitude dropped from the vertex. That's little $a = A/\sqrt{2}$, or $A = a$ times $\sqrt{2}$, either way.

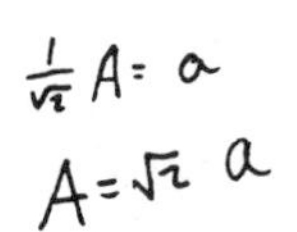

10. And I think I can do things like that because if I remember I take these 45 degree angle things and I go 1, 1, $\sqrt{2}$.
11. And if I want to have a times $\sqrt{2}$ . . . then I do that . . . mmm . . . wait a minute . . . I can try and figure out how to construct $\frac{1}{\sqrt{2}}$.
12. O.K. So I just got to remember how to make this construction. So I want to draw this line through this point and I want this animal to be . . . $\frac{1}{\sqrt{2}}$ times A. I know what A is, that's given. So all I

got to do is figure out how to multiply $\sqrt{\frac{1}{2}}$ times it.

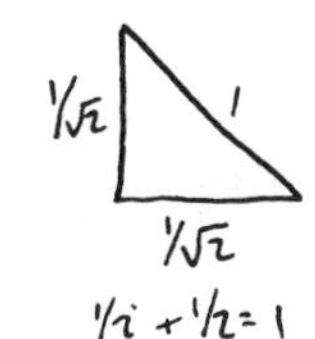

13. Let me think of it. Ah huh! Ah huh! Ah huh! $\frac{1}{\sqrt{2}}$ . . . let me see here . . . ummm . . . that's $\frac{1}{2}$ plus $\frac{1}{2}$ is 1 . . .
14. So of course if I have a hypotenuse of 1 . . .
15. Wait a minute: $\frac{1}{\sqrt{2}} \times \frac{\sqrt{2}}{\sqrt{2}} = \frac{\sqrt{2}}{2}$. . . that's dumb!

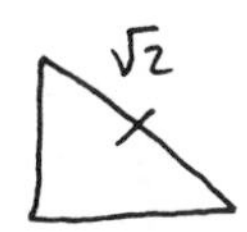

16. Yeah, so I construct $\sqrt{2}$ from a 45, 45 90. O.K. so that's an easier way. Right?
17. I bisect it. That gives me $\frac{\sqrt{2}}{2}$ I multiply it by A . . . now how did I used to do that?

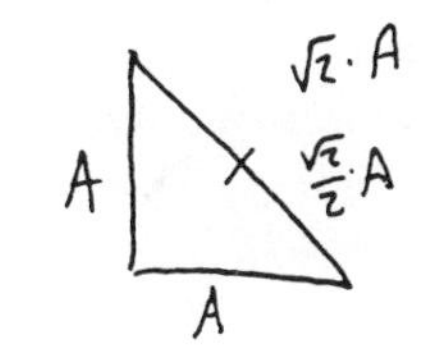

18. Oh heavens! How did we used to multiply times A. That . . . the best way to do that is to construct A . . . A . . . then we get $\sqrt{2}$ times A, and then we just bisect that and we get A times $\frac{\sqrt{2}}{2}$. O.K.
19. That will be . . . what! . . . mmm . . . that will be the length . . . now I drop a perpendicular from here to here. O.K. . . . and that will be . . . ta, ta . . . little a.

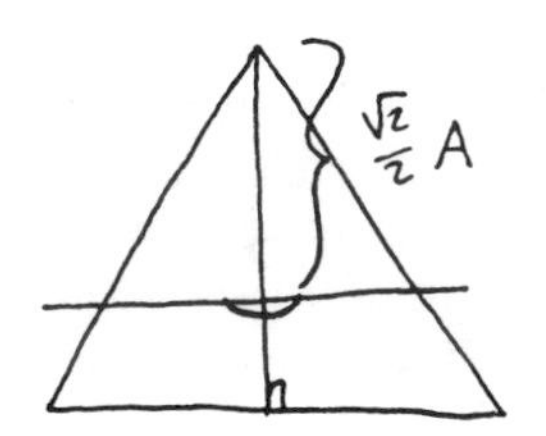

20. So that I will mark off little a as being A times $\frac{\sqrt{2}}{2}$. O.K. and automatically when I draw a line through that point . . . I'd better get $\frac{\sqrt{2}}{2}$ times big B. O.K.
21. And when I multiply those guys together I get $\frac{2}{4}AB$. So I get half the area . . . what? . . . yeah . . . times $\frac{1}{2}$ . . . so I get exactly $\frac{1}{2}$ the area in the top triangle so I better have half the area left in the bottom one. O.K.
22. O.K., now can I do it with 5 parts?
23. Assuming 4 lines.
24. Now this is going to be interesting since these lines are going to have to be graduated . . . that . . .
25. I think, I think, that rather than get a whole lot of triangles here, I think the idea, the essential question is can I slice off . . . $\frac{1}{5}$ of the area . . . mmm . . .
26. Now wait a minute! This is interesting. Let's get a . . . how about four lines instead of . . .

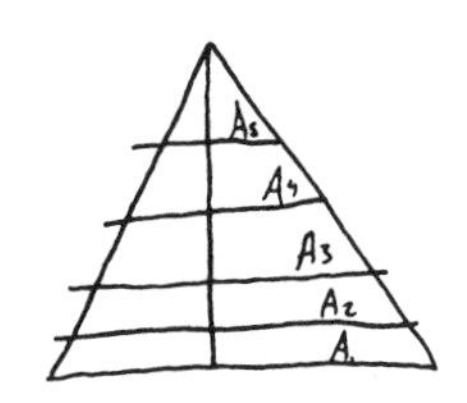

27. I want these to be . . . all equal areas . . . right? A_1, A_2,A_3, A_4, A_5, right?
28. Sneak! I can . . . I can do it for a power of 2 . . . that's easy because I can just do what I did at the beginning and keep slicing it in half all the time.
29. Now cán I use that kind of induction thought.
30. I want that to be $\frac{2}{5}$. And that to be $\frac{3}{5}$.
31. So let's make a little simpler one here.
32. If you could do that then you can construct $\sqrt{5}$. But I can construct $\sqrt{5}$ to 1 . . . square root of 5, right?
33. So I can construct . . . O.K. So that certainly isn't going to do it. No contradiction . . .
34. Now, I do want to see, therefore, what I have here.

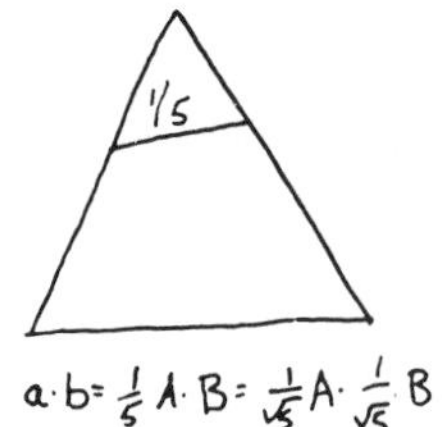

35. I'm essentially saying is it possible for me to construct it in such a way that that is 1, 2, 3, 4, 5, $\frac{1}{5}$ the area . . . O.K.
36. So little a times little b has got to equal $\frac{1}{5}AB$. So I can certainly chop the top piece off and have it be $\frac{1}{5}$ of the area. Right? Right?

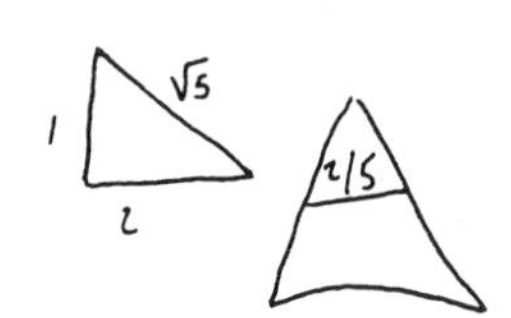

37. Now, from the first part of the problem . . . I know the ratio of the next base to draw . . . because it is going to be $\sqrt{2}$ times this base. So I can certainly chop off the top $\frac{2}{5}$.
38. Now, from the first part of the problem I know the ratio of the top . . . uh, O.K. now this is $\frac{2}{5}$ here, so top $\frac{4}{5}$ . . . O.K. . . . all right . . . so all I got to be able to do is chop off the top $\frac{3}{5}$ and I'm done . . .

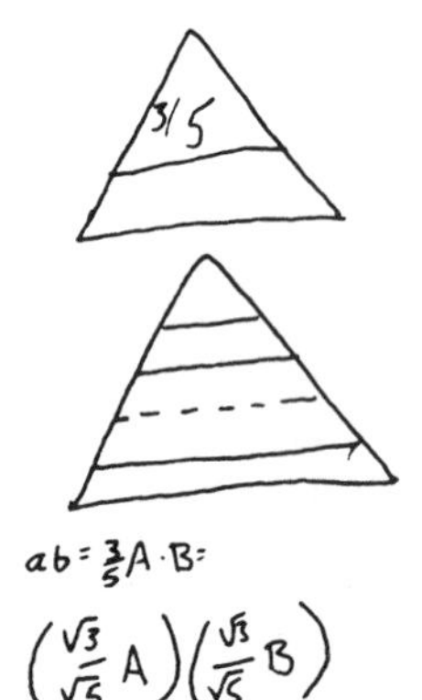

39. It would seem now that it seems more possible . . . let's see . . .
40. We want to make a base here such that little a times little b is equal to . . . the area of this thing is going to be $\frac{3}{5}$ . . . $\frac{3}{5}AB$ . . . in areas, right! . . . and that means little a times little b is $(\frac{\sqrt{3}}{\sqrt{5}}A)(\frac{\sqrt{3}}{\sqrt{5}}B)$. O.K. then can I construct $\frac{\sqrt{3}}{\sqrt{5}}$. If so then this can be done in one shot.
41. Well let's see. Can I construct $(\frac{\sqrt{3}}{\sqrt{5}})$. That's the question. $\frac{\sqrt{3}}{\sqrt{5}} \times \frac{\sqrt{5}}{\sqrt{3}} = \frac{\sqrt{15}}{5}$.

42. $\sqrt{15}$, $\sqrt{15}$. Wait a minute! $\frac{\sqrt{15}}{5}$. Is $\sqrt{15}$ constructable? $\sqrt{15}$ is . . .

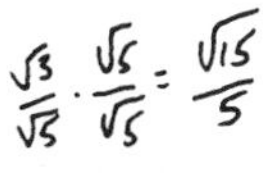

43. It is $\sqrt{16-1}$. But I don't like that. It doesn't seem the way to go.

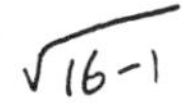

44. $16^2 - 1^2$ equals . . . (expletive deleted)
45. Somehow it rests on that.
46. (expletive) If I can do $\sqrt{15}$. Can I divide things and get this?
47. Yeah, there is a trick! What you do is you lay off 5 things. 1, 2, 3, 4, 5. And then you draw these parallel lines by dividing them into fifths. So I can divide things into fifths so that's not a problem.

48. So it's just constructing $\sqrt{15}$ then I can answer the whole problem.
49. I got to think of a better way to construct $\sqrt{15}$ then what I'm thinking of . . . or I got to think of a way to convince myself that I can't . . .- ummm . . . $x^2 - 15$.

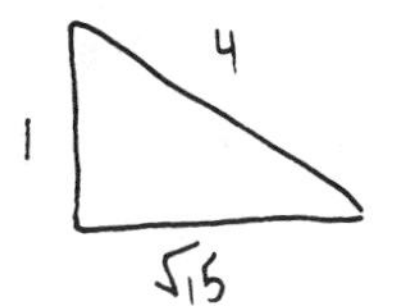

50. Trying to remember my algebra to knock this off with a sledgehammer.
51. It's been so many years since I taught that course. It's 5 years. I can't remember it.
52. Wait a minute! Wait a minute!
53. I seem to have in my head somewhere a memory about quadratic extension.
54. Try it differently here. mmm . . .
55. So if I take a line of length one and a line of length . . . And I erect a perpendicular and swing a 16 (transcriber's note: for mathematical clarity he really means 4 instead of 16) here . . . then I'll get $\sqrt{15}$ here, won't I?
56. I'll have to, so that I can construct $\sqrt{15}$ times anything because I'll just multiply this by A and this by A and this gets multiplied by A divided by 5 using that trick. Which means that I should be able to construct this length and if I can construct this length then I can mark it off on here and I can draw this line and so I will answer the question as YES!!

Appendix 10.E:Protocol 5

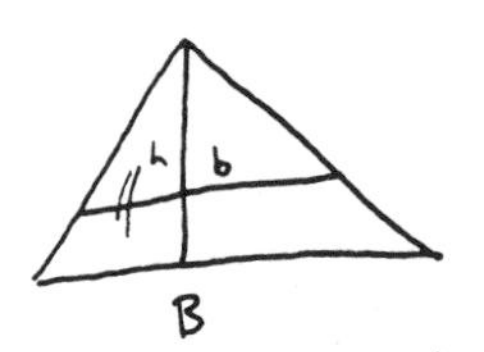

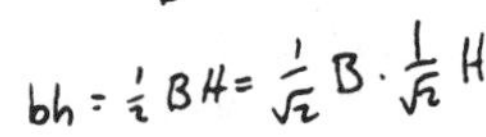

1. (Reads problem) Same as Protocol 4.
2. The first thought is that the two triangles for the first question will be similar.
3. And since we'll want the area to be one half. And area is related to the product of the altitude and the base we want the area of the smaller triangle to be one half.
4. And corresponding parts of similar triangles are proportional. We want the ratio of proportionality between the altitudes and the bases both to be $1/\sqrt{2}$.
5. So I will draw a diagram . . . and I'm drawing that parallel and checking that algebra.
6. I hope you can hear the pencil moving because that's what's happening at this point.
7. And now I'm writing a bunch of letters on my diagram and multiplying them together . . . leaving the one half out, of course . . . and I want that to be one half of that.
8. So, that certainly seems like a reasonable solution. So all I have to be able to do is construct $\sqrt{2}$. And I can do that with a 45 right triangle, and then given a certain length, namely the altitude, to the base B, which I can find by dropping a perpendicular. I want to construct a length which is $\sqrt{\frac{1}{2}}$ times that, and I can do that with the ordinary construction for multiplication of numbers.
9. So, I can do the problem.
   I: You can do all the constructions?
10. Yeah, I do them in the winter term. This line, this line, here's one, you want to multiply P times q, you draw these parallels and it's pq. (The solution of Part 2 is omitted)

Reference Notes

1. Kantowski, M. G. *The use of heuristics in problem solving: An exploratory study* (Appendix D, Final Tech. Rep. NSF project SED 77–18543).

2. Neches, R. *Promoting self-discovery of improved strategies.* C.I.P. working Paper No. 398, 1979, Department of Psychology, Carnegie–Mellon University.

References

Anzai, Y., & Simon, H. A. The theory of learning by doing. *Psychological Review,* 1979, *86*(2), 124–140.

Bobrow, D., & Collins, A. *Representation and understanding.* New York: Academic Press, 1975.

Brown, A. Knowing when, where, and how to remember: A problem of meta-cognition. In R. Glaser (Ed.), *Advances in instructional psychology* (Vol. 1). Hillsdale, New Jersey: Erlbaum, 1978.

Ericsson, K. A., & Simon, H. A. Verbal reports as data. *Psychological Review,* 1980, *87*(3), 215–251.

Kulm, G., Campbell, P., Frank, M., & Talsma, G. *Analysis and synthesis of mathematical problem solving processes.* Paper presented at the Annual Meeting of the National Council of Teachers of Mathematics, St. Louis, April, 1981.

Larkin, J., McDermott, J., Simon, D., & Simon, H. Expert and novice performance in solving physics problems. *Science,* 1980, *208,* 1335–1342.

Lucas, J. F., Branca, N., Goldberg, D., Kantowski, M. G., Kellogg, H., & Smith, J. P. A process-sequence coding system for behavioral analysis of mathematical problem solving. In G. A. Goldin & C. E. McClintock (Eds.), *Task variables to mathematical problem solving.* Columbus, Ohio: ERIC/SMEAC, 1979.

McDermott, J., & Forgy, C. Production system conflict resolution strategies. In D. A. Waterman & F. Hayes (Eds.), *Pattern-directed inference systems.* New York: Academic Press, 1978.

Neves, D. M. A computer program that learns algebraic procedures by examining examples and by working test problems in a textbook. In *Proceedings of the Second Annual Conference of the Canadian Society for Computational Studies of Intelligence,* Toronto, 1978.

Newell, A. *On the analysis of human problem solving protocols.* Paper presented at the International Symposium on Mathematical and Computational Methods in the Social Sciences, Rome, July 1966.

Polya, G. *Mathematical discovery* (Vol. 2). New York: Wiley, 1965.

Schoenfeld, A. H. Teaching problem solving in college mathematics: The elements of a theory and a report on the teaching of general mathematical problem-solving skills. In R. Lesh, D. Mierkiewicz & M. Kantowski (Eds.), *Applied mathematical problem solving.* Columbus, Ohio: ERIC/-SMEAC, 1979.

Schoenfeld, A. H. Teaching problem-solving skills. *American Mathematical Monthly,* 1980, *87*(10), 794–805.

Schoenfeld, A. H. Measures of problem solving performance and of problem solving instruction. *Journal for Research in Mathematics Education,* 1982, *13*(1), 31–49.

Simon, D., & Simon, H. Individual differences in solving physics problems. In R. Siegler (Ed.), *Children's thinking: What develops?* Hillsdale, New Jersey: LEA, 1978.

Simon, H. Problem solving and education. In D. T. Tuma & F. Relf (Eds.), *Problem solving and education: Issues in teaching and research.* Hillsdale, New Jersey: Erlbaum, 1980.

Stewart, P. *Jacobellis* v. *Ohio.* Decision of United States Supreme Court, 1964.

Author Index

The numbers in italics indicate the pages on which the complete references appear.

A

Abramowitz, S, 47, 48, *86*
Adams, V. M., 239, 253, *261*
Adi, H., 47, 63, 75, *86*, *88*
Adler, M. J., *226*
Ames, L. B., 14, 19, *42*
Anderson, B., 206, 226, *226*
Angelo, S., 184, *201*
Anick, C. M., 13, 26, 27, 28, *40*
Anzai, Y., 350, *395*
Appel, M., 81, *88*
Arnett, L. D., 18, *40*
Asher, J. J., 231, *258*
Asso, D., 190, *200*
Ausubel, D. P., 179, *200*

B

Banerji, R., 236, *261*
Bang, V., 46, *89*, 162, *173*
Banwell, C. S., 175, *200*
Barakat, M. K., 181, *200*
Barnett, J., 234, *258*
Bart, W. M., 63, *86*
Beardsley, L. M., 91, *125*
Beattle, I. D., 10, *40*
Behr, M., 121, *124*, 263, 307, *342*
Behrens, M. S., 9, 18, *42*
Bell, M. S., 251, *258*
Bender, L., 188, *200*
Benhadj, J., 169, *174*
Bentley, W., 63, *86*
Bernoff, R., 81, *88*
Berry, J. W., 195, *200*
Bishop, A. J., 180, 184, 185, 186, 197, *200*
Blade, M. F., 185, *200*
Blaut, J. M., 178, 194, *200*
Blume, G., 13, 26, 33, 37, *40*
Bobrow, D., 370, *395*
Boelke, W. W., 80, *88*
Bolduc, E. J., Jr., 10, *40*
Bradbard, D. A., 230, *259*
Branca, N. A., 233, 237, 239, *260*, *261*, 348, 369, *395*
Brennan, M. L., 233, *261*
Brian, D., 9, *44*
Briars, D. J., 35, *40*
Brinkmann, E. H., 186, *200*
Bronfenbrenner, U., 240, *258*
Brousseau, G., 161, *173*
Brown, A., *258*, 369, *395*
Brown, J. S., 17, 19, *40*, 237, 239, 252, *258*
Brown, S. I., 226, *226*
Browne, C. E., 9, *40*
Brownell, W. A., 9, 11, 14, 19, *40*, 232, *258*
Bruner, J. S., 101, *124*

Bundy, A., 256, *258*
Burger, W. F., 212, *226*
Burns, P. C., 28, *43*
Burtis, J., 64, *89*
Burton, R. R., 237, 239, 252, *258*
Buswell, G. T., 7, 9, *40*
Butterworth, I. P., 46, *88*

C

Caldwell, J., 232, *258*
Campbell, P. F., 233, 237, *260*, 348, *395*
Cardone, I. P., 266, *343*
Carpenter, T. P., 8, 10, 13, 15, 18, 19, 22, 23, 24, 28, 29, 32, 35, 37, 38, 39, *40*, *41*, *42*, 91, 107, *124*, *125*, 256, *258*
Carper, D. V., 9, *40*
Case, R., 39, *41*, 50, 72, 73, 74, 75, *86*
Charles, R. I., 239, 247, 253, *258*
Chase, W. G., 250, *258*
Chi, M., 236, *257*
Clement, C. A., 65, *89*
Clements, M. A., 181, 183, 186, 199, *202*
Cloutier, R., 63, 75, *88*
Cobrun, T. G., 91, 107, *124*, *125*
Cohen, J., 68, *88*
Cohen, L., 186, *201*
Cohen, P., 68, *88*
Collea, F. P., 63, *89*
Collins, A., 370, *395*
Collis, K. A., 39, *43*
Connor, J. M., 185, *201*
Corbitt, M. K., 91, *125*, 256, *258*
Cox, M. V., 177, *201*
Coxford, A. F., 121, *125*, 212, *226*
Cronbach, L. J., 7, *41*

D

Das, J. P., 250, *258*
Davidov, V. V., 17, *41*
Davis, G., 231, *258*
Day, M. C., 72, *89*
de Avila, E. A., 64, *88*
de Benedictis, T., 65, *89*
DeGuire, L., 250, *258*
de Ribaupierre, A., 63, 72, 74, *88*
Deichmann, J. W., 10, *40*
Deregowski, J. B., 185, 189, *201*
Dienes, Z. P., 100, 101, 123, *125*, 206, 221, 225, *226*
Docherty, E. M., 63, *88*
Dodson, J. W., 235, *258*
Donaldson, M., 177, *201*
Douady, R., 161, *173*
Drauden, G. M., 185, *201*
Duncan, C. P., 231, *258*
Dussouet, A., 169, *174*

E

Eakin, J., 81, *88*
Eggleston, V., 13, *43*
Eliot, J., 177, *201*
Ellerbruch, L. W., 94, *125*
Ericsson, K. A., 238, *259*, 351, *395*
Ernst, G. W., 236, *261*
Errecalde, P., 169, *174*

F

Feltovich, P. J., 236, *257*
Fennema, E. H., 63, 65, 75, *88*, 108, *125*
Fielker, D. S., 181, 200, *201*
Fischbein, E., 180, *201*
Fishbein, H. D., 182, *201*
Flavell, J. H., 11, *41*, 72, *88*, 239, *259*
Forgy, C., 350, *395*
Formisano, M., 95, *125*
Frank, M., 233, 237, *260*, 348, *395*
Freeman, M., 185, *201*
Freudenthal, H., 48, *88*, 148, 165, *173*, 207, 210, *227*
Fruchter, B., 181, *202*
Furman, I., 50, 63, 74, *88*
Fuson, K. C., 13, 17, 20, 31, 39, *41*, *43*, 177, 180, *201*

G

Gagne, L., 73, *89*
Gagne, R. M., 104, *125*
Gallistel, C. R., 13, *41*
Ganson, R. E., 96, 97, *125*
Garofalo, J., 250, *259*, *260*
Geddes, D., 213, *226*
Geeslin, W. E., 178, *201*
Gelman, R., 12, 13, *41*, *43*
Genkins, E. F., 180, *201*
Gerling, M., 108, *125*
Giacobbe, J., 142, 160, *174*
Gibb, E., 10, 14, 15, 19, 28, *41*
Ginbayashi, H., 17, *41*

Ginsburg, H., 19, *41*
Glaser, R., 236, *257*
Glocker, D., 224, *227*
Gold, A. P., 80, 82, *88*
Goldberg, D. J., 233, 237, *259*, *260*, 348, 369, *395*
Goldin, G. A., 231, 232, 233, 234, 235, *259*
Golding, G. A., 232, *258*
Goldschmid, M. L., 63, 75, *88*
Goodenough, D., 64
Goodnow, J., 180, *201*
Goodstein, M. P., 80, *88*
Gorman, C. J., 233, *259*
Greeno, J. G., 15, 19, 26, 27, 35, 38, *43*, 231, 256, *259*
Greenwood, J., 206, 226, *226*
Grize, J. B., 46, *89*, 162, *173*
Groen, G., 18, 20, *44*
Groen, G. J., 18, 19, 20, 21, 33, *41*, *44*
Grouws, D. A., 10, 15, *41*
Guay, R. B., 184, *201*
Guilford, J. P., 181, *202*
Gunderson, A. G., 122, *125*
Gunderson, E., 122, *125*

H

Hadamard, J., 183, *201*
Hall, J. W., 13, 31, *41*, *43*
Hamilton, E., 102, *126*, 268, 307, *343*
Hamrick, A. K., 121, *125*
Hamza, M., 182, *201*
Harik, F., 232, 237, 248, 249, *259*
Harris, L. T., 183, 197, *201*
Harris, P., *201*
Hart, K., 47, 48, *88*, 149, 167, *173*
Hart, R. A., 179, *201*
Harvey, J. G., 17, 26, *43*
Hatano, G., 38, *41*
Hatfield, L. F., 230, *259*
Havassy, B., 64, *88*
Hawkins, P., 92, *125*
Hebbeler, K., 10, *41*
Heller, J. I., 15, 19, 26, 27, 35, 38, *43*
Hermann, D. J., 236, *257*
Herron, J. D., 80, *88*
Hess, P., 180, *201*
Hiebert, J., 13, 15, 26, 27, 29, 37, 38, *40*, *41*, *42*, 94, *125*
Higgins, J. L., 108, *126*
Hirstein, J. J., 10, 11, *42*, *43*
Hoffer, A., 211, 224, *227*
Horn, J. L., 64, *88*
Howe, A., 81, *88*
Hoy, C., 307, *343*
Hudson, T., 38, *42*
Hughes, F. P., 182, *203*
Hughes, M., 177, *201*
Hutko, P., 182, *203*

I

Ibarra, C. G., 10, 15, 19, 27, *42*
Ilg, F., 14, 19, *42*
Inhelder, B., 46, 47, 49, 50, 63, 65, 71, 75, *88*, *89*, 94, *126*, 176, 182, *202*

J

Jacobson, M., 244, 253, *258*
Jarman, R. F., 250, *258*
Jerman, M. E., 9, 11, *42*, *44*, 234, *259*
Johnson, D. C., 10, *43*
Jones, J., 195, 197, *201*
Judd, H., 7, 9, *40*

K

Kantowski, M. G., 1, *6*, 230, 231, 233, 237, 239, 240, 251, *259*, *260*, 348, 369, *394*, *395*
Karplus, E. F., 47, 63, 79, *88*, 95, *125*
Karplus, R., 47, 48, 50, 63, 73, 75, 79, 80, 81, *88*, 95, 107, *125*, 162, *173*
Katriel, T., 15, *43*
Kearins, J., 191, *201*
Keating, D. P., 63, *88*
Keiffer, K., 182, *201*
Kellogg, H., 233, 237, *260*, 348, 369, *395*
Kelly, G. A., 197, *202*
Kennedy, J. M., 193, *202*
Kepner, H., 256, *258*
Kepner, H. S., Jr., 91, *125*
Kerr, D. R., 241, *257*
Kester, F. K., 230, 233, 234, 238, *260*
Kieren, T. E., 47, *88*, 92, 93, 96, 97, 99, 101, 107, 108, 117, 122, *125*, 165, *173*, 288, 301, *343*
Kilpatrick, J., 209, *227*, 230, 231, 233, 237, 240, *259*, *260*
Kirby, J., 250, *258*

Kirsch, A., 169, *173*
Klahr, D., 5, *6*, 13, *42*, 107, *125*
Knight, F. B., 9, 18, *42*
Kogan, N., 65
Kouba, V., *42*
Krutetskii, V. A., 182, 199, *202*, 231, 232, 233, 235, *260*
Kuchemann, D. E., 179, *202*
Kulm, G., 233, 234, 237, *260*, 348, *395*
Kurtz, B., 49, 80, 81, 82, *88*, 95, *125*
Kyllonen, P., 64, *88*

L

Lancy, D. F., *202*
Landau, M., 102, *126*, 268, 297, 307, *343*
Lankford, F. G., 31, *42*
Lankford, F. G., Jr., 91, *126*
Larkin, J. H., 35, *40*, 236, 252, *260*, 369, *395*
Laurendeau, M., 178, *202*
Lawson, A. E., 47, 63, 75, 80, 81, *88*
LeBlanc, J. F., 10, 15, *42*, 241, 244, 253, *257*, *258*
Leach, M. L., 185, *202*
Lean, G. A., 183, 185, 199, *202*
Lesh, R., 99, 101, 102, 106, *126*, 177, *202*, 225, *227*, 230, 251, *260*, 261, 263, 266, 268, 269, 270, 307, *342*, *343*
Lester, F. K., 244, 247, 253, *258*
Leutzinger, L. D., 32, *42*
Lewis, D., 197, *202*
Lewis, S., 182, *201*
Lindquist, M. M., 91, *125*, 256, *258*
Lindvall, C. M., 10, 15, 19, 27, *42*
Linn, M. C., 64, 65, *88*, *89*
Loftus, E. F., 11, *44*
Lovell, K., 46, *88*
Lucas, J. F., 231, 233, 237, *260*, 348, 369, *395*
Luger, G. F., 234, *260*
Lunzer, A. E., 46, *88*
Luria, A. R., 250, *260*
Lybeck, L., 49, 74, *88*, 148, 162, 165, *173*

M

McClintock, C. E., 230, 234, 250, *259*, *260*
McCloskey, P., 185, *202*
McDaniel, E. D., 184, *201*
McDermott, J., 236, 252, *260*, 350, 369, *395*
MacFarlane, Smith I., 181, 183, 184, *202*
McGee, M. G., 181, *202*
MacLane, S., 219, 221, 222, *227*
MacLatchy, J. H., 10, *42*
Manes, E., 222, *227*
Marshall, G. G., 10, *42*
Marthe, P., 140, 142, 160, *174*
Martin, J. L., 177, *202*, 221, *227*
Martorano, S. C., 63, *88*
Matz, M., 237, 252, *257*
Metregiste, R., 140, 142, 160, *174*
Meyer, R. A., 232, *260*
Michael, W. P., 181, *202*
Mierkiewicz, D., 230, 251, *260*, 266, 268, *343*
Mirman, S., 11, *42*
Mitchelmore, M. C., 178, 179, 180, 194, 197, *202*
Montgomery, M. E., 17, 26, *43*
Moore, G. T., 179, *201*
Morningstar, M., 18, *44*
Moser, J., 13, *42*
Moser, J. M., 10, 15, 17, 18, 19, 23, 24, 26, 28, 29, 32, 35, 37, 38, *40*, *41*, *43*
Moses, B. E., 183, 186, *202*, 247, *260*
Murray, C., 177, 180, *201*
Murray, J. E., 9, *42*, 181, *202*

N

Neches, R., 350, *394*
Nelson, D., 47, *88*, 96, 97, *125*, *173*
Nesher, P., 15, 28, *42*, *43*
Neves, D. M., 350, *395*
Newell, A., 349, *395*
Noelting, G., 47, 48, 49, 50, 51, 52, 73, 74, 75, *89*, 95, 96, 97, 107, *126*, 148, 162, 165, 166, 167, *173*
Novak, J. D., 179, *202*
Novillis, C., 92, *126*
Novillis-Larson, C., 94, 113, 118, *126*
Nummedal, S. G., 63, *89*
Nussbaum, J., 179, *202*

O

Oehmke, T., 233, 240, 255, *257*
Oltman, P. K., 64, *89*
Owens, D. T., 93, *126*

P

Parkman, J. M., 18, 19, 20, 21, *41*

Pascual-Leone, J., 50, 63, 64, 72, 74, 82, *88*, *89*
Paulsen, A. C., 95, *125*
Payne, J. N., 91, 94, *125*
Perham, F., 177, *202*
Peterson, R. W., 47, 48, *88*, 162, *173*
Piaget, J., 11, 12, 13, *43*, 46, 47, 49, 50, 63, 65, 71, 75, *88*, *89*, 94, *126*, 162, *173*, 176, 182, *202*
Pimm, D., 13, *40*
Pinard, A., 178, *202*
Pinchback, Carolyn L., 121, *126*
Polkinghorne, A. R., 93, *126*
Poll, M., 18, 33, *41*
Polya, G., 5, *6*, 206, *227*, 239, 255, *260*, 349, *395*
Post, T. R., 99, 100, *126*, 233, *261*, 263, 307, *342*
Pottle, H. L., 9, *43*
Prigge, G. E., 180, *203*
Proudfit, L., 233, 240, 247, 248, 255, *257*, *261*
Pulos, S., 63, 64, 65, *86*, *89*, 107, *125*
Pumfrey, P. D., 46, *88*
Putt, I. J., 233, 237, 239, 247, *261*
Pyshkalo, A. M., 209, 210, *227*

R

Rappaport, D., 92, *126*
Raskin, E., 64, *89*
Rathmell, E. C., 32, *43*
Reed, S. K., 236, *261*
Rees, R., 11, *42*, 234, *259*
Resnick, L. B., 15, 17, 18, 19, 27, 32, *41*, *43*, *44*
Reys, R. E., 91, 100, 107, *124*, *125*, 256, *258*
Rhonheimer, M., 47, 49, *89*
Ricco, G., 140, 142, 160, *173*, *174*
Richards, J., 18, 31, 39, *44*
Richardson, A., 183, *203*
Riess, A. P., 92, *126*
Riley, M. S., 15, 19, 26, 27, 35, 38, *43*
Robinson, E., 175, *203*
Robinson, M., 19, 37, 39, *43*
Rogalski, J., 154, *174*
Romberg, T. A., 17, 26, 39, *43*, 233, 240, 255, *257*
Rosenthal, D. J. A., 15, 18, 27, *43*
Ross, A. S., 193, *202*
Rosskopf, M. F., 177, *203*
Rouchier, A., 140, 142, 160, 169, 170, 173, *174*
Rupley, W. H., 47, 60, 73, 79, 82, *89*
Rusch, J. J., 81, *88*

S

Salim, M., 167, *174*
Sambo, Abdussalami A., 93, *126*
Saunders, K. D., 175, *200*
Schaefer, R. A., 63, *88*
Schaeffer, B., 13, *43*
Schell, L. M., 28, *43*
Schoen, H. L., 233, 240, 255, *257*
Schoenfeld, A. H., 233, 236, 239, 244, 253, *257*, *261*, 349, 367, *395*
Schultz, K., 178, *203*
Scott, J. L., 13, *43*
Secada, W. G., 31, *43*
Serbin, L. A., 185, *201*
Shar, A. O., 178, *201*
Sherman, J. A., 63, 65, 75, *88*
Shores, J. H., 10, 28, *43*
Shumway, R. J., 175, *203*
Siegler, R. S., 5, *6*, 19, 37, 39, *43*, 46, *89*, 107, *125*
Silver, E. A., 232, 236, 239, 251, 253, *261*
Simon, D. P., 146, 236, 252, *260*, *261*, 369, *395*
Simon, H. A., 146, 236, 238, 250, 252, *258*, *259*, *260*, *261*, 350, 351, 369, *395*
Smith, J. H., 18, *43*
Smith, J. P., 233, 237, *260*, 348, 369, *395*
Smock, C. D., 176, *203*
Smothergill, D. W., 182, *203*
Southwell, B., 96, 97, *125*, *173*, 288, *343*
Spearman, C. E., 181, *203*
Spikes, W. C., 11, *43*
Stage, E. K., 107, *125*
Starkey, P., 12, *43*
Stea, D., 178, 194, *200*
Steffe, L. P., 10, 11, 15, 18, 31, 32, 39, *43*, *44*
Stengel, A., 244, 253, *258*
Sternberg, R. J., 250, *261*
Stewart, P., 357, *395*
Stone, C. A., 72, *89*
Suarez, A., 47, 49, 50, 75, 79, *89*
Sullivan, F., 81, *88*
Sullivan, P. A., 65, *89*
Suppes, P., 7, 9, 11, 18, 20, *41*, *44*
Suydam, M. N., 108, *126*, 167, *174*, 230, *261*

Swenson, E. J., 32, *44*
Szeminska, A., 46, *89*, 94, *126*, 162, *173*, 176, *202*

T

Tahta, D. G., 175, *200*
Talsma, G., 233, 237, *260*, 348, *395*
Taylor, C. C., 183, *203*
Theile, C. L., 32, *44*
Thier, H. D., 83, *89*
Thompson, D. W., 18, 31, 39, *44*
Thornton, C. A., 32, *44*
Thurstone, L. L., 64, *89*, 181, *203*
Timmons, S. A., 182, *203*
Tirosh, D., 180, *201*
Tolman, E. C., 178, *203*
Tonnessen, L. H., 94, *125*
Torgerson, W. S., 61, *90*
Tulving, E., 105, *126*

U

Underhill, R. G., 10, 28, *43*
Usiskin, Z. P., 92, *126*, 213, 214, *226*, 266, *343*

V

van Hiele, P. M., 205, 207, 208, 209, 211, 224, *226*, *227*
van Hiele-Geldof, D., 207, *227*
Van Lehn, K., 17, 19, *40*
Vergnaud, G., 8, 16, *44*, 74, *90*, 128, 140, 142, 160, 169, *174*
Vladimirskii, G. A., 185, *203*
Vollrath, H. J., 177, 179, 197, *203*
Vos, K., *261*
Vygotsky, L. S., 4, *6*, 99, *126*

W

Wagman, H. G., 178, *203*
Wallace, J. G., 13, *42*
Wallach, M., 65, *90*
Walter, M., 226, *226*
Waters, W., 234, *261*
Watson, W. S., 185, *200*
Wearne, D., 233, 240, 255, *257*
Weaver, J. F., 10, 20, *44*
Webb, N., 250, *259*
Webb, N. L., 231, 233, 234, 236, 237, *257*, *261*
Weil, E. M., 250, *261*
Weinzweig, A. I., 221, *227*
Werdelin, I., *203*
Werner, H., 179, *203*
Wheatley, G. H., 80, *88*
Wheeler, N., 211, *226*
White, R. T., 104, *125*
Willis, R., 83, *89*
Wilson, J. W., 91, 107, *124*
Winch, W. H., 45, *90*
Wirszup, I., 205, 209, *227*, 231, 240, *260*
Witkin, H., 64, *90*
Witkin, H. A., 64, *89*
Wohlwill, J., 72, *88*
Wolfe, L. R., 186, *203*
Wollman, W., 47, 80, *88*, *90*, 95, *125*
Wollman, W. T., 63, *88*
Wood, R., 183, *203*
Wood, S., 108, *125*
Woods, S. S., 18, 19, 33, *44*
Wrigley, J., 181, *203*
Wyke, N., 190, *200*

Z

Zimmerman, W. S., 181, *202*
Zykova, V. I., 180, 181, *203*

Subject Index

A

Ability, 64, 235, 242
  cognitive, 250
  mathematical, 181
  spatial, 177, 181, 184, 247
Abstract, 220
Abstraction, 30–34, 123, 222; levels of, 32, 34
Achievement, 235
Addition, 28, 180
Addition and subtraction, 7–39
  Change, 15, 16
  Combine, 15, 16
  Compare, 15, 16
  Equalize, 15, 16
  Joining, 12, 13, 15, 16
  Separate, 15–16
Adjustment, 53, 58, 59, 72
Adolescent, 45–85
Affective, 243
Age-level
  college, 81, 217, 239, 350–351
  elementary school, 81, 168, 209–210, 211, 215, 225, 241
  grade level differences, 59, 66, 67, 71
  high school, 80–81, 211, 214–217
  junior high school, 80, 211
  middle school, 51, 96, 215, 219, 239, 241, 307
Algebra, 80, 93
Algorithm(s), 97, 236, 280, 281, 345
American educational system, 224
Anxiety, 191, 193
Approximation, *see* Estimation
Area, 93–94, 128, 135, 137, 151, 159, 177, 178, 211, 276, 291, 308; lateral, 158
Arithmetic, 7–39, 243, 252, 267
Artificial intelligence, 256, 346, 347, 348, 349, 350, 369
Assessment, 240, 254–255, 257, 297–306, 307, 351, 352, 353, 355, 357, 360
Attitude, 63, 65, 70, 71, 72, 75, 79, 83, 235, 238; correlates of, 62–72
Attribute, 215–217
Ausubel, 179
Axiom, 217, 222

B

Basic skill, 211, 230, 266
Binary Law of Composition, 129–130
Brooklyn Project, 213, 214, 218, 224
Brownell, 7, 9

C

Calculation, 149–153
Calculator, 84
Cartesian product, 134, 135, 138
Chemistry, 80, 220
Chicago Project, 213, 214, 215, 218, 224
Class inclusion, 12, 13
Coefficient, 160

Cognitive adaptability, 64, 79
Cognitive characteristics, 3
Cognitive conflict, 158
Cognitive correlates, 62–72
Cognitive development, 213
Cognitive disequilibrium, 124
Cognitive elements, 60–63, 67–68, 70–74, 76, 78–79, 84–86
Cognitive functioning, 94, 109
Cognitive tasks, 63
Communication, levels of, 218
Comparison, 53, 56, 59, 61, 155
  problems, 48, 162
  tasks, 50
Computation, 280
Concept, 92, 172, 231, 264
  acquisition of, 266
  addition and subtraction, 30
  application, 81
  development, 100, 104, 169
  formation, 120, 265
  fraction, 297
  introduction, 81
  number, 14
  ratio, 297
  rational number, 91–124, 162, 219, 263–308
  spatial, 196
Conceptual field, 127–128
Conceptual model, 2, 263–308
  acquisition of, 306–308
  between-concept systems, 269–270
  within-concept networks, 268–269
Concrete materials, *see* Manipulative aids
Conservation, 12, 13, 116, 222–223
Consumption, 143
Continuous Embodiment Task, 114–116
Coordination, 177
Counting, 18, 25, 38
  process, 291
  strategies, 32, 35
  systems, 195
Cultural amplifier, 4
Cuisenaire rod, 103
Curriculum, 92–98, 218

D

Decalage, 3, 168; horizontal, 92
Decimal, 84, 100, 267
Decision making, 345–394
  strategic, 345–347, 349, 350, 352, 353–360, 365, 367, 369, 370
  tactical, 345–348, 350, 353, 354, 365, 369
Deductive system, 207, 211, 217
Development, 179, 199
  cognitive, 213
  conceptual, 100
  geometric, 210
  spatial, 198
Developmental level, 256
Developmental theories, 75, 197
Diagram, 180–181, 187–189, 192, 193, 198, 199, 297, 299–301
Didactic experiment, 154, 207, 240, 244–246
Dienes, 221
Discrete Embodiment Task, 116–117
Discrimination, 120
Distance, 138
Divergent thinking, 64, 71, 238
Division, 130–132, 139–140, 143, 150, 180, 269; indicated, 95
Drawing; *see* Diagram

E

Early number-research, 3
Economics, 220
Educational background, 218
Egocentrism, 182
Elementary school level, 81, 168, 209–210, 211, 215, 225, 241
Embedded figures, 192
Environment, 198, 244; factor, 253
Episode, 347, 354–369
  analysis, 356, 357–358, 366
  exploration, 356, 358–359, 361, 365, 366, 367
  implementation, 356, 359–360, 363, 364, 367
  local assessment, 359, 363, 366, 369, 370
  new information, 359, 363
  planning, 356, 359–360, 361, 364
  reading, 356, 357, 361, 365, 366, 367
  transition, 357, 358, 360, 361, 363, 364
  verification, 357, 360, 364, 366
Equivalence, 109, 223; class, 97
Error analysis, 19, 107, 237, 280, 284, 287–288, 289, 292; patterns, 19, 219
Estimation, 286, 296, 301
Executive decisions, *see* Decision making

F

Field dependence–independence, 64
Flexibility, 33, 34, 114, 115, 248
Formula, 149, 150, 157, 159
Fraction, 47, 70, 76, 79, 80, 83, 84, 113, 148, 160–171, 183, 219, 223, 308
  addition of, 103, 165, 273, 274, 278–291, 295–297
  concept, 91–124
  equivalence of, 96
  function operators, 164–167
  measurement, 99
  model, 94–95, 101
  multiplication of, 165, 273, 277–278, 279, 291–295, 295–297, 303
  operations of, 302, 328–336
  part–part, 288, 289, 290
  part–whole, 93–94, 97, 99, 118, 288, 289, 290, 296, 301
  ratio, 288, 289
  scalar operator, 162–164, 165–169
  subtraction of, 165
  unit, 120, 122–123, 162–164, 168
Functions, 46

G

Gender-related difference, 59, 63, 66, 70, 75, 183, 235
Geography, 178, 179
Geometry, 175, 207–226, 267, 269, 347
  abilities of, 181–186, 198–199
  concept, 176–181, 194–195
  education, 175–176
  Euclidean, 175, 177
  idealization, 222
  invariance, 222
  problem, 361–395
  pseudoequivalence, 223
  transformation, 175, 177, 211, 220–221
Gestaltist, 184
Gravity, 179
Guttman scale, 61, 67

H

Height, 137, 138, 159
Heuristic, 236–237, 238–239, 241, 242, 253, 345, 346, 347, 348, 349, 365
  looking back, 255–257
  trial and error, 27, 131

I

Idea
  analysis, 3, 5, 104, 265–266, 307
  mathematical, 109
  spatial, 197
Implementation decisions, *see* Decision making
Impulsivity, 243
Individual difference, 39, 182
Information processing, 39, 63, 71, 74, 79, 250
Insight, 205
Instruction, 4, 5, 32, 37, 98, 104, 109, 119–120, 154–160, 165, 172–173, 179, 184, 186, 196–197, 199, 205–206, 213, 214, 218, 224, 238–239, 241, 244–247, 253, 254, 257, 302
  intervention, 1
  research, 2
Integer, 267
Integration–differentiation, 93, 154, 158
Intelligence, 64, 70, 71, 72, 83, 127, 250, 369
Interpreting Figural Information (IFI), 184–186, 193, 198–199
Intervention, 107
Interview, 18, 19, 51, 106, 149, 213, 218, 224, 231, 241, 264, 272–298, 307
  clinical, 2
  protocol, 347–394
Intuition, 180
Inverse relationship, 131, 132
Isomorphism of measure, 50, 129–133, 136–138, 140–142, 148, 161, 305
Iterative strategy, 46
  subtraction, 21, 23–26, 30, 33–34, 36
    Adding On, 22
    Choice, 22
    Counting Down From, 21
    Counting Down To, 21
    Counting Up, 22
    Matching, 22
    Separation From, 21
    Separation To, 21

K

Knowledge, 127, 154, 198; mathematical, 93
Krutetskii, 182–183, 199, 235–236

L

Language, 120, 185, 189, 190, 193, 195, 198, 215, 264, 271, 293, 294–295, 296, 302; written, 299–301, 307
Learning,
cycle, 221
directed orientation, 206, 208
expliciting, 206, 208
free orientation, 206, 208
inquiry, 206, 208
integration, 206, 208
phases of, 206, 225
Linear function, 133, 140, 169–171, 172
Logic, 217, 220
Logical deduction, 175
Logico-mathematical concept, 46
Longitudinal study, 23, 24, 37

M

Management decision, *see* Decision making
Managerial behavior, 355,360
Managerial strategy, *see* Metacognition
Manipulative aid, 10, 19, 21, 23, 36, 82, 83, 101–103, 105, 108, 120, 121–122, 123, 150, 180, 238, 271, 273, 280–291, 293–296
Mass, 138
Mathematical Problem Solving Project, 239, 240–247
Mathematics education, 1, 230–231, 238, 250, 349, 369; research, 3, 121, 347
Meaningful learning, 179
Measurement, 149–154, 155, 164, 209, 267, 269, 308
Memory, 186, 192, 194, 207, 267, 291, 348
structure, 104–105
visual, 191
Metacognition, 239, 243–244, 253, 257, 346, 350, 367, 369, 370
Misconception, 237
Missing factor, 131
Missing-value problem, 48, 51
Model, 78, 82, 178, 236, 237, 265, 268, 271, 370
competency, 252–255
computer simulation, 35, 36
concrete, 101
performance, 252–255
simulation, 37
Modeling, 18, 20, 25, 27, 30, 38, 121, 198, 264, 270, 369; modeling strategies, 35
Morphism, 219, 220, 221, 222–223, 224, 225
Move, 249
Multiple regression analysis, 66–68
Multiplication, 129–130, 134–138, 139, 143, 180

N

National Assessment of Educational Progress, 256, 299
National Council of Supervisors of Mathematics, 230
National Council of Teachers of Mathematics, 241
National Science Foundation, 230, 241, 348
Noelting, 48, 50, 73, 74, 166–167
Number fact, 8, 19, 23, 25, 31, 32, 34; derived, 34
Number line, 94, 111, 112, 113, 117–119, 269, 308
Numerical relationship, 81, 82, 83

O

Observation, 106
Open Sentence, 10
Operation, 12, 13, 128, 130, 243, 267
Operator, 132, 136, 137, 349; rational number, 96, 100
Order, 109
Oregon Project, 212, 214, 224

P

Partitioning, 97, 109
Percent, 308
Perception, 182, 215–217
Perceptual cues, 108–120, 282
Perceptual distractor, 110, 111–120
Perceptual features, 123
Perspective, 177; multiple, 115
Physics, 80
Piaget, 11, 30, 46, 49, 73, 74, 92, 127, 176–177, 178, 182, 206
Piagetian research, 11–13, 46–47
Polya, 239, 255, 349
Postulate, 217, 221
Probability, 46, 183
Problem-solving, 127, 169–170, 172, 211, 306
applied, 251, 265–272, 270, 307
assessment of, 237

definition of, 231–232
environmental factors, 233, 238–239
expertise, 346–347, 349
factors of, 232, 236–238
instrumentation and research methodology, 233, 239–240
realistic situation, 251
research, 5, 229–257, 263–308
search-spaces, 249
solution process, 17, 22–29, 34, 117
solution stage, 267
strategy, 14
task factors, 232, 234–235, 235–236, 237, 248, 251, 266–268, 303
verbal, 35
word problem, 11, 16, 22–29, 36, 79, 81, 252, 264, 266–267, 272–273, 295, 303
Problem structure, 8–11, 14, 22–29, 36–37
Procedure, 142, 144, 145–149, 289
erroneous function, 147, 148
erroneous product, 147, 148
erroneous quotient, 147, 148
erroneous scalar, 147, 148
erroneous scalar and function, 147, 148
function, 142, 145, 148, 149, 156
function decomposition, 148
inverse, 147, 148
rule-of-three, 146, 148, 149
scalar, 142, 145, 148, 149, 156
scalar decomposition, 146, 148
unit value, 145, 148
Process, 278, 282, 287, 289, 292, 293
addition and subtraction, 17, 29
cognitive, 14, 17, 237, 243, 252, 256, 268
mental, 97
modeling of, 307
of solution, 17
thought, 216
visual (VP), 184, 194, 195, 198–199
Product of measures, 134–138
Proficiency, 347
Proof, 211, 213, 214, 217, 218
Property, 146, 171, 206, 207, 215, 216, 269, 270, 303
Proportion, 45, 79, 80, 95, 129, 131, 134, 135, 136, 138, 140, 159, 162, 165, 183, 297, 308
multiple, 138–142
within–between distinction, 48–50, 54–58, 60, 71, 73–75, 78, 82
Protocol, 360–364, 371–394
Psychogenetic approach, 128, 144

Q

Qualitative correspondence, 46
Qualitative reasoning, 58, 72
Quantitative notion, 122
Quantitative reasoning, 73
Quantitative skill, 13

R

Ratio, 45, 52, 61, 73, 76, 79, 95, 99, 148, 152, 153, 157, 158, 160–171 269, 296, 308
equal, 47, 56, 59
integer, 47, 56, 60, 75
integral, 82
noninteger, 60, 70, 75, 83
nonintegral, 47
unequal, 83
unit, 48, 73, 80, 82, 84
Rational number, 160–172, 267, 272–306; *see also* Fractions
assessment of, 309–343
linear coordinate, 100
quotient, 95, 99
rate, 99
subconstruct, 93–97, 99–100
Rational Number Project, 98, 105, 109
Rational-Number Test, 105
Real number, 221, 306
Reasoning, 78, 84, 216–218
additive, 72
formal, 65, 68, 71, 75
patterns, 58, 225
proportional, 5, 45–85
qualitative, 58, 72
Representation, 101–105, 119, 121, 123, 142, 149, 151–152, 169, 170, 178, 180, 182, 187, 193, 194, 195, 223, 252, 264, 265, 270, 271, 280, 281, 286–288, 291, 294–295, 296, 299, 300, 303, 304, 305, 307, 365; visual, 83
Research
methodologies, 1
paradigms, 8
Response categories, 54
additive, 54, 55, 57

illogical, 54, 56
incomplete, 54
proportional, 54, 55, 57
qualitative, 54, 56
Response latency, 18
Restructuring
adaptive, 48
cognitive, 64, 70–71, 79
Rule-of-three problem, 132–133, 144, 148

S

Schema, 99, 116, 357, 370
Schemata, 36, 347
Scheme, 63, 100, 220, 237, 239, 244, 348, 358, 369
Science, 187
Semantic information, 105
Semantic structure, 14, 15, 38
Sequence, 154, 205, 207, 214, 224
Seriation, 12, 13, 46
Sex-related differences; *see* Gender-related differences
Small group, 198
Soviet studies, 209–210, 223
Space, 149, 155, 175–200
Spatial orientation, 198
Speed, 46, 138
Stability, 142, 292
Stages, 91, 178
Stages of proportional reasoning, 48
Standardized question, 2
Static relationships, 15
Strategy, 37, 50, 75, 78, 79, 82, 286
addition, 19, 24–26, 30, 33–34, 36
Counting All, 19
Counting All with Models, 20
Counting On, 20, 30–31
keeping track procedure, 20
MIN strategy, 21
SUM strategy, 20
choice of, 33
counting, 32, 35
derived, 32
integral, 60
iterative, *see* Iterative strategy
Structure, 172, 208, 210, 220, 235, 236, 257, 268, 269, 350, 359
additive, 8, 128, 158
cognitive, 103
mathematical, 2
multiplicative, 127–173
Subtraction, 22, 180; *see also* Addition and subtraction
Symbol, 264, 272, 280, 283, 285, 286, 295, 299, 302, 303, 305, 307
Symmetry, 177
Syntax, 241, 242–243
Syntactic variable, 14
System, 269, 270; physical, 46

T

Task environment, *see* Problem solving; Task factors
Task variable, *see* Problem solving; Task factors
Teaching experiments, *see* Didactic experiments
Text, 219
Theorem, 217
Theory, 250; cognitive, 75
Thorndike, 7
Thought level, 205, 219, 224
Time, 138, 141
Topology, 177
Transfer, 186
Translation, 102, 103, 106, 119, 121, 264, 270, 271, 287, 295, 299, 300, 305, 306, 307

U

Unit, 156, 274, 290
cubic, 157
fraction, 282, 289, 290
liquid, 156
recognition of, 94, 109, 118

V

Van Hiele–Based Research, 205–226
Variable, 81
Vector-space, 171, 172
Vergnaud, 17
Visual cue, 112, 215
Visual images, 82
Visualization, 184
Vocabulary, 206; spatial, 198
Volume, 135, 137, 138, 140, 141, 142, 149–154, 155–160, 211

W

Within–between distinction, 48, 49, 50, 54–58, 60, 71, 73, 74–75, 78, 82
Word problem, *see* Problem solving

DEVELOPMENTAL PSYCHOLOGY SERIES

Continued from page ii

EUGENE S. GOLLIN. (Editor). *Developmental Plasticity: Behavioral and Biological Aspects of Variations in Development*

W. PATRICK DICKSON. (Editor). *Children's Oral Communication Skills*

LYNN S. LIBEN, ARTHUR H. PATTERSON, and NORA NEWCOMBE. (Editors). *Spatial Representation and Behavior across the Life Span: Theory and Application*

SARAH L. FRIEDMAN and MARIAN SIGMAN. (Editors). *Preterm Birth and Psychological Development*

HARBEN BOUTOURLINE YOUNG and LUCY RAU FERGUSON. *Puberty to Manhood in Italy and America*

RAINER H. KLUWE and HANS SPADA. (Editors). *Developmental Models of Thinking*

ROBERT L. SELMAN. *The Growth of Interpersonal Understanding: Developmental and Clinical Analyses*

BARRY GHOLSON. *The Cognitive-Developmental Basis of Human Learning: Studies in Hypothesis Testing*

TIFFANY MARTINI FIELD, SUSAN GOLDBERG, DANIEL STERN, and ANITA MILLER SOSTEK. (Editors). *High-Risk Infants and Children: Adult and Peer Interactions*

GILBERTE PIERAUT-LE BONNIEC. *The Development of Modal Reasoning: Genesis of Necessity and Possibility Notions*

JONAS LANGER. *The Origins of Logic: Six to Twelve Months*

LYNN S. LIBEN. *Deaf Children: Developmental Perspectives*